Post-Genomic Perspectives in Modeling and Control of Breathing

ADVANCES IN EXPERIMENTAL MEDICINE AND BIOLOGY

Recent Volumes in this Series

Volume 539
BLADDER DISEASE, Part A and Part B: Research Concepts and Clinical Applications
Edited by Anthony Atala and Debra Slade

Volume 540
OXYGEN TRANSPORT TO TISSUE, VOLUME XXV
Edited by Maureen Thorniley, David K. Harrison, and Philip E. James

Volume 541
FRONTIERS IN CLINICAL NEUROSCIENCE: Neurodegeneration and Neuroprotection
Edited by László Vécsei

Volume 542
QUALITY OF FRESH AND PROCESSED FOODS
Edited by Fereidoon Shahidi, Arthur M. Spanier, Chi-Tang Ho, and Terry Braggins

Volume 543
HYPOXIA: Through the Lifecycle
Edited by Robert C. Roach, Peter D. Wagner, and Peter H. Hackett

Volume 544
PEROXISOMAL DISORDERS AND REGULATION OF GENES
Edited by Frank Roels, Myriam Baes, and Sylvia De Bie

Volume 545
HYPOSPADIAS AND GENITAL DEVELOPMENT
Edited by Laurence S. Baskin

Volume 546
COMPLEMENTARY AND ALTERNATIVE APPROACHES TO BIOMEDICINE
Edited by Edwin L. Cooper and Nobuo Yamaguchi

Volume 547
ADVANCES IN SYSTEMS BIOLOGY
Edited by Lee K. Opresko, Julie M. Gephart, and Michaela B. Mann

Volume 548
RECENT ADVANCES IN EPILEPSY RESEARCH
Edited by Devin K. Binder and Helen E. Scharfman

Volume 549
HOT TOPICS IN INFECTION AND IMMUNITY IN CHILDREN
Edited by Andrew J. Pollard, George H. McCracken, Jr., and Adam Finn

Volume 550
BRAIN DEATH AND DISORDERS OF CONSCIOUSNESS
Edited by Calixto Machado and D. Alan Shewmon

Volume 551
POST-GENOMIC PERSPECTIVES IN MODELING AND CONTROL OF BREATHING
Edited by Jean Champagnat, Monique Denavit-Saubié, Gilles Fortin, Arthur S. Foutz, Muriel Thoby-Brisson

---

POST-GENOMIC PERSPECTIVES IN MODELING AND CONTROL OF BREATHING

Edited by

Jean Champagnat
Monique Denavit-Saubié
Gilles Fortin
Arthur S. Foutz
Muriel Thoby-Brisson
CNRS, UPR2216, Genetic & Integrative Neurobiology,
Gif sur Yvette, France

Kluwer Academic / Plenum Publishers
New York, Boston, Dordrecht, London, Moscow

Proceedings of the IXth Oxford Conference on Modeling and Control of Breathing,
held September 13–16 in Paris, France

ISSN 0065-2598

ISBN 0-306-48507-9

233 Spring Street, New York, New York 10013

http://www.wkap.nl

10 9 8 7 6 5 4 3 2 1

A C.I.P. record for this book is available from the Library of Congress

Printed in the United States of America

Preface

Since 1978, meetings of the Oxford Conference on Modeling and Control of Breathing are composed primarily of voluntary contributions from respiratory physiologists and modelers in order to promote interactions between these two domains and to stimulate interests of young researchers in modeling and physiological research in general. The first Oxford Conference was in September 1978 at the University Laboratory of Physiology in Oxford by the late Dr. D.J.C. Cunningham (Oxford), R. Hercynski (Warsaw) and others, who saw a need to bring physiologists and mathematicians together in order to address the critical issues in understanding the control of breathing. Four years later, a group at UCLA including respiratory/exercise physiologists, anesthesiologists and system engineers led by Drs. B.J. Whipp, S.A. Ward, J.W. Bellville and D.M. Wiberg organized a sequel meeting at Lake Arrowhead, California, USA, which proved to be as much a success as the first one. The next Conference was the first to be hosted in France: it was held in Solignac in 1985 and was organized by G. Benchetrit, P. Baconnier and J. Demongeot, from the University of Grenoble. Since then, the Oxford Conference has been continued every 3 years in the USA, Japan, U.K. and Canada with the most recent event in 2000 held at North Falmouth (Cape Cod, Massachusetts, USA) and co-chaired by Chi-Sang Poon (MIT) and H. Kazemi (Harvard).

This volume is dedicated to the memory of our colleague and member of the International Oxford Committee, Yoshiyuki Honda, MD, PhD, from the Department of Physiology at Chiba University School of Medicine (Chiba, Japan). Dr. Y. Honda passed away of acute heart failure on August 1st 2003, a little more than one month before the opening of the 9th Conference in Paris. This was just after he finished three of the seven days of an experiment on dyspnea induced by combination of hypercapnea and hypoxia with several co-workers. One could say that he fell in the battle which he had pursued to the very end of his 77 years. Dr. Yoshiyuki Honda was not only the scientist we know, working on many aspects of physiology, and particularly on the impact of life at high altitude, was also the organizer of the unforgettable 1991 Conference, at the foot of majestic Mont Fuji, the first of the series to be officially titled "Oxford Conference". We, his colleagues, express our deepest sympathy to his family and friends. The first section of this volume is specially dedicated to his memory. This section includes his chapter on ventilatory vs. respiratory sensation responses together with a tribute to Yoshiyuki Honda by John W. Severinghaus, and chapters on the breathing behaviour in humans. The second section of the volume is on central and peripheral chemoreceptive responses, to recognize the first *in vitro* studies identifying cells responsive to acidic perfusates in medullary slices (Fukuda, Y. & Honda, Y. pH-sensitive cells at ventro-lateral surface of rat medulla oblongata, Nature, 1975, 256: 317–318).

Close to the historical center of Paris, the 9th Oxford Conference presented a superb forum for formal and informal discussions. More than 140 participants from 16 countries attended this conference. The scientific program was comprised both of oral and poster contributions. The hundred presented papers covered a wide spectrum of modeling and experimental studies of respiratory control ranging from genetics and ion channels to respiratory disease and respiratory perception. One of the highlights of the meeting was a research competition for pre- and post-doctoral trainees, in which four finalists made oral presentations in a dedicated session of the conference: Kevin J. Cummings from Calgary (Canada), Efstratios Kosmidis from Yale (USA), Philip N Ainslie from Calgary and Laura Guimarães from Porto (Portugal) who was congratulated by the International Organizing Committee for the best trainee presentation. The Committee stressed the difficulty of selecting these finalists in view of the high scientific quality demonstrated by all other trainee participants, whose work fill many chapters of this volume. At the business meeting of the Conference, the International Committee also decided on the venue of the next Oxford Conference which is to be organized by Marc Poulin from Calgary (Canada). We believe the long history of the Conference series and its past accomplishments will ensure its continued success in the upcoming meeting. We are very grateful to every participant who contributed scientifically and constructively to the fruitful discussions and congenial atmosphere of the Paris Conference.

The Editors

Acknowledgments

The Organizing Committee gratefully acknowledges support from:

NATIONAL SCIENCE FOUNDATION
CNRS (DEPARTEMENT DES SCIENCES DE LA VIE)

THE IXth OXFORD CONFERENCE: INTERNATIONAL ORGANIZING COMMITTEE

Yoshiyuki Honda *deceased 8/1/2003*
Department of Physiology
Chiba University,
260-8672, Japan

Gila Benchetrit
Laboratoire de Physiologie Respiratoire Expérimentale, Théorique et Appliquée (PRETA-TIMC, UMR CNRS 5525),
Faculté de Médecine de Grenoble,
Université Joseph Fourier,
38700, La Tronche, France

Ikuo Homma
Department of Physiology,
Showa University School of Medicine
Hatanodai 1-5-8, Shinagawa-Ku,
Tokyo 142-8555, Japan

Richard L. Hughson
Cardiorespiratory and Vascular Dynamics Laboratory,
Faculty of Applied Health Sciences,
University of Waterloo, Waterloo,
Ontario, Canada

Homayoun Kazemi
Pulmonary and Critical Care Unit
Bulfinch 148, Massachusetts General Hospital
Harvard Medical School,
Boston, MA 02114, USA

Chi-Sang Poon
Harvard-MIT Division of Health Sciences & Technology
Bldg. 56-046,
Massachusetts Institute of Technology
77 Massachusetts Avenue
Cambridge, MA 02139, USA

Peter A. Robbins
University Laboratory of Physiology,
University of Oxford, Parks Road,
Oxford, OX1 3PT, United Kindom

J.W. Severinghaus
Department of Anesthesiology,
University of California Medical School,
San Francisco, Cal 94143-0542, USA

Susan A. Ward, Director,
Centre for Exercise Science and Medicine, West Medical Building,
University of Glasgow,
Glasgow G12 8QQ, United Kingdom

Brian J. Whipp
Division of Respiratory & Critical Care Medicine
Research & Education Institute,
Harbor-UCLA Medical Center
1124W. Carson Street, RB-2 Torrance,
California 90502, USA

THE IXth OXFORD CONFERENCE: LOCAL ORGANIZING COMMITTEE

Neurobiologie Génétique et Intégrative (NGI), CNRS UPR 2216
Institut de Neurobiologie Alfred Fessard
91198, Gif-sur-Yvette, France

General Organization:

Jean Champagnat
Annette Mont-Reynaud
Sandra Autran
Eliane Boudinot
Lydie Collet
Isabelle Germon
Valérie Mézière-Robert
Alain Pérignon

Local Scientific Committee:

Jean Champagnat
Caroline Borday
Fabrice Chatonnet
Monique Denavit-Saubié
Gilles Fortin
Arthur S. Foutz
Muriel Thoby-Brisson
Christelle Thaëron

Contents

5. RESPIRATORY ACTIVITY DURING SLEEP AND ANESTHESIA

6. CARDIO-RESPIRATORY REGULATIONS AND CEREBRAL BLOOD FLOW

7. VENTILATORY RESPONSE TO EXERCISE

8. VARIABILITY AND PLASTICITY OF BREATHING

9. CONCLUSIONS AND PERSPECTIVES

1

Views on the Human Breathing Behavior

Memories of Yoshiyuki Honda

Different Profile in Ventilatory vs. Respiratory Sensation Responses to CO_2 with Varying Po_2

Atsuko Masuda, Yoshikazu Sakakibara, Toshio Kobayashi, Michiko Tanaka and Yoshiyuki Honda

1. Introduction

Recently, we[1] reported that the slope of the respiratory sensation response curve to CO_2 assessed by visual analog scale (VAS), exhibited a parallel leftward shift when combined with hypoxic stimulation. However, further analysis additionally elucidated the presence of significant upward shift of this VAS response curve in the same experimental condition. On the other hand, the CO_2-ventilation response curve increased its slope with increasing hypoxic stimulation, and the extrapolated response lines converged at the horizontal axis known as the so-called Oxford fan[2]. These contrasting change between ventilatory vs. VAS response curves led us to certain speculations and assumptions about the different control mechanisms and anatomical regions possibly responsible for our experimental findings.

2. Methods

Details of the experimental design and procedure were described in the preceding publication[1]. Briefly, 29 healthy young college students (11 males and 18 females) participated in the study. Their age, height and weight were 23 ± 5 yrs, 162.1 ± 9.0 cm and 54.2 ± 8.6 kg (mean ± SD), respectively. All subjects gave their informed consent before the experiment, and the study protocol was approved by the local ethics committee. The subjects were exposed to progressive hypercapnia by modified Read's rebreathing method using four different gas mixtures: A. 7% CO_2 + 93% O_2 or hyperoxia run, B. 7% CO_2 + 19% O_2 for normoxia run, C. 7% CO_2 + 13% O_2 for mild hypoxia run and D. 7% CO_2 +

Atsuko Masuda • Tokyo Med & Dent Univ, Tokyo, 113-8519, Japan. **Yoshikazu Sakakibara** • Kanazawa Inst Technol, Kanazawa. **Toshio Kobayashi** • Hiroshima Univ, Hiroshima. **Michiko Tanaka** • Miyazaki Pref Nurs Univ, Miyazaki. **Yoshiyuki Honda** • Chiba Univ, Chiba.

Post-Genomic Perspectives in Modeling and Control of Breathing, edited by Jean Champagnat, Monique Denavit-Saubié, Gilles Fortin, Arthur S. Foutz, Muriel Thoby-Brisson. Kluwer Academic/Plenum Publishers, 2004.

11% O_2 for moderate hypoxia run. Amounts of 6–8 L of the above gas mixtures were put into a rubber bag and used for the rebreathing test. During rebreathing, by introducing a small amount of O_2 or N_2 gas to the inspiratory portion of the respiratory valve, the P_{ETO_2} level was consecutively adjusted to >300, 100, 80 and 60 Torr in the A, B, C and D runs, respectively. To avoid psychological intervention, a paper screen was placed between the subject and the experimenter. The magnitude of VAS was continuously recorded in a range of 0 to 10 degrees. Breath-by-breath V_T and P_{O_2} and P_{CO_2} signals in the airway gas were also detected and recorded by using a hot wire flowmeter (Minato, RF-H, Osaka, Japan) and a rapid response O_2 and CO_2 analyzer (NEC-Sanei, 1H21, Tokyo, Japan). Minute ventilation (V_E) was calculated from V_T and respiratory frequency (f) and normalized body surface area. Arterial O_2 saturation (S_{pO_2}) was continuously recorded by pulse oximeter (Ohmeda, BIOX·, USA). For each subject, four runs were randomly performed twice on the same day, once in the morning and once in the afternoon. Hypercapnic ventilatory and VAS response curves were obtained by linear regression analysis between P_{ETCO_2} and normalized minute volume or VAS, respectively.

3. Results

Figure 1 shows examples of CO_2-ventilation and CO_2-VAS responses. The CO_2-V_E response curve became steeper as the oxygen level decreased (left), the VAS response curves are not only shift in parallel to the left with increasing hypoxic stimulation (right). Figure 2 presents the averaged mean P_{ETCO_2}-V_E (left) and PET_{CO_2}-VAS (right) response curves of all 29 subjects. The slopes of the ventilatory response curves were augmented by advancing hypoxia and the extrapolated response lines converged at the horizontal axis. On the other hand, the VAS response curve shifted to the left in parallel and simultaneously moved upward with intensifying hypoxia.

4. Discussion

The major finding in this study is the different profile in ventilatory vs. respiratory sensation responses to progressive hypercapnia with varying P_{O_2}. The former exhibited

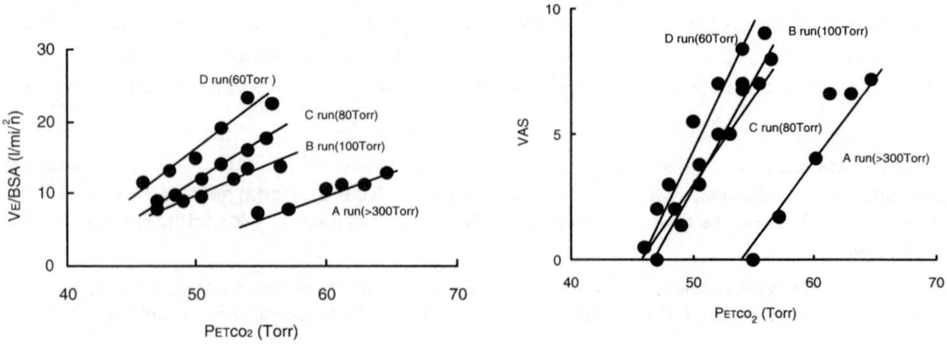

Figure 1. CO_2-ventilatory (left) and CO_2-VAS (right) responses obtained in one subject.

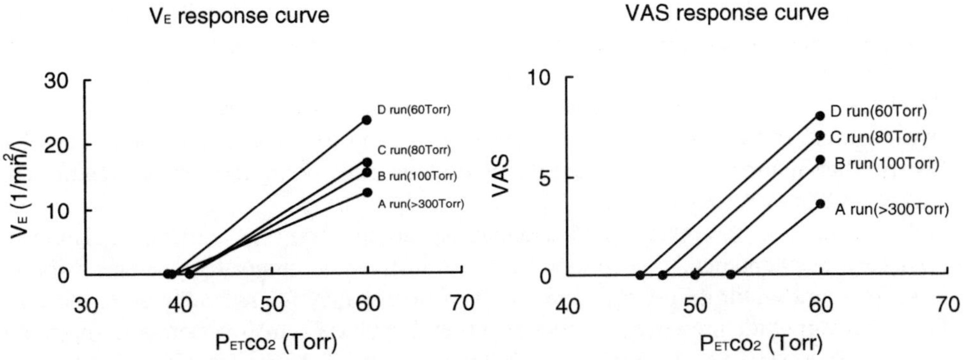

Figure 2. Average ventilatory and VAS responses of all subjects (n = 29).

increased response slopes, with extrapolated response lines converging at the horizontal axis. On the other hand, the latter showed parallel left and upward shifts with increasing hypoxic stimulation. We could verify that there was positive interaction between CO_2 and hypoxic stimulation in ventilatory response. Such strong positive interaction was not only seen in ventilatory response, but was also found in the discharges of the carotid body located in the brainstem[3]. On the other hand, no such positive interaction was detected in the present VAS response.

Ventilatory response is assumed to be mainly elicited by augmented peripheral as well as central chemoreceptor activities in the brainstem respiratory control system, but the VAS response likely originated from a somewhat different mechanism.

Figure 3 describes the possible implication of our results in schematic form. We speculate that the main respiratory sensation is produced in the sensory cortex as a response to the activated behavioral respiratory control system consisting of such emotion and behavior related systems as the limbic system. From the profile of the P_{ETCO_2}-VAS response

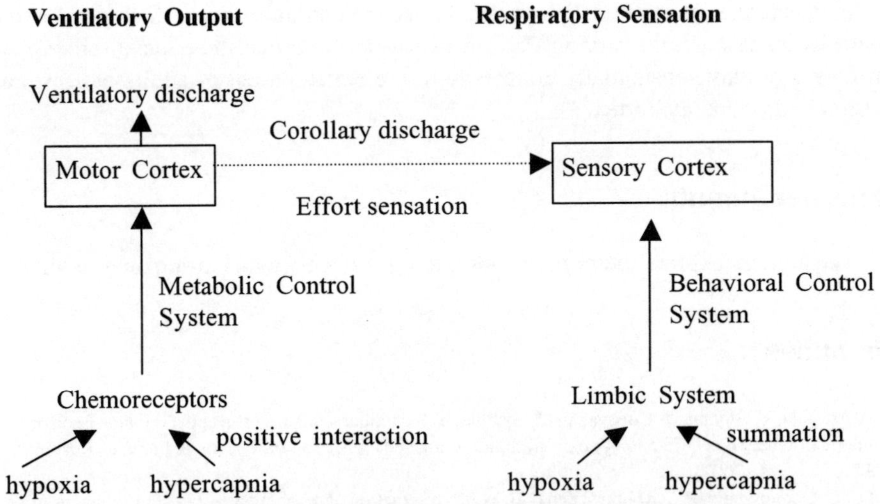

Figure 3. Possible implication derived from our experiment.

curve, we propose that respiratory sensation may have been induced additively by hypoxic and hypercapnic stimulation. Corollary discharge[4], which conveys the sense of ventilatory effort from the motor cortex, may have played a subsidiary role in our experiments. On the other hand, we surmise that stimulation of hypoxia and/or hypercapnia not only strongly drives ventilation by augmenting peripheral and/or central chemoreceptor activities[2], but also specifically arouses suprapontine CNS structures including the limbic system, thus eliciting additional VAS elevation.

A number of studies have indicated that the suprapontine brain structures elicit weak ventilatory, but appreciable behavioral response to hypoxic and/or hypercapnic stimulations. Horn and Waldrop[5] described in their review that they found studies from the early 90's that reported the presence of specific CO_2 and hypoxia-sensitive neural discharges in the caudal hypothalamus, both in vitro and in vivo. Chen et al.[6] reported in unanesthetized cats in 1991 that respiratory-associated rhythmic firing found in the mesencephalon, which is vulnerable to light anesthesia and is synchronous with EEG activities, was presumed to convey the dyspnea sensation. Corfield et al.[7] observed in awake humans in 1995 that in response to CO_2-stimulated breathing, dyspnea and regional brain blood flow in the limbic system were proportionally augmented. They used positron emission tomography (PET) to detect increased regional brain blood flows in their study, considering to indicate augmented local neural activities. More recently, similar PET studies observing CO_2 loading were also conducted by Brannan et al.[8]. They found not only activation, but also deactivation in some restricted areas. Since we found different VAS response feature when hypoxia was combined with hypercapnia, further exploration using PET deserves to be conducted. Furthermore, Curran et al.[9] recently demonstrated moderate tachpneic hyperventilation in unanesthetized dogs, in which the carotid body was maintained as normocapnic, normoxic and normohydric by separate extracorporeal perfusion and by specifically induced CNS hypoxia. They considered that this indicated the presence of hypoxia-sensitive respiratory-related neurons in the CNS, i.e., outside of the brainstem chemosensitive-ventilatory control system.

All these findings may support our speculation that VAS-related respiratory activity is present in the behavioral respiratory control system described above.

In conclusion, the contrasting profile between ventilatory vs. VAS responses to progressive hypercapnia with varying Po_2 led us to speculate that the behavioral respiratory control system may substantially contribute to the development of additional respiratory sensation in the sensory cortex.

Acknowledgments

The authors express their appreciation to the subjects who participated in this study.

References

1. A. Masuda, Y. Ohyabu, T. Kobayashi, C. Yoshino, Y. Sakakibara, T. Komatsu, and Y. Honda, Lack of positive interaction between CO_2 and hypoxic stimulation for Pco_2-VAS response slope in humans, Respir. Physiol. 121, 173–181 (2001).
2. D.J.C. Cunningham, P.A. Robbins, and C.B. Wolff, Interaction of respiratory responses to changes in alveolar partial pressure of CO_2 and O_2 in arterial pH, in: Handbook of Physiology, Sec 3, The respiratory System

Vol.·, Control of breathing, Part 2, Bethesda, MD, edited by A.P. Fishman, N. S. Cherniack, J. D. Widdicombe, and S. T. Geiger (Am. Physiol. Soc, Berthesda, MD, 1986) pp. 475–528.

3. R.S. Fitzgerald, D.C. Parks, Effect of hypoxia on carotid chemoreceptor response to carbon dioxide in cats, Respir. Physiol. 12, 218–229 (1971).
4. D.I. McClosky, Corollary discharges: motor commands and perception, in: Handbook of Physiology. The nervous system. Sec 1, Vol 2, part 2, edited by J. K. Brookhart, V.B. Mountcatsle (Am. Physiol. Soc, Berthesda, MD, 1981), pp. 1415–1447.
5. E.M. Horn and T.G. Waldrop, Frontiers review. Suprapontine control of respiration, Respir. Physiol. 114, 201–211 (1998).
6. Z. Chen, F.L. Eldridge, and P.G. Wagner, Respiratory-associated rhythmic firing of midbrain neurons in cats: relation to level of respiratory drive, J. Physiol. 437, 305–325 (1991).
7. D.R. Corfield, G.R. Fink, S.C. Ramsay, K. Murphy, H.R. Harty, J.D.G. Watson, L. Adams, B.S.J. Frackowiak, and A. Guz, Evidence for limbic system activation during CO_2-stimulated breathing in man, J. Physiol. 488, 77–84 (1995).
8. S. Brannan, M. Liotti, G. Egan, R. Shade, L. Madden, R. Robillard B. Abplanalp, K. Stofer, D. Denton, and P. T. Fox, Neuroimaging of cerebral activations and deactivations associated with hypercapnia and hunger for air, Proc. Natl. Acad. Sci. USA 98(4), 2029–2034 (2001).
9. A.K. Curran, T.R. Rodman, P.R. Eastwood, K.S. Henderson, J.A. Dempsey, and C.A. Smith, Ventilatory response to specific CNS hypoxia in sleeping dogs. J. Appl. Physiol. 88, 1840–1852 (2000).

Amygdala and Emotional Breathing in Humans

Yuri Masaoka and Ikuo Homma

1. Introduction

According to many reports the amygdala plays a role in fear, attention and anxiety[1]. In addition to these emotional roles, the amygdala which is involved in the conditioning process projects to many anatomical areas to elicit physiological responses such as blood pressure elevation, skin conductance response and respiration as well as behavioral responses such as freezing and facial expression of fear. In an awake state, the amygadala evaluates a variety of environmental stimuli to determine whether the stimulation is harmful or safe; if it is harmful, the amygdala immediately elicits emotions of fear and anxiety simultaneously with physiological changes. In other words, measuring physiological responses could be an index to determine the level of emotion occurred in a situation.

Respiratory psychophysiology studies have reported that respiratory patterns are affected by fear and anxiety in humans[2]. The brainstem regulates respiration to adjust for a metabolic requirement but final respiratory output appears to be from interactions between maintaining homeostasis and input from many types of sensory information and emotions from the higher cortical and limbic structures. In animal studies, there have been reports on electrical stimulation of the amygdala which altered respiratory pattern[3]. Investigation of the relationship between respiratory output and the amygdala in humans is limited, but in this chapter we reports our studies investigating anticipatory anxiety in normal subjects and in two patients with epilepsy who had lesions of the left amygdala. We also give direct evidence concerning the effect of electrical stimulation of the amygdala on the total respiratory time, inspiratory time and expiratory time in a patient with epilepsy.

1.1. Measuring Respiratory Patterns and Metabolism in Anticipatory Anxiety

Our previous research on personality differences in breathing patterns during mental stress and physical load found that levels of individual anxiety affect respiratory frequency[4].

Yuri Masaoka and Ikuo Homma • Department of Physiology II, Showa University School of Medicine, 1-5-8 Hatanodai, Shinagawa-ku, Tokyo, Japan, 142-8555.

Post-Genomic Perspectives in Modeling and Control of Breathing, edited by Jean Champagnat, Monique Denavit-Saubié, Gilles Fortin, Arthur S. Foutz, Muriel Thoby-Brisson. Kluwer Academic/Plenum Publishers, 2004.

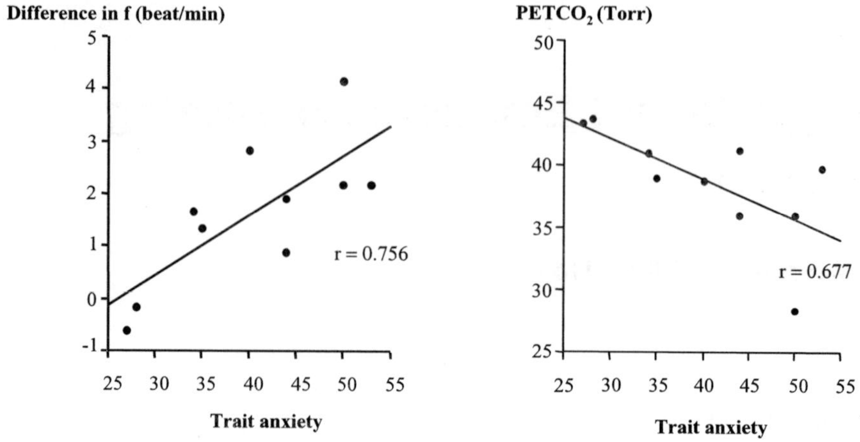

Figure 1. Relation between fR and trait anxiety scores (left), and between $PETCO_2$ and trait anxiety scores (right).

From this study we hypothesized that an increase of respiratory frequency might not be caused by metabolic changes but by input from higher centers related to anxiety. In a laboratory situation, anticipatory anxiety was produced in subjects while measuring V_E, V_T, fR, T_I, T_E and P_{ETCO_2}[5]. The subjects were informed that electrical stimulation attached to the forefinger would be delivered within two minutes after the onset of a warning light. "Anticipatory anxiety" was defined as the time between warning onset and stimulation. This increase of respiratory frequency correlated with trait anxiety scores, and P_{ETCO_2} had a negative correlation with trait anxiety scores (Figure 1). There was no difference in VO_2 and VCO_2. The trait anxiety scores are used to evaluate how people generally feel, asking their tendency in various situations. Large increases of respiratory frequency in high trait anxiety subjects means that trait anxiety influences behavioral breathing independently from metabolic demands. What area is related to anxiety? That question prompted us to look into the amygdala.

1.2. EEG Dipole Tracing in Anticipatory Anxiety

Recent research on emotions has been investigated by neuro-psychologists using PET and fMRI. Among these methods, we have developed the EEG dipole method (DT) utilizing a realistic scalp-skull-brain head model (BS-navi, Brain research and Development, Japan)[6]. BS-navi is a method for estimating the source of brain activity from potentials recorded by electroencephalogram (EEG). The method has been used to evaluate patients with epilepsy showing the foci of the epileptic spike measured by deep electrodes corresponding to foci estimated by SSB/DT[7]. One feature of this method is that it can determine the source of the brain from averaged potentials which are triggered by physiological responses. Our hypothesis was that if the subjective feeling of anxiety enhances the respiratory rate, electric current sources synchronized with this onset of inspiration could be found somewhere in the limbic areas. During anticipatory anxiety, significant increases in anxiety state and respiratory rate were observed. The onset of inspiration during increased respiratory frequency was used as a trigger for averaging EEG. From 350 ms to 400 ms after the onset of inspiration,

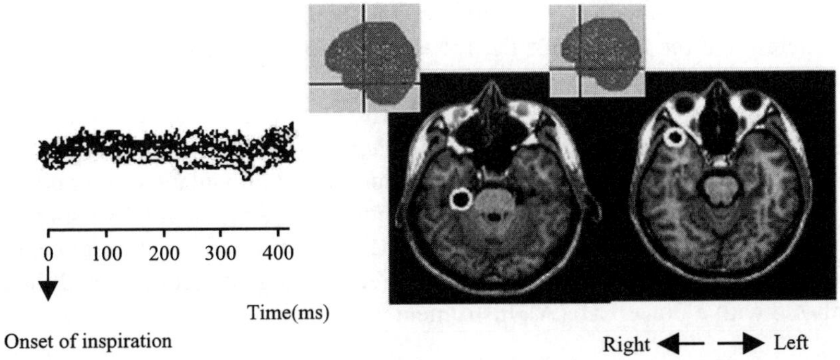

Figure 2. Respiratory-related Anxiety Potential (RAP) (right) and dipole location during RAP (left).

a positive wave was observed in the averaged potentials. This positive wave is referred to as "respiratory-related anxiety potentials" (RAP) (Figure 2 left)[8]. SSB/DT estimated the location of the sources from RAP which were found in the right temporal pole, and in the right temporal pole and the left amygdala in the most anxious subjects (Figure 2 right).

Anticipatory anxiety has been tested in patients with epilepsy who had a lesion of the left amygdala. The left amygdala lesion decreased state anxiety, respiratory frequency, skin conductance response during anticipatory anxiety, as well as trait anxiety [9].

In this chapter we give direct evidence of electrical stimulation on the left amygdala in patient who had deep electrodes installed to evaluate the location of the epileptic spike. All of the experiments reported here were approved by the Ethical Committee of Showa University School of Medicine and Tokyo Women's University.

2. Method

A patient (female, aged 26) had experienced epileptic seizures for more than ten years and had been diagnosed as having typical temporal seizures. Focus of the epileptic spike was located in the left amygdala confirmed with a intracerebral depth electrodes composed of six contact points, four electrodes from the tip placed in the amygdala had a 5 mm interval and the other two had a 10 mm interval (Figure 3) (Unique Medical Co., LTD). The location of

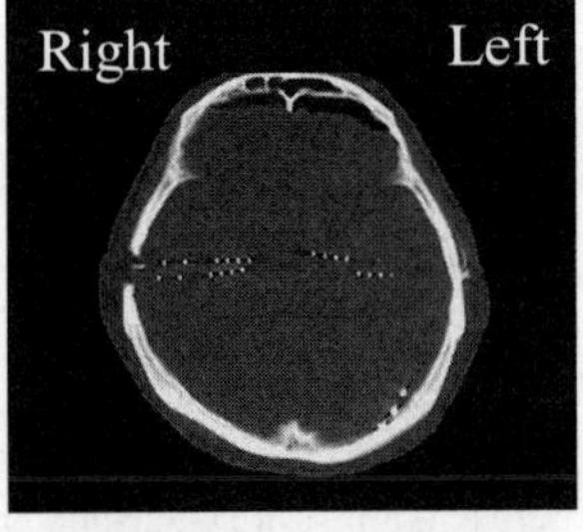

Figure 3. CT image with depth electrodes from a patient with epilepsy.

the focus also confirmed with an estimation of the SSB/DT[7]. Epileptic patients are routinely tested to determine the location of the focus of the spike by electrical stimulation on the amygdala and by asking the patient whether or not this stimulation causes a feeling which is similar to that before an epileptic seizure. The stimulation was delivered through the depth electrode by a cortical stimulator (Ojemann, Radionics). Stimulation of 0.5 mmA was delivered for 60s. During the stimulation on the left amygdala, chest and abdominal movements were measured with a respiratory induction plethymography using Respitrace transducer (Amburatory Monitoring) and volume was measured by a transducer connected to a flow meter (Minato Medical). Data was stored on a digital recorder (PC208AX, Sony) and analyzed with a PowerLab (ADInstruments).

3. Results

Figure 4 (top) shows Ttot, TI and TE during baseline, 0.5 mA of stimulation and 1 mA of stimulation. Rib and abdominal movements, and volume during the 0.5 mA of stimulation on the left amygdala are shown in Figure 4 (bottom).

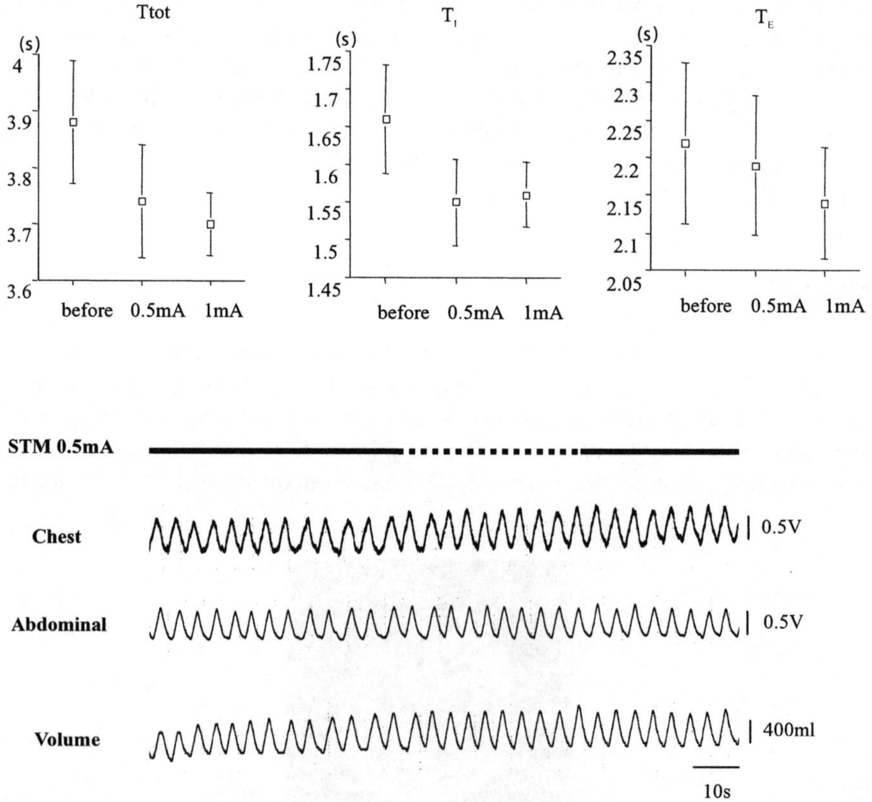

Figure 4. Ttot, TI and TE during baseline, 0.5 mA of stimulation and 1mA of stimulation (top). Chest and abdominal movements and volume during stimulation of the left amygdala (bottom).

4. Discussion

In this chapter, we have focused on the role of the amygdala during anxiety and anxiety related respiratory changes in humans. A number of investigators have studied neuroanatomical correlates of anxiety using various modality in humans[10]. In animals, many reports show the effect of the electrical stimulation of the amygdala on respiratory timing. These studies prove that the amygdala has a role for fear and anxiety, and the physiological outputs related to these emotions. From these studies and our data, our interest focuses on three questions. The first two are related to the behavioral aspect. What does an increase in respiratoy frequency during anxiety mean, and is this tendency determined by individual trait? These respiratory timing changes were dominantly affected by anxiety more than by metabolism regulation. As we mentioned, the amygdala has a role for fear conditioning. The amygdala immediately evaluates the outer stimulation and produces various physiological changes to protect ourselves. In respiration, conditioning was observed in ventilatory responses to auditory stimuli after pairing with hypoxic stimulus[11] and in an anticipatory of physical exercise[12], and these responses involved respiratory feed forward mechanism. However, it is unknown whether the conditioning process is determined by individual trait or whether the personality trait dominantly affect respiratory frequency before conditioning.

Second, if activity of the amygdala elicits frequency changes in respiration, what is the relationship between the amygdala and other respiratory related areas? One study found that the central nucleus of the amygdala (ACE) contributes to excitation of the inspiratory cycle, possibly through the projection to the parabrachial nucleus; this study also demonstrated that the respiratory entrainment to ACE stimulation occurs during the alert state[13]. The lateral part of the ACE projects to the medial part of the central nucleus, restricted parts of the bed nucleus of the stria terminals, and the parabrachial nucleus in the pons[14]. From these studies it is reasonable to assume that the respiratory outputs modulated by either direct electrical stimulation or emotional stimulation of the amygdala is conducted through the route of the amygdala-parabrachial nucleus.

In this chapter, we report the results from only one case of amygdala stimulation on respiration; however, emotional breathing mediated from the amygdala might be related to the movement of the chest rather than the abdomen which would explain the tightness of the chest when one feels anxiety.

References

1. The amygdala, edited by J. P. Aggleton (Oxford University Press, New York, 2001).
2. F. A. Boiten, N. H. Frijda, and C. J. E. Wientjes, Emotions and respiration, Int. J. Psychophysiol. 17, 103–128 (1994).
3. C. D. Applegate, B. S. Kaap, M. D. Underwood, and C. L. McNall, Autonomic and somatomotor effects of amygdala central n. stimuation in awake rabbits, Physiol. Behav. 31, 353–360.
4. Y. Masaoka and I. Homma, Expiratory time determined by individual anxiety levels in humans, J. Appl. Physiol. 86(4), 1329–1336 (1997).
5. Y. Masaoka and I. Homma, The effect of anticipatory anxiety on breathing and metabolism in humans, Respir. Physiol. 128, 171–177 (2001).
6. S. Homma, T. Musha, Y. Nakajima, and T. Sato, Location of electric current sources in the human brain estimated by the dipole tracing method of the scalp-skull-brain (SSB) head model, Clin. Neurophysiol. 91, 374–382 (1994).

7. I. Homma, Y. Masaoka, K. Hirasawa, F. Yamane, T. Hori and Y. Okamoto, Comparison of source localization of interictal epileptic spike potentials in patients estimated by the dipole tracing method with the focus derectly recorded by the depth electrodes, Neurosci. Lett. 304, 1–4 (2001).
8. Y. Masaoka and I. Homma, The source generator of respiratory-related anxiety potential in the human brain, Neurosci. Lett. 283, 21–24 (2000).
9. Y. Masaoka, K. Hirasawa, F. Yamane, T. Hori, and I. Homma, Effect of left amygdala lesions on respiration, skin conductance, heart rate, anxiety, and activity of the right amygdala during anticipation of negative stimulus, Behavior Modification, 27, 607–619 (2003).
10. Cognitive neuroscience of emotion, edited by R. D. Lane and L. Nadel (Oxford University press, New York, 2000).
11. J. Gallego and P. Perruchet, Classical conditioning of ventilatory responses in humans, J. Appl. Physiol. 70, 676–682, 1991.
12. M. J. Tobin, W. Perez, S. M. Guenther, G. D'Alonzo, and R. Dantzker, Breathing pattern and metabolic behavior during anticipation of exercise, J. Appl. Physiol. 60, 1306–1312.
13. R. M. Harper, R. C. Frysinger, R. B. Trelease, and J. D. Marks, State-dependent alternation of respiratory cycle timing by stimulation of the central nucleus of the amygdala, Brain Research, 306, 1–8 (1984).
14. G. D. Petrovich and L. W. Swanson, Projection from the lateral part of the central amygdalar nucleus to the postulated fear conditioning circuit, Brain Research, 763, 247–254 (1997).

Immediate Effects of Bilateral Carotid Body Resection on Total Respiratory Resistance and Compliance in Humans

Benjamin Winter and Brian J. Whipp

1. Introduction

In 1962, Nadel and Widdicombe[1] conclusively demonstrated a reflex increase in airways resistance (Raw) in response to carotid body stimulation in the dog. This effect could be eliminated by blocking either the glossopharyngeal or vagus nerves, i.e. the afferent or efferent limbs of the reflex respectively. That the peripheral chemoreceptors not only act reflexly on respiration but also cause reflex bronchoconstriction was also stressed by Widdicombe[2] in his review of "Chemoreceptor Control of the Airways." This is among the reasons often cited as a basis for potential benefits that might accrue from bilateral carotid body resection (BCBR) in patients with severe chronic obstructive pulmonary disease (COPD).[3]

However, there is no convincing evidence that this reflex is actually operative in humans and, even were it to be so, whether it is sufficiently potent to change Raw in the face of the numerous other factors which lead to airflow impairment in such patients. What evidence is available, however, suggests that this might be the case. For example, Vermeire et al.[4] have demonstrated that Raw (determined plethysmographically) was reduced significantly in a group of COPD patients following BCBR (although this was not apparent in their flow-volume analysis). In addition, Whipp and Ward[5] reported a small but significant increase in forced expiratory volume in 1 second (FEV_1) after BCBR, without an increase in total lung capacity (TLC), in a group of 146 patients with severe COPD, confirming the previous report of Winter.[6] Although such results are suggestive of improved Raw as a result of BCBR, they are not conclusive evidence because of other secondary factors that could influence Raw, such as altered blood and alveolar gas tensions[7] and increased mucus expectoration.[6]

Benjamin Winter • (deceased), Garfield Hospital, Monterey Park, California, 91754. **Brian J. Whipp** • Department of Physiology, St George's Hospital Medical School, London SW17 0RE, U.K.

Post-Genomic Perspectives in Modeling and Control of Breathing, edited by Jean Champagnat, Monique Denavit-Saubié, Gilles Fortin, Arthur S. Foutz, Muriel Thoby-Brisson. Kluwer Academic/Plenum Publishers, 2004.

Table 1. Pulmonary function and arterial blood gases and pH pre- and post-BCBR.

	FEV_1 % pred	MVV % pred	RV % pred	TLC % pred	PaO_2 Mm Hg	$PaCO_2$ mm Hg	pHa
Pre	31.9 ± 18	31.0 ± 11	260 ± 71	128 ± 19	57.0 ± 12.4	40.3 ± 4.5	7.43 ± 0.03
Post	38.6 ± 20	33.2 ± 16	230 ± 61	125 ± 21	50.1 ± 11.9	42.6 ± 3.2	7.40 ± 0.04

In an attempt to clarify this issue, we therefore examined total respiratory resistance (Rrs) and compliance (Crs) breath-by-breath in 32 consecutive patients immediately prior to and immediately following BCBR, using a customized ventilator.

2. Methods

Thirty-two patients with severe COPD were studied (approved by the Surgical Committee of Garfield Hospital): 23 males, 9 females (63 ± 10 yr, S.D.) (Table 1). Prior to electing to undergo BCBR surgery, all patients had been under the care of their physicians and consequently had a range of prior treatment regimens; in each case, their disease had failed to respond to conventional medical therapy. On admission, patients underwent a detailed history, physical examination, clinical laboratory tests, chest X-ray and ECG. Patients with evidence of cardiac failure (e.g. dependent edema, enlarged liver, ascites) were immediately digitalized and rapidly diuresed with appropriate dosage of diuretics. Pulmonary function tests and arterial blood-gas analyses were performed, and routinely repeated on the second post-operative day, at which time patients were discharged.

BCBR was performed as described by Winter:[8] this technique preserves barostasis by avoiding the carotid sinus baroreceptor afferents which predominantly course in the lateral aspect of the intercarotid nerve plexus in humans.[8] After induction (sodium pentothal and anectine; no pre-medication), endotracheal intubation was undertaken with general anesthesia being maintained with nitrous oxygen and oxygen (range: 30–60%) at light planes of anesthesia, so that patients could be awakened quickly on conclusion of BCBR.

Once the patient was stable, the endotracheal tube was attached to a specially-designed servo-ventilator (Elema-Schonander 900) for ventilation during BCBR surgery. The ventilator[9] provided a constant inspiratory flow ($\dot{v}_I$) profile that allowed the necessary total respiratory system pressure (Ptot[rs]) to overcome flow resistance (Rrs) and elastic recoil (Crs) to be determined. Respired flow and pressure were displayed throughout (Siemens Mingograph). At intervals, $\dot{v}_I$ was abruptly stopped, but with the thoracic volume maintained (i.e. expiration was not allowed) so that airway pressure fell as the flow-resistive component (Pres[rs]) decreased to zero, leaving only the pressure necessary to overcome the system's elastic recoil pressure (Pst[rs]) (Fig. 1). Inspiratory (t_I) and expiratory (t_E) durations were 40% and 60% of the total (t_T), respectively; the constant-flow component of inspiration was 25% of t_T, and the zero-flow component 15%.

In healthy lungs, as Ptot[rs] drops very quickly to [Pst(rs)] when $\dot{v}_I$ ceases, the Pres[rs] component is easily determined (Fig. 2). In COPD patients, however, the degree of obstruction often varies in different lung compartments, and the resulting difference in time

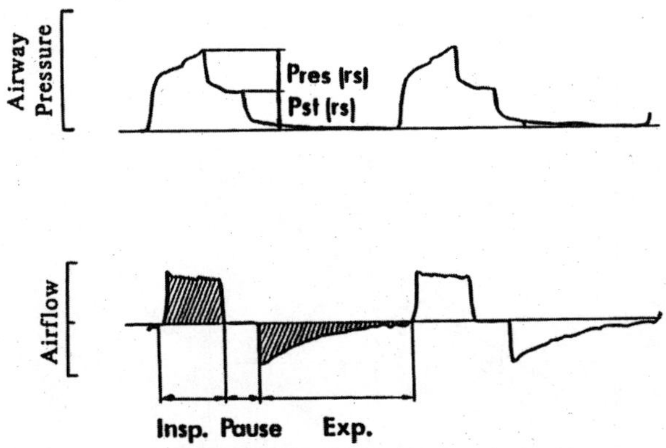

Figure 1. Schematic of the airway pressure and flow generated by the ventilator. With cessation of flow, pressure decreases to a nadir (Pres[rs]), reflecting the thoracic static recoil characteristics at that lung volume.[9]

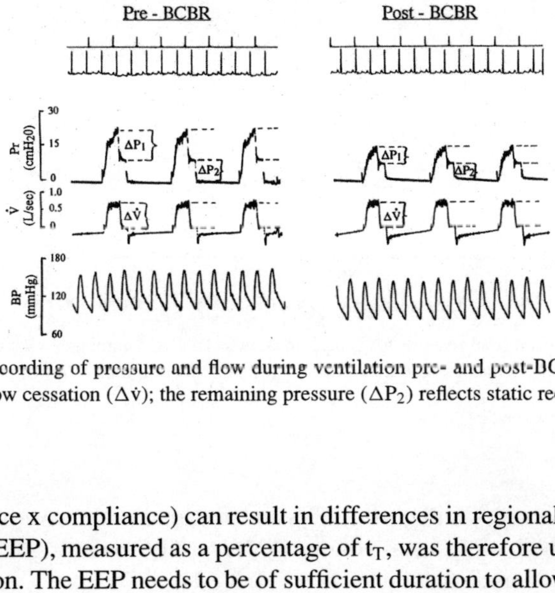

Figure 2. On line recording of pressure and flow during ventilation pre- and post-BCBR. Note the decline in pressure (ΔP_1) with flow cessation ($\Delta \dot{v}$); the remaining pressure (ΔP_2) reflects static recoil.

constants (resistance x compliance) can result in differences in regional emptying. An end-expiratory pause (EEP), measured as a percentage of t_T, was therefore used as a standard to mark end-expiration. The EEP needs to be of sufficient duration to allow a pressure plateau to be established (Fig. 2) and to avoid pressure gradients due to regional flow within the system.

Pre-BCBR measurements during BCBR surgery were standardly initiated within 15 min following induction, i.e. after the carotid bifurcations had been exposed but before any interference with the region. The surgical procedure to remove both carotid bodies took ~15 min, and the post-BCBR measurements were made ~5 min after the resection.

Each subject served as his/her own control—a sham-operated control group was naturally not feasible. Statistical significance was established using a paired-differences Student's t-test, using a level of $p < 0.05$ for discrimination of significant changes.

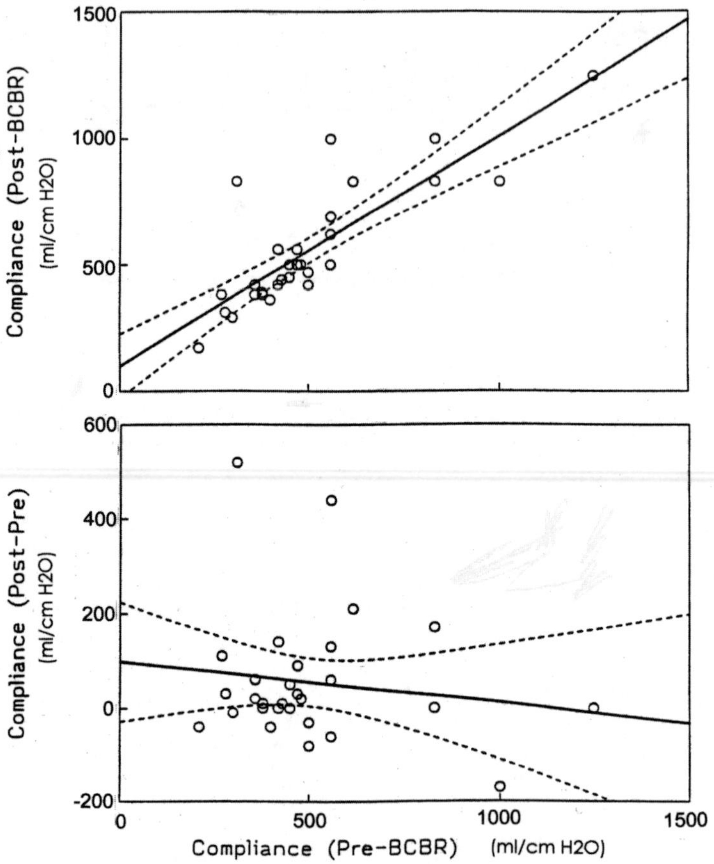

Figure 3. Effect of BCBR on total respiratory compliance: post-BCBR compliance vs. pre-BCBR compliance (upper panel); change in compliance (post- minus pre-) vs. pre-BCBR compliance (lower panel).

3. Results

All patients had severe COPD, with FEV_1 and maximum voluntary ventilation (MVV) ~30% of predicted, and both residual volume (RV) and TLC being significantly increased (~200% and ~130% of predicted, on average) (Table 1).

The influence of BCBR on Crs was highly variable among the subjects, with the changes not being significant for the group (Fig. 3).

Pre-operatively, Raw was 20.7 ± 7.9 cm H_2O/L/s. However, it should be noted that this measurement included the series resistance of the endotracheal tube and its attachment to the ventilator (which, unfortunately, were not measured at the time of the study) and hence does not represent airways resistance *per se*. Nevertheless, there was a reduction in mean Raw of some 25% following BCBR (to 16.8 ± 6.8 cm H_2O/L/s).

The change in Raw among subjects was highly variable, however (Fig. 4). In fact, for subjects in whom Rrs was less than 20 cm H_2O/L/s pre-BCBR, there was no significant change. However, for subjects with levels greater than 20 cm H_2O/L/s pre-BCBR, the change

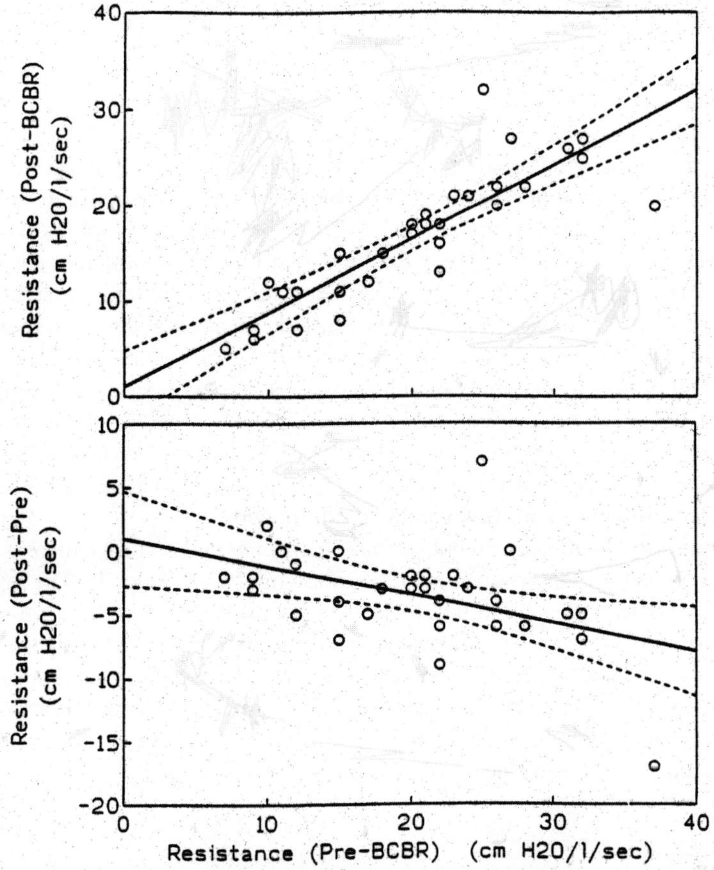

Figure 4. Effect of BCBR on total respiratory resistance: post BCBR resistance vs. pre BCBR resistance (upper panel); change in resistance (post- minus pre-) vs. pre-BCBR resistance (lower panel).

in Rrs consequent to BCBR was highly significant. The least-squares linear regression between pre- and post-BCBR Rrs (Fig. 4) resulted in a slope of 0.77, which was significantly less than 1.0 (95% confidence interval: 0.61–0.94) and a Y-intercept of 1.0, which was not significantly different from zero (95% confidence interval: −2.7 to 4.7).

4. Discussion

The carotid bodies in humans subserve an important role in both ventilatory and cardiovascular control. They are responsible for the ventilatory response to acute hypoxia,[10,11,12] some 20–30% of the ventilatory response to acute hypercapnia in euoxia.[12,13] have a significant role in the kinetics of the ventilatory response to moderate exercise[14,15] and play a dominant role in regulating pHa in response to acute metabolic acidemia during exercise.[15,16] The carotid bodies also play a significant role in the tolerance to breath-holding[17] and in the heart rate and blood pressure response to acute hypoxia.[12,18]

It has been suggested that the carotid bodies may also be the afferent source of a reflex which results in bronchoconstriction.[6] This reflex has been demonstrated in the dog.[1] When the carotid bodies were stimulated by hypoxia or nicotine bitartrate, reflex bronchoconstriction ensued which could be abolished by either sectioning the glossopharyngeal or cooling the vagus nerves that serve as the afferent and efferent pathways of this reflex, respectively. Whether such a reflex is functional in humans, however, is uncertain. Suggestive evidence, from experiments on pulmonary mechanics in subjects who have had both carotid bodies surgically resected, demonstrates improvement in FEV_1[5,6] or plethysmographically-determined Raw.[4] However, while suggestive, these findings do not provide sufficient support for the operation of a reflex directly influencing Raw. Several other factors, such as blood and alveolar gas tensions, acid-base status and the amount, location and consistency of airway mucus could also be involved.

The design of the current study obviates many of these concerns. The magnitude and pattern of ventilation and the inspired gas concentrations remained constant throughout the experiment (obviating offsetting bronchodilator effects of increased tidal volume[2,19]), and the acute changes in Rrs were computed immediately prior to and after BCBR. That is, any slower secondary influences on Raw would be minimized.

While our results naturally do not include sham-operated controls, there was a significant reduction in Rrs for the group as a whole following BCBR (Fig. 4). The magnitude of the decrease in Rrs was greatest in the subjects in whom Rrs was highest pre-operatively. In contrast, respiratory system compliance was not significantly affected by the procedure, suggesting that the tissue-flow component of the total resistance was not responsible for the decrease in system resistance. Our results are therefore most consistent with the operation of a reflex involving afferent signals from the carotid bodies leading to a component of bronchoconstriction in these subjects with severe COPD.

References

1. J. A. Nadel, and J. G. Widdicombe, Effects of changes in blood gas tensions ans carotid sinus pressure on tracheal volume and total lung resistance to air flow. *J. Physiol. (Lond.)* **163**, 13–33 (1962).
2. J. G. Widdicombe, Chemoreceptor control of the airways. *Respir. Physiol.* **87**, 373–381 (1992).
3. B. Winter, Carotid body resection: Controversy-confusion-conflict. *Ann. Thorac. Surg.* **16**, 648–659 (1973).
4. P. Vermeire, W. de Backer, R. van Maele, J. Bal, and W. van Kerckhoven, Carotid body resection in patients with severe chronic airflow limitation. *Clin. Respir. Physiol.* **23**(suppl. 11), 165S–166S (1987).
5. B. J. Whipp, and S. A. Ward, Physiologic changes following bilateral carotid-body resection in patients with chronic obstructive pulmonary disease. *Chest* **101**, 656–661 (1992).
6. B. Winter, Surgical treatment of asthma, chronic bronchitis and emphysema by bilateral carotid body resection. *Respir. Ther.* **5**, 8–28 (1975).
7. Y. Honda, S. Watanabe, I. Hashizume, Y. Satomura, N. Hata, Y. Sakibara, and J. W. Severinghaus, Hypoxic chemosensitivity in asthmatic patients two decades after carotid body resection. *J. Appl. Physiol.* **46**, 632–638 (1979).
8. B. Winter, Bilateral carotid body resection for asthma and emphysema: a new surgical approach without hypoventilation or baroreceptor dysfunction. *Int. Surg.* **57**, 458–464 (1972).
9. L. Nordstrom, On automatic ventilation. *Acta Anaesth. Scand.* **47**(suppl.), 1–56 (1972).
10. R. Lugliani, B. J. Whipp, C. Seard, and K. Wasserman, Effect of bilateral carotid-body resection on ventilatory control at rest and during exercise in man. *New Engl. J. Med.* **285**, 1105–1111 (1971).
11. Y. Honda, Respiratory and circulatory activities in carotid body-resected humans. *J. Appl. Physiol.* **73**, 1–8 (1992).

12. R. Lugliani, B. J. Whipp, and K. Wasserman, A role for the carotid body in cardiovascular control in man. *Chest* **63**, 744–750 (1973).
13. K. Wasserman, and B. J. Whipp, in: *Morphology and Mechanisms of Chemoreceptors*, edited by A. S. Paintal (V.P.C.I., Delhi, 1976), pp. 156–175.
14. T. L. Griffiths, L. C. Henson, and B. J. Whipp, Influence of peripheral chemoreceptors on the dynamics of the exercise hyperpnoea in man. *J. Physiol. (Lond.)* **380**, 387–403 (1986).
15. K. Wasserman, B. J. Whipp, S. N. Koyal, and M. G. Cleary, Effect of carotid body resection on ventilatory and acid-base control during exercise. *J. Appl. Physiol.* **39**, 354–358 (1975).
16. B. J. Whipp, Peripheral chemoreceptor control of the exercise hyperpnea in humans. *Med. Sci. Sports Exer.* **26**, 337–347 (1994).
17. J. T. Davidson, B. J. Whipp, K. Wasserman, S. N. Koyal, and R. Lugliani, Role of the carotid bodies in the sensation of breathlessness during breath-holding. *New Engl. J. Med.* **290**, 819–822 (1974).
18. P. M. Gross, B. J. Whipp, J. T. Davidson, S. N. Koyal, and K. Wasserman, Role of the carotid bodies in the heart rate response to breath holding in man. *J. Appl. Physiol.* **41**, 336–340 (1976).
19. E. H. Vidruk, Hypoxia potentiates, oxygen attenuates deflation-induced reflex tracheal constriction. *J. Appl. Physiol.* **59**, 941–946 (1985).

Memories of Yoshiyuki Honda, MD, PhD

John W Severinghaus

In 1962, in connection with a World Congress of Anesthesia in Kyoto, my wife Elinor and I visited Yoshi Honda in Kanazawa, with an invitation to speak on blood gas analysis. He patiently translated each sentence of my lecture for the audience, and then showed us his department and laboratory, followed by a tour of the ancient city and its amazing aqueduct system. Even at that time he was the respiratory physiologist in Japan best known in the US from his publications in English.

A few years later I helped recommend him for the Professorship in Chiba, based on my review of his publications and our discussions in Kanazawa. His special reason to move was to have access to the patients who had undergone bilateral carotid body denervation in attempts to relieve asthma and chronic respiratory distress. He became the world's authority on the various pathophysiologic effects of those people.

I participated in studies of the spinal fluid acid base balance of that group in the 1970s, bringing my portable blood gas electrode system with me. Spinal fluid pH proved to be normal on the average, even in those with chronic hypercapnia.

Benjamin Winters, a surgeon in southern California, had been denervating carotid bodies on patients with severe dyspnea. He reported relief and gratitude of many patients, and no evidence of prompt death from respiratory failure. Julius Comroe proposed that the NIH support a study of these patients, before and after denervation, My associates and I organized a small conference to plan the study, to which Dr. Honda came. The proposal met an enthusiastic response from the NIH study section but was vetoed by the Heart Lung Council as an unethical study. The result was that Winters was expelled from the California Medical Society, and his license revoked, without any contact with the group of respiratory physiologists who had begun to investigate those patients. Ignorance and unjustified criticism and fear blocked good research. Many of the bilaterally denervated patients in Japan survived for many years, and Yoshi and associates conducted valuable studies of their performance and regulation of respiration. There remains no doubt that humans can survive for years despite pulmonary obstructive dyspnea that, before resection, proved incapacitating.

In 1978, Honda invited me again to speak in Japan with support from industry. I was living in the Netherlands on sabbatical, and arranged to fly from Amsterdam to Tokyo, and then to South Africa for another professional visit. At that time the Dutch did not need a visa

Post-Genomic Perspectives in Modeling and Control of Breathing, edited by Jean Champagnat, Monique Denavit-Saubié, Gilles Fortin, Arthur S. Foutz, Muriel Thoby-Brisson. Kluwer Academic/Plenum Publishers, 2004.

Tokyo 1987, J. Severinghaus, Y. Honda, T. Aoyagi.

to enter Japan, but Americans did. My Dutch travel agent failed to warn me, and I arrived at Narita without visa. Yoshi had come to meet my plane, but found me at the airport passport barrier, where I was detained. I was told I would have to fly to some other country, perhaps Seoul, get a visa and then come back. However, Yoshi went to the legislature and arranged within 24 hrs to get a special law passed to admit me. I stayed a day at the airport hotel, not permitted to leave, and was warned never to do that again.

The conference, which Dr. Honda had organized, was unique. Some 20 Japanese investigators presented papers in English at a Japanese Respiratory Physiologic society

Ayoagi explains his pulse oximeter theory to Honda and me, 1987.

meeting in Tokyo, at which only two non-Japanese physiologists were present, Professor Hans Loeschcke from Germany and I. Our role was to be the critics for each paper, both for its science and English presentation. The best of these papers were then to be presented at international meetings at a later date.

In 1985 the first international conference on pulse oximetry was organized by Prof Jimmy Payne of the Faculty of Anaesthetists, supported by Ohmeda Co, in a London suburb, and I was invited to discuss the history of the discovery. The published literature suggested that it had been invented by a Japanese surgeon, a Dr. Nakayama, and I failed to obtain direct information. When my incorrect history was published, Yoshi took the trouble to obtain and send me the real story, identifying the actual inventor, Takuo Aoyagi, a bioengineer working for Nihon Kohden Co in Tokyo. The later editions of this history, thanks to Honda, revealing the inventor, resulted in Aoyagi being permitted to return to his research from an unproductive desk job. Aoyagi eventually obtained his PhD. He then invented what he called the Pulse Spectrophotometer, a multi wavelength version of a pulse oximeter which can use an IV injection of indocyanine dye to measure cardiac output, plasma volume and liver blood flow. Aoyagi told me that none of these career changes would have been possible had not Yoshi Honda identified him, notified me, and thus made Aoyagi's name famous to the world and especially to his boss who had not appreciated his revolutionary discovery.

Memories of Dr. Yoshiyuki Honda

Yoshikazu Sakakibara

Dr. Yoshiyuki Honda died of acute heart failure on the morning of August 1, 2003 in the middle of his successive seven-day experiment on dyspnea induced by various kinds of combined administrations of hypercapnia and hypoxia. In fact, he fell in the battle that he had been pursuing his whole scientific life, for half a century—52 years.

Dr. Honda graduated from Kanazawa Medical College in 1950, in the old capital city in the northwest region of Japan. He obtained his medical license in 1951 and then started his clinical work as a pediatrician in a nearby small town. He soon left this job and came back to the physiology laboratory of his home college to devote himself to his endless study of physiology. His first task was to directly measure arterial blood pH. To that purpose he constructed a pH meter with an electrode made of pH sensitive glass tubing that he purchased[1] himself. The dynamic profile of the pH changes in concert with respiration attracted and impressed him greatly, and from this he drew motivation to become a respiratory physiologist. The first paper on his experimental results was published in 1957[1].

He started a collaboration between Kanazawa and Chiba Universities on systematic investigations of patients with resected carotid bodies. During the course of this collaboration, Dr. Honda received the chair of Professor of Physiology at Chiba University in 1974. The idea for this work was originally proposed by Dr. John W. Severinghaus in 1967. Dr. Honda built the experimental setup for the project in his laboratory room. The patient was reclined on a second hand dental chair. Rubber tubing connected the mouth-piece, spirometer, Douglas bag, an electric cleaner to wash out the inside of the system, and cylinders of CO_2, O_2, N_2 and several calibration gases. The set-up also involved advanced O_2 and CO_2 sensors and other electric devices such as amplifiers, calculators, and a recorder. All of these took up substantial space in the cramped experimental room and, in total, presented the appearance of a factory. Many patients and subjects first looked astonished when entering this room, but almost all were then brave enough to perform the subject's role in the chair. Experiments were repeated and repeated. I can see clearly even now in my mind's eye the scene of Dr. Honda paying careful attention to the O_2 sensor on the days of experimentation. Throughout the studies he documented on such patients[2], he was particularly impressed by the fact that the carotid body functions were often different from those reported in other animal species. Dr. Honda always insisted in a convincing manner that the systematic investigation on humans as a whole is indispensable and that active and frank communications

---

Yoshikazu Sakakibara • Yoshikazu Sakakibara, Kanazawa Inst Technol, Kanazawa.

Post-Genomic Perspectives in Modeling and Control of Breathing, edited by Jean Champagnat, Monique Denavit-Saubié, Gilles Fortin, Arthur S. Foutz, Muriel Thoby-Brisson. Kluwer Academic/Plenum Publishers, 2004.

must be established among different laboratories and departments. Since then, he devoted almost all of his scientific life to the unraveling of the mechanisms of the chemical control of respiration. His attention was directed at every aspect of the ventilatory control system on the basis of experimental studies as well as clinical cases, from high mountains to the deep sea, and from exercise to sleep. As a consequence, he published a number of new findings and presented them at many international symposia. Thus, many Japanese researchers might well have viewed him the leader of this field in Japan. He, himself pointed out[2] that all of these findings were the result of the kind and generous collaboration of many people from different areas and departments.

After he retired from Chiba University in 1992, he was engaged in publishing for the several years, but then he resumed his research activity in order to examine the relationship between respiratory sensation and ventilatory responses during hypoxic and hypercapnic challenges. He improved the visual analog scale (VAS) to better assess the dyspneic sensation, and he modified Read's rebreathing method in order to facilitate progressive hypoxia under steady and different levels of hypercapnia, and vice versa. In 1998, as soon as he started the new experiments, he noted that the hypercapnic VAS dyspnea slope showed only a leftward shift with decreasing SaO_2, whereas the ventilatory responses became steeper. This result was submitted to Respir. Physiol. Although this journal finally accepted his paper (2001), there were hard and long discussions between Dr. Honda and the referees. In addition, in the summer of 2000 during the time of these debates, and on the evening of the day that his summer experiments were just completed, he suddenly lost his wife due to brain hemorrhage. Regret and grief struck him acutely. Still, even in the middle of such tragic circumstances he persevered. Finally, several months after his wife's death he received the letter of acceptance from the journal.

Last summer his colleagues from different universities gathered for experiments, as was the usual case for several years. Dr. Honda designed the experiments to elicit a confirmation of the behavior of the respiratory sensation rated with VAS with relation to ventilatory responses during the combined stimuli of O_2 and CO_2. But the experiments suddenly terminated in the morning of the fourth day.

As these brief memories suggest, Dr. Honda was a real pioneer. It must also be stressed that he never once begrudged spending time and effort to offer educational and moral support to graduate students and young researchers or doctors. Thanks to his support and encouragement, many young researchers got the chance to study in Bochum, Pennsylvania, Strasbourg, Lhasa and Oxford, and also obtained important positions at institutes and universities.

References

1. Y. Honda, H. Nomura, and M. Minoguchi, Effects of vagotomy on the excitability of the respiratory center to blood CO_2, *Jpn. J. Physiol*, 7, 137–146 (1957).
2. Y. Honda, Preface in 'Regulation of blood gas and ventilation—Collected papers of Dr. Honda (1957–1992)', edited by Y. Fukuda (1992).

2

Mechanisms of Central and Carotid Bodies Chemoreception

Chemosensory Control of the Respiratory Function

Working Towards Understanding the Role of ATP-Mediated Purinergic Signalling

Alexander V. Gourine, Nicholas Dale, K. Michael Spyer

1. Introduction

Respiratory drive is sensitive to small changes in arterial PO_2 and PCO_2 which are sensed by the peripheral chemoreceptors located within the carotid and aortic bodies and by central chemoreceptors localised within the brain (Gonzalez *et al.*, 1994; Daly, 1997; Nattie, 1999; Prabhakar, 2000). In adult mammals type I (glomus) cells of the carotid body are the primary peripheral chemosensitive elements which detect changes in arterial PO_2 (Gonzalez *et al.*, 1994; Prabhakar, 2000) and transmit this information to the afferent nerve fibres of the carotid sinus nerve, which in turn relays to the brainstem respiratory centres to produce adaptive changes in ventilation. Levels of PCO_2 are monitored predominantly by the central chemoreceptors localised within the medulla oblongata (primarily at, or in close proximity to, the ventral medullary surface [Loeschcke, 1982]), as up to 80% of the CO_2-evoked ventilatory response is mediated by the action of CO_2 at the brainstem chemosensitive sites (Heeringa *et al.*, 1979).

There has been growing interest in the role of ATP-mediated purinergic signalling in the nervous system. Extracellular ATP acting through ionotropic P2X and metabotropic P2Y receptors has numerous physiological functions in the brain and periphery (North, 2002). Over the last several years we have been studying extensively the role of purinergic signalling in central and peripheral mechanisms of chemosensitivity. As a result of these studies a unifying hypothesis of central and peripheral chemosensory transduction which involves ATP as a common key mediator has been proposed. We suggest that in the ventrolateral medulla (VLM) ATP is produced following CO_2/H^+-induced activation

Alexander V. Gourine and K. Michael Spyer • Department of Physiology, Royal Free and University College London Medical School, Rowland Hill Street, London NW3 2PF, UK. **Nicholas Dale** • Department of Biological Sciences, University of Warwick, Coventry CV4 7AL, United Kingdom.

Post-Genomic Perspectives in Modeling and Control of Breathing, edited by Jean Champagnat, Monique Denavit-Saubié, Gilles Fortin, Arthur S. Foutz, Muriel Thoby-Brisson. Kluwer Academic/Plenum Publishers, 2004.

of central chemosensory elements (neuronal and/or glial) and acts within the respiratory network to produce physiologically relevant changes in ventilation. In the carotid body, ATP contributes in a significant manner to the transmission of the sensitivity of the carotid body to changes in arterial PO_2 and may be considered as a key transmitter released by chemoreceptor cells to activate endings of the sinus nerve afferent fibres. This short review discusses the data obtained in experiments *in vivo* and *in vitro* supporting this hypothesis.

2. Evidence

2.1. ATP Induces Changes in Central Respiratory Drive When Applied into the VLM or onto the Ventral Surface of the Medulla

The effects of P2 receptor agonists and antagonists on central respiratory drive have been examined in anaesthetised and artificially ventilated rats. ATP, the stable ATP analogue α,β-methylene ATP (α,β-meATP) and P2 receptor blocker suramin were injected into the rostral VLM, an area of the medulla which plays a key role in the generation and patterning of respiration. The effect of this treatment on resting phrenic nerve discharge was investigated (Thomas *et al.*, 2001). Unilateral microinjection of ATP into the VLM was shown to reduce in a dose-dependent manner, or to abolish totally, phrenic nerve activity. This effect of ATP was mimicked by microinjection of α,β-meATP into the same brainstem site. Whilst suramin had no effect on resting respiration it blocked the effects of ATP microinjected into the VLM.

By contrast, application of ATP onto the ventral surface of the medulla resulted in an immediate increase in phrenic nerve discharge in anaesthetised rats which were either artificially ventilated or spontaneously breathing (Gourine, Dale and Spyer, unpublished observations). This effect of ATP was reduced by suramin and was augmented by the adenosine receptor blocker 8-phenyltheophylline.

These data indicated that ATP-mediated purinergic signalling may play a role in the medullary mechanisms of respiratory control. An important observation was that CO_2-evoked augmentation of respiration can be mimicked by application of exogenous ATP to the ventral medullary surface—the primary central chemosensitive site. However, studies involving microinjection or application of ATP are not sufficient to determine its functional role. The latter can be revealed by experiments in which P2 receptors in the VLM are blocked by specific antagonists applied either by microinjection or ionophoretically and the effect of this treatment on basal respiration and respiratory responses to chemosensory stimulation are monitored.

2.2. Blockade of P2 Receptors Within the VLM Attenuates the Respiratory Response Evoked by Hypercapnia

In a series of experiments performed in anaesthetised and artificially ventilated rats the effect of P2 receptor blockade within the VLM on changes in phrenic nerve activity evoked by rising levels of inspired CO_2 have been studied. It was found that P2 receptor antagonist suramin, when microinjected into the VLM, attenuates the increase in phrenic nerve discharge evoked by rising levels of inspired CO_2 (Thomas *et al.*, 1999). A similar effect was observed after desensitization of certain P2X receptors within the VLM with α,β-meATP.

These results provided the first evidence that ATP-mediated purinergic signalling may play an important role in the VLM mechanisms responsible for the increase in ventilation during hypercapnia. However, more detailed electrophysiological studies are required to define this role more precisely.

2.3. Blockade of P2 Receptors Reduces CO_2-Evoked Activation of the VLM Pre-Inspiratory and Inspiratory Neurones

In anaesthetised and artificially ventilated rats the activity of the respiratory neurones within the VLM has been recorded and the effect of P2 receptor antagonists suramin and pyridoxal-5′-phosphate-6-azophenyl-2′,4′-disulphonic acid (PPADS) on changes in the activity of VLM respiratory neurones during hypercapnia has been determined. Microionophoretic application of suramin or PPADS completely blocked hypercapnia-evoked increases in the activity of pre-inspiratory and inspiratory VLM neurones (Thomas and Spyer, 2000).

It was concluded from these data that during hypercapnia, ATP acting on certain P2 receptors is responsible for the increases in activity of medullary pre-inspiratory and inspiratory neurones and, therefore, mediates CO_2-evoked increase in central respiratory output. The subtypes of the P2 receptor responsible for the increases in activity of VLM inspiratory neurones during hypercapnia remain to be determined.

2.4. VLM Respiratory Neurones Express P2X Receptors

Expression of $P2X_1$, $P2X_2$, $P2X_5$ and $P2X_6$ receptor subunits has been demonstrated in the VLM (Kanjhan *et al.*, 1999; Yao *et al.*, 2000; Thomas *et al.*, 2001; Yao *et al.*, 2003).

To characterise the profile of P2X receptor subunits on individual VLM respiratory neurones we determined in particular whether VLM respiratory neurones express $P2X_1$ and $P2X_2$ receptor subunits. $P2X_2$ receptor subunit immunoreactivity was detected in ~50% of expiratory neurones and in ~20% of neurones with inspiratory-related discharge: pre-inspiratory and inspiratory (Gourine *et al.*, 2003). In contrast, no identified VLM respiratory neurones were detectably immunoreactive for the $P2X_1$ receptor subunit (Gourine *et al.*, 2003). Microionophoretic application of ATP increased the activity of ~80% of expiratory neurones and of ~30% of VLM neurones with inspiratory-related discharge and these effects were abolished by the P2 receptor blocker suramin (Gourine *et al.*, 2003).

These data indicate that a significant proportion of VLM neurones that display rhythmic respiratory-related activity express $P2X_2$ receptor subunits and increase their discharge in response to exogenously-applied ATP. These observations also suggested that activation of the $P2X_2$ receptors during hypercapnia may contribute to the increases in activity of the VLM respiratory neurones that express these receptors.

However, the unexpected finding of that study was that in the VLM only about 20% of inspiratory neurones contain $P2X_2$ immunoreactivity, indicating that the proportion of VLM neurones with inspiratory-related discharge that express $P2X_2$ receptor subunit is smaller than the proportion of these cells that are excited during hypercapnia. An earlier study showed that CO_2 excited 85% of inspiratory and 66% of pre-inspiratory neurones in the area of rostral VLM (Thomas and Spyer, 2000). These data taken together led us to an obvious conclusion that P2X receptors that contain $P2X_2$ subunits are unlikely to be the sole factor responsible for CO_2-induced excitation of the VLM pre-inspiratory and inspiratory neurones. We suggest that P2X receptors other than, or in addition to, $P2X_2$, or P2Y receptors may be involved in mediating hypercapnia-induced changes in the activity of VLM neurones with inspiratory-related discharge.

2.5. ATP is Released Rapidly on the Ventral Surface of the Medulla Oblongata During Hypercapnia

The hypothesis of purinergic signalling involvement in chemosensory transduction within the VLM has recently received its strongest supporting evidence with our direct real-time measurement of ATP release during hypercapnia.

In anaesthetised and artificially ventilated rats we have demonstrated, using amperometric enzymatic ATP biosensors (Llaudet *et al.*, 2003), that hypercapnia induced an immediate and profound increase in the concentration of ATP on the surface of the VLM (Gourine, Llaudet, Dale and Spyer, unpublished observations). This increase in ATP concentration coincided with the enhancement in the amplitude of the phrenic nerve discharge. Interestingly, ATP release from the ventral surface was also observed during hypercapnia in peripherally chemodenervated animals (vagi, aortic and carotid sinus nerve sectioned). Further *in vitro* experiments revealed that CO_2-induced acidification (decrease in pH from 7.4 to 7.0) evoked ATP release from horizontal slices that contained the ventral surface of the medulla (Gourine, Llaudet, Dale and Spyer, unpublished observations).

2.6. Hypothesis: A Role for ATP in Central Chemosensory Transduction

The evidence obtained to date in our laboratories suggest that ATP is released into the extracellular fluid in the VLM during hypercapnia due to activation of central chemoreceptors. We propose that during hypercapnia a rapid increase in extracellular concentration of ATP and its action on P2 receptors localised in the close proximity to the VLM surface is responsible for the increases in activity of medullary inspiratory neurones and, therefore, for the overall augmentation of the central respiratory output. The cellular sources of ATP released during hypercapnia as yet remain to be identified.

2.7. ATP Induces Rapid Activation of the Carotid Sinus Nerve Chemosensory Afferents

The effects of ATP and its stable analogue α,β-meATP, on sinus nerve activity in the *in vitro* mouse carotid body/sinus nerve preparations have been determined (Rong *et al.*, 2003). It was found that both ATP and α,β-meATP evoked an immediate and marked increase in the carotid sinus nerve discharge. This effect of ATP was blocked by P2 receptor antagonist PPADS. Analysis of the chemosensory afferent responses evoked by ATP and

α,β-meATP in preparations taken from the wildtype mice and from the mice deficient in either $P2X_2$, $P2X_3$, or both receptor subunits, suggested that sinus nerve terminals contain functional homomeric $P2X_2$ and $P2X_3$ receptors as well as heteromeric $P2X_{2/3}$ receptors (Rong *et al.*, 2003).

2.8. $P2X_2$ Receptor Subunit Deficiency Results in an Attenuation of the Ventilatory Response to Hypoxia

Adult conscious mice lacking $P2X_2$, $P2X_3$ or both P2X subunits (Cockayne *et al.*, 2000; 2002) were exposed to graded levels of hypoxia and the respiratory rate and tidal volume were monitored by whole body plethysmography. Whilst the resting ventilation was identical in all groups of animals, during hypoxia $P2X_2$ subunit deficient mice showed markedly diminished ventilation as compared to their wild type counterparts (Rong *et al.*, 2003). Interestingly the ventilatory response to hypoxia was not affected by the deficiency of the $P2X_3$ receptor subunit.

These data provide clear evidence that P2X receptors that contain $P2X_2$ receptor subunit, with or without $P2X_3$ subunit, play a crucial role in the mechanisms responsible for the increase in ventilation during hypoxia. As in adult mammals arterial PO_2 levels are monitored primarily by the carotid bodies it was imperative to determine whether carotid body function is affected by the $P2X_2$ receptor subunit deficiency.

2.9. $P2X_2$ Receptor Subunit Deficiency Impairs Carotid Body Function

To investigate the functional role of P2X receptors in the carotid body, we recorded electrophysiologically the activity of the carotid sinus nerve in the carotid body/sinus nerve preparations taken from mice deficient in $P2X_2$, $P2X_3$ or both subunits and from their respective wild type control mice (Rong *et al.*, 2003). These experiments revealed that hypoxia-induced increases in the carotid sinus nerve discharge were markedly reduced in mice lacking $P2X_2$ subunit and was even smaller in mice deficient in both $P2X_2$ and $P2X_3$ receptor subunits. However, the afferent responses of the sinus nerve to hypoxia were not affected in mice lacking $P2X_3$ receptor subunits.

These results were consistent with the ventilatory observations described above and clearly indicated that in the carotid body ATP acting via P2X receptors that contain $P2X_2$ receptor subunit is a key mediator responsible for the hypoxia-induced increase in the carotid sinus nerve activity.

2.10. Blockade of P2 Receptors Abolishes Hypoxia-Induced Activation of the Carotid Sinus Nerve Afferent Fibres

We further suggested that if oxygen-sensitive carotid body elements release ATP as the main transmitter to stimulate sinus nerve afferent terminals via interactions with P2X receptors, then P2X receptor antagonists should attenuate the increase in sinus nerve activity evoked by hypoxia. This was indeed the case. The P2 receptor antagonist PPADS reversibly reduced sinus nerve background discharge, and markedly attenuated increase in firing induced by hypoxia. At a high concentration (100 μM), PPADS induced profound and irreversible reduction in the background activity, and abolished the hypoxia-induced increase in afferent discharge (Rong *et al.*, 2003).

These results were in accord with the data obtained using carotid body/sinus nerve preparations taken from the $P2X_2$ receptor subunit deficient mice and supported the hypothesis that in the carotid body ATP may indeed act as a main transmitter to activate sinus nerve afferent fibres during hypoxia. However, the presence of the P2X receptors in the mouse carotid body remained to be documented immunohistochemically.

2.11. $P2X_2$ and $P2X_3$ Receptor Subunits are Present on the Afferent Terminals of the Carotid Sinus Nerve

Consistent with previous studies in rats (Prasad *et al.*, 2001), we have found that in mice both $P2X_2$ and $P2X_3$ receptor subunit immunoreactivities are present in the carotid body (Rong *et al.*, 2003). Interestingly, the patterns of staining for $P2X_2$ and $P2X_3$ receptor subunits were similar, and when observed under the confocal microscope both appeared to be on the afferent terminals surrounding individual glomus cells or their clusters (Rong *et al.*, 2003).

2.12. ATP is Released Rapidly in the Carotid Body During Hypoxia

In the isolated carotid body/sinus nerve preparation of the rat and mouse we have shown using amperometric enzymatic biosensors (Llaudet *et al.*, 2003) that a decrease in PO_2 in the perfusate induces rapid release of ATP from the carotid body (Gourine, Rong, Llaudet, Dale and Spyer, unpublished observations).

2.13. Hypothesis: A Role for ATP in Peripheral Chemosensory Transduction

Our data indicate that in the carotid body, P2X receptors that contain the $P2X_2$ subunit play a crucial role in transmitting information about PO_2 levels in the arterial blood, and are therefore essential for a normal ventilatory response to hypoxia. Given that $P2X_2$ and $P2X_3$ immunoreactivities were detected on sinus nerve terminals rather than on the glomus cells, we suggest that during hypoxia, oxygen-sensing glomus cells release ATP as the key mediator to activate afferent terminals of the sinus nerve via interaction with P2X receptors that contain the $P2X_2$ subunit, with or without $P2X_3$ subunit.

3. Conclusion: A Unifying Hypothesis

The results obtained in our laboratories over the last years and discussed above provide convincing evidence in favour of ATP-mediated purinergic signalling having a pivotal role in chemosensory control of the respiratory function. Given the available evidence we propose a unifying hypothesis of central and peripheral chemosensory transduction which involves ATP as a common key mediator.

Indeed, ATP applied to the ventral surface of the medulla (the primary central chemosensitive site) evokes a rapid increase in phrenic nerve activity, while ATP applied to the carotid body (the primary peripheral chemosensitive site) evokes rapid excitation of the carotid sinus nerve afferents. Blockade of the P2 receptors within the VLM attenuates the respiratory response evoked by hypercapnia, while blockade of the P2X receptors in the carotid body (or their elimination in the knockout mice) greatly diminishes the ventilatory response to hypoxia and impairs carotid body function. Ionotropic, P2X receptors for ATP are expressed by the VLM respiratory neurones as well as by the peripheral chemosensory afferent neurones, which relay information to the brainstem. Finally, ATP is released during chemosensory stimulation: from the ventral surface of the medulla during hypercapnia and from the carotid body during hypoxia. ATP is released at these sites at the right time and in sufficient quantities to evoke adaptive changes in ventilation. The cellular sources and the mechanisms underlying release of ATP in response to a decrease in PO_2 and/or increase in $PCO_2/[H^+]$ as yet remain to be investigated in detail.

4. Acknowledgements

The experimental work described in this paper was supported by the Biotechnology and Biological Sciences Research Council (UK) and the Wellcome Trust (UK).

References

Cockayne, D., Dunn, P.M., Burnstock, G., and Ford, A., 2002, Generation and electrophysiological characterization of $P2X_2$ and $P2X_2/P2X_3$ knockout (KO) mice. *Neurosci. Abstr.* **52**: 12.

Cockayne, D., Hamilton, S.G., Zhu, Q.M., Dunn, P.M., Zhong, Y., Novakovic, S., Malmberg, A.B., Cain, G., Berson, A., Kassotakis, L., Hedley, L., Lachnit, W.G., Burnstock, G., McMahon, S.B., and Ford, A.P., 2000, Urinary bladder hyporeflexia and reduced pain-related behaviour in $P2X_3$-deficient mice. *Nature* **407**: 1011–1015.

Daly, M. DeB., 1997, Peripheral arterial chemoreception and respiratory-cardiovascular integration. In *Monograph for the Physiological Society*. Oxford University Press, Oxford.

Gonzalez, C., Almaraz, L., Obeso, A., and Rigual, R., 1994, Carotid body chemoreceptors: from natural stimuli to sensory discharges. *Physiol. Rev.* **74**: 829–898.

Gourine, A.V., Atkinson, L., Deuchars, J., and Spyer, K.M., 2003, Purinergic signalling in the medullary mechanisms of respiratory control in the rat: respiratory neurones express the $P2X_2$ receptor subunit. *J. Physiol.* **552**: 197–211.

Heeringa, J., Berkenbosch, A., de Goede, J., and Olievier, C.N., 1979, Relative contribution of central and peripheral chemoreceptors to the ventilatory response to CO_2 during hyperoxia. *Respir. Physiol.* **37**: 365–379.

Kanjhan, R., Housley, G.D., Burton, L.D., Christie, D.L., Kippenberger, A., Thorne, P.R., Luo, L., and Ryan, A.F., 1999, Distribution of the $P2X_2$ receptor subunit of the ATP-gated ion channels in the rat central nervous system. *J. Comp. Neurol.* **407**: 11–32.

Llaudet, E., Botting, N.P., Crayston, J.A., and Dale, N., 2003, A three-enzyme microelectrode sensor for detecting purine release from central nervous system. *Biosens. Bioelectron.* **18**: 43–52.

Loeschcke, H.H., 1982, Central chemosensitivity and the reaction theory. *J. Physiol.* **332**: 1–24.

Nattie, E., 1999, CO_2, brainstem chemoreceptors and breathing. *Prog. Neurobiol.* **59**: 299–331.

North, R.A., 2002, Molecular physiology of P2X receptors. *Physiol. Rev.* **82**: 1013–1067.

Prabhakar, N.R., 2000, Oxygen sensing by the carotid body chemoreceptors. *J. Appl. Physiol.* **88**: 2287–2295.

Prasad, M., Fearon, I.M., Zhang, M., Laing, M., Vollmer, C., and Nurse, C.A., 2001, Expression of $P2X_2$ and $P2X_3$ receptor subunits in rat carotid body afferent neurones: role in chemosensory signalling. *J. Physiol.* **537**: 667–677.

Rong, W., Gourine, A.V., Cockayne, D.A., Xiang, Z., Ford, A.P.D.W., Spyer, K.M., and Burnstock, G., 2003, Pivotal role of nucleotide P2X_2 receptor subunit of the ATP-gated ion channel mediating ventilatory responses to hypoxia. *J. Neurosci.* 23: 11315–11321.

Thomas, T., Ralevic, V., Bardini, M., Burnstock, G., and Spyer, K.M., 2001, Evidence for the involvement of purinergic signalling in the control of respiration. *Neuroscience* **107**: 481–490.

Thomas, T., Ralevic, V., Gadd, C.A., and Spyer, K.M., 1999, Central CO_2 chemoreception: a mechanism involving P2 purinoceptors localized in the ventrolateral medulla of the anaesthetized rat. *J. Physiol.* **517**: 899–905.

Thomas, T., and Spyer, K.M., 2000, ATP as a mediator of mammalian central CO_2 chemoreception. *J. Physiol.* **523**: 441–447.

Yao, S.T., Barden, J.A., Finkelstein, D.I., Bennett, M.R., and Lawrence, A.J., 2000, Comparative study on the distribution patterns of P2X_1-P2X_6 receptor immunoreactivity in the brainstem of the rat and the common marmoset (*Callithrix jacchus*): association with catecholamine cell groups. *J. Comp. Neurol.* **427**: 485–507.

Yao, S.T., Gourine, A.V., Spyer, K.M., Barden, J.A., and Lawrence, A.J., 2003, Localisation of P2X_2 receptor subunit immunoreactivity on nitric oxide synthase expressing neurones in the brain stem and hypothalamus of the rat: a fluorescence immunohistochemical study. *Neuroscience* **121**: 411–419.

Brainstem NHE-3 Expression and Control of Breathing

Martin Wiemann, Heidrun Kiwull-Schöne, Stilla Frede, Dieter Bingmann, and Peter Kiwull

1. Introduction

Intracellular acidification by selective inhibition of the Na^+/H^+ exchanger type 3 (NHE-3) has been shown to enhance the bioelectric activity of CO_2/H^+ sensitive neurons cultured *in vitro* from the ventrolateral medulla oblongata of newborn rats (Wiemann et al., 1998, 1999; Wiemann and Bingmann, 2001). Recently, we demonstrated NHE-3 immunoreactive neurons in brainstem areas with prevalence for central chemosensitivity in adult rabbits (Kiwull-Schöne et al., 2001). During anaesthesia, NHE-3 inhibition by the brain-permeant substance S8218 (Aventis Pharma) significantly lowered the arterial threshold PCO_2 for central apnea upon mechanical hyperventilation, both under normal blood gas conditions and after prolonged respiratory acidosis (Kiwull-Schöne et al., 2001, 2003). This could be of clinical importance, since there is evidence that increased Na^+/H^+ antiporter (NHE) activity may predispose patients to sleep apnea (Tepel et al., 2000).

Now, we studied medullary NHE-3 mRNA in rabbits by RT-PCR, to explore whether brainstem NHE-3 plays a role for the adjustment of individual arterial base-line PCO_2 ($PaCO_2$) during wakefulness.

2. Methods

Experimental procedures *in vitro* concerning the conditions for organotypic culturing, measurement of intracellular pH (pHi), recording from ventrolateral neurons under normo- and hypercapnic conditions, as well as the characteristics of S1611 have all been described (Wiemann et al., 1999). To stain NHE-3 expressing cells a mouse monoclonal anti rat NHE-3 antibody (Chemicon, MAB3134, diluted 1:100) was applied to paraformaldehyde (PFA) fixed cultures (4% PFA in 0.1 M sodium phosphate buffer). A biotinylated anti-mouse IgG

Martin Wiemann, Stilla Frede, Dieter Bingmann • Department of Physiology, University of Essen, Essen, Germany D-45122. **Heidrun Kiwull-Schöne, Peter Kiwull** • Department of Physiology, Ruhr-University, Bochum, Germany D-44780.

Post-Genomic Perspectives in Modeling and Control of Breathing, edited by Jean Champagnat, Monique Denavit-Saubié, Gilles Fortin, Arthur S. Foutz, Muriel Thoby-Brisson. Kluwer Academic/Plenum Publishers, 2004.

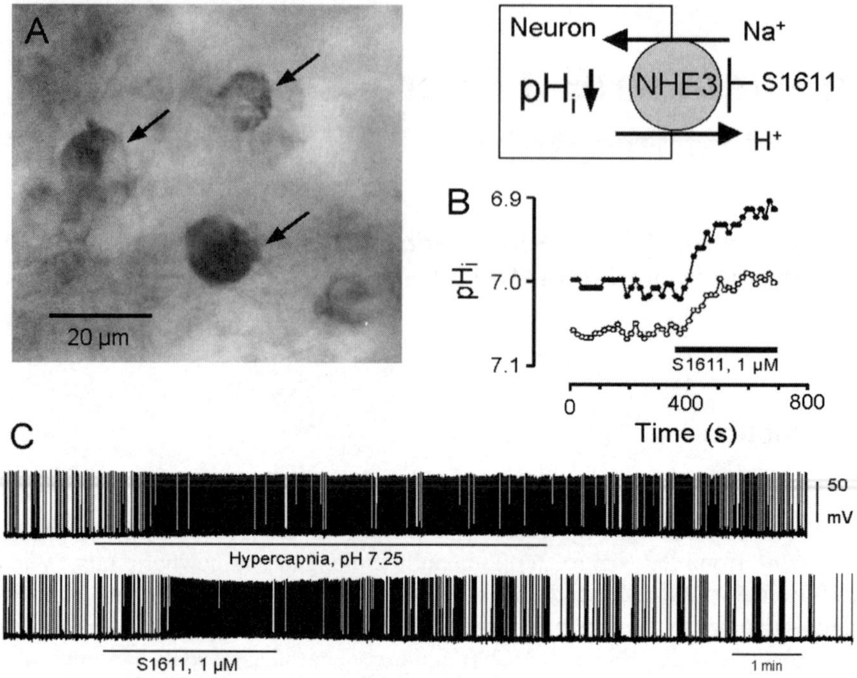

Figure 1. Prevalence and function of the Na^+/H^+ exchanger type 3 (NHE-3) *in vitro*. A drop of pHi is to be expected upon inhibition of ongoing NHE-3 activity (inset upper right). (A) NHE-3 immunoreactive neurons of the ventrolateral medulla oblongata (arrows) detected in organotypic cell cultures from new-born rats. (B) Intracellular acidification of two chemosensitive neurons by the NHE-3 inhibitor S1611. (C) Similar responses of neuronal discharge to either hypercapnia or pharmacological NHE-3 inhibition.

and the ABC kit (Vector Laboratories) were used to obtain a dark color reaction based on peroxidase and 3,3′-diaminobenzidine as a chromogene (Fig. 1A).

Brainstem NHE-3 expression was assessed by standard RT-PCR, using primers specific for rabbit NHE-3 mRNA to obtain a 645-bp DNA fragment (Kiwull-Schöne et al., 2001). PCR products were visualized under UV light and bands were quantified from digital images after background subtraction. By this, optical signals were linearly converted into arbitrary units. NHE-3 expression was normalized to the expression of glycerine-aldehyde-phosphate-dehydrogenase (GAPDH) which was determined in parallel according to published protocols (Platzer et al., 1994).

Healthy adult male rabbits (2.4–3.9 kg) were kept in a metabolic cage on standard alkali-rich pellet feed. Food and water were given ad libitum. Daily food consumption, water intake and urine excretion were supervised. Using local skin anaesthesia, blood was taken from the central ear artery while the animals inhaled oxygen enriched air ($F_IO_2 \sim 0.60$). Samples were analyzed for oxygen and CO_2 partial pressures (PaO_2, $PaCO_2$) and pH (pHa) by conventional equipment (ABL 5 Radiometer, Denmark). Anaesthesia was performed by intravenous sodium pentobarbital (about 60 mg/kg) before animals were sacrificed to quickly remove and snap freeze brainstem tissue from the obex region.

3. Results

3.1. Prevalence and Function of NHE-3 in Newborn Rat Cell Cultures

Figure 1A demonstrates the incidence of NHE-3 in the ventrolateral brainstem of new-born rats, namely NHE-3 immunoreactive neurons. In this ventrolateral area of the brainstem many neurons show a sustained increase of their bioelectric activity in response to CO_2. Likewise, these neurons can be acidified by specific NHE-3 inhibition by 0.05–0.1 pH (Fig. 1B). This is within the physiological range and leads to an increase in action potential frequency which strongly resembles the response to hypercapnia (Fig. 1C).

3.2. Prevalence and Function of NHE-3 in Adult Rabbits

Since the described *in vitro* findings suggest a significant role for NHE-3 in central respiratory chemosensitivity, the NHE-3 expression was studied in the brainstem of rabbits, whose blood gases were determined during wakefulness. Figure 2 shows the inter-individual variability of NHE-3 mRNA, compared to the uniform results for the reference marker. Mean values and ranges are given by Table 1. Within these ranges, the ratio NHE-3/GAPDH was significantly proportional to $PaCO_2$ and inversely correlated with PaO_2 (Fig. 3). No significant correlations were found between brainstem NHE-3 expression and either arterial pH or (calculated) actual bicarbonate (HCO_3^-a).

4. Discussion

This study shows for the first time that brainstem NHE-3 expression (assessed by RT-PCR) not only varies among different animal individuals but also that NHE-3 levels correlate with arterial PCO_2 and PO_2 values. Because measurements of breathing parameters

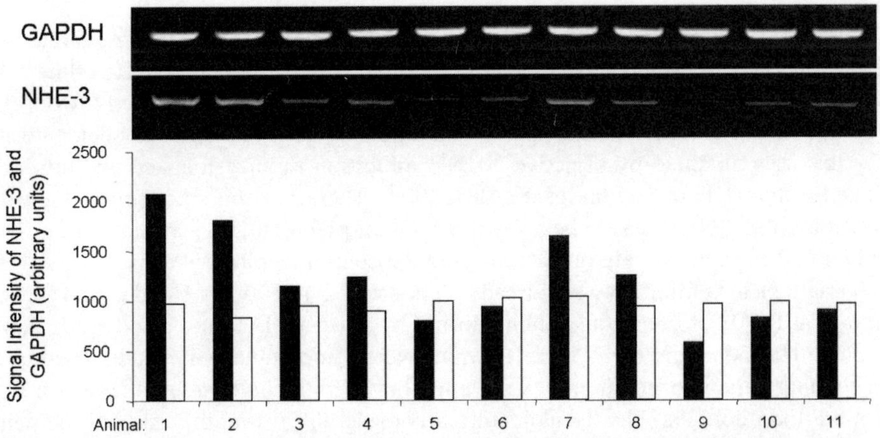

Figure 2. Variability of the expression of the Na^+/H^+ exchanger type 3 (NHE-3) in the brainstem of rabbits. For 11 animals original RT-PCR products and their quantifications are shown of NHE-3 (filled bars) and of the reference gene GAPDH (empty bars). For printing reasons, the optical signal of GAPDH was amplified.

Table 1. Means and ranges for NHE-3, blood gases and acid-base data.

Variable	Means ± SEM	Range
PaO_2 [mmHg]	211.4 ± 16.2	122.2–306.0
$PaCO_2$ [mmHg]	36.4 ± 1.3	30.3–45.8
pHa	7.428 ± 0.011	7.350–7.470
HCO_3^-a [mM]	23.5 ± 0.9	17.4–27.5
NHE-3 [units]	1202 ± 141	573–2076
GAPDH [units]	960 ± 17	836–1033
NHE-3/GAPDH	1.27 ± 0.16	0.60–2.17

Data for 11 rabbits during wakefulness breathing oxygen enriched air ($F_IO_2 \sim 0.60$).

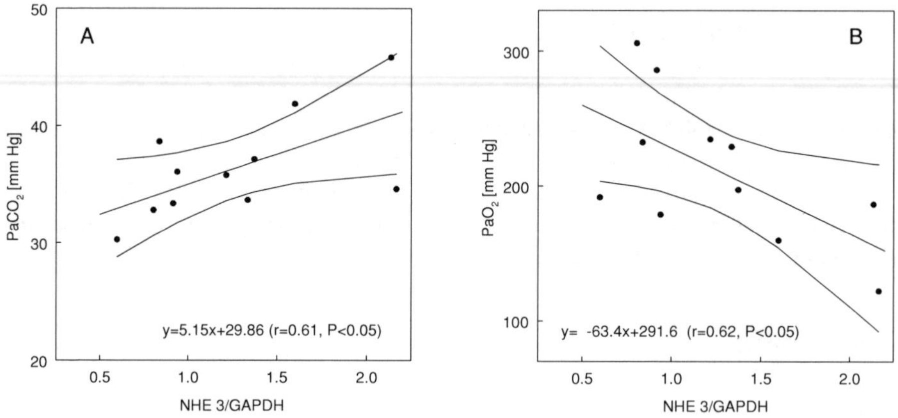

Figure 3. Significant correlation between brainstem NHE-3 expression and arterial blood gases in awake rabbits. Arterial CO_2 (A) and O_2 (B) partial pressure as linear functions of the NHE-3 over GAPDH ratio.

may suffer from the use of anaesthetics, it is of major importance to note that the variations of NHE-3 expression measured here could be correlated with the spontaneous base-line values of $PaCO_2$ during wakefulness: Animals with low NHE-3 expression showed low levels of $PaCO_2$ and vice versa. This finding appears analogous to the reduction in apnoeic threshold $PaCO_2$ that was obtained by selective NHE-3 inhibition in anaesthetized and artificially ventilated rabbits (Kiwull-Schöne et al., 2001, 2003). Therefore, this study further highlights the possible role of NHE-3 as to be a key molecule of pHi regulation within neurons being actively involved in the setting or in modifying the central respiratory drive.

Variations in ventilation were already demonstrated in 1968 by Honda, who analysed the end-tidal PCO_2 of conscious rabbits during hyperoxia. He found standard deviations of ~4.9 mmHg and ranges of ~15 mmHg which were comparable to our study. Thereby, the corresponding values for pulmonary ventilation between the highest and the lowest PCO_2 were more than doubled, clearly indicating a considerable variability also of the central respiratory controller. Inter-individual variations of end-tidal and arterialized capillary PCO_2 have been investigated also in humans breathing normal room air (Crosby and Robbins, 2003). After considering a great number of possible causal factors, such as diurnal changes in vigilance, metabolic rate, acid-base condition or variations in respiratory and renal controller functions, the authors concluded that renal and respiratory controllers for pH were not independent. This may be of special importance with respect to NHE-3 which is the main sodium-proton-exchanger of the proximal tubulus and, thus, fundamentally involved in acid-base regulation of the kidney (Schultheis et al., 1998). Further studies should clarify whether NHE-3 is equivalently regulated in brainstem and kidney.

There is growing evidence from studies *in vitro* and *in vivo* that the intracellular pH of ventrolateral and other brainstem neurons is the adequate signal to elicit respiratory responses (Ritucci et al., 1998; Wiemann et al., 1998, 1999, 2001; Nattie et al., 2002). Accordingly, pHi has been termed the proximate stimulus of central CO_2/H^+ receptors (Filosa et al., 2002) to express its central transmitting role. There are still unknown signal transduction processes comprising cytoplasmic factors as well as potassium and/or calcium channels which are certainly important for the firing behaviour of neurons (Wiemann and Bingmann, 2001). It appears plausible that even a partly individual orchestration of all these constituents may underlie inter-individual differences in respiratory drive and ventilation. However, the finding that a single gene product such as NHE-3 clearly correlates with baseline ventilation, suggests a causal relationship between steady state pHi and pHi regulation of central NHE-3 expressing neurons and ventilation.

In our study, a contribution of the peripheral chemoreflex controller can be excluded, since the animals inhaled oxygen enriched air, eliciting arterial PO_2 values considerably above the normoxic level of 100 mmHg. Nevertheless, also these hyperoxic values of PaO_2 showed a significant inverse dependence of brainstem NHE-3 expression. Although the possible role of metabolic rate and acid base regulation of the kidney should not be neglected, it is much more likely that the NHE-3 expression in central chemosensitive areas primarily determines pulmonary ventilation with corresponding changes in $PaCO_2$ and PaO_2.

Together, our findings suggest that the sodium/proton exchanger subtype 3 in brainstem chemosensitive neurons is at least one of the factors determining the set point of the respiratory control system and thus causally related to the inter-individual variation of ventilation in awake rabbits.

5. Acknowledgment

This article is dedicated to Professor Dr. Yoshiyuki Honda (1926–2003) with deep respect in memory of his life-work on the control of breathing.

References

Crosby, A., and P. A. Robbins, 2003, Variability in end-tidal PCO_2 and blood gas values in humans. *Exp. Physiol.* **88**:603–610.

Filosa, J. A., Dean, J. B., and R. W. Putnam, 2002, Role of intracellular and extracellular pH in chemosensitive response of rat locus coeruleus neurones. *J. Physiol.* **541**:493–509.

Honda, Y., 1968, Ventilatory response to CO_2 during hypoxia and hyperoxia in awake and anaesthetized rabbits. *Respir. Physiol.* **5**:279–287.

Kiwull-Schöne, H., Wiemann, M., Frede, S., Bingmann, D., Wirth, K. J., Heinelt, U., Lang, H. J., and P. Kiwull, 2001, A novel inhibitor of the Na^+/H^+ exchanger type 3 activates the central respiratory CO_2 response and lowers the apneic threshold. *Am. J. Resp. Crit. Care Med.* **164**:1303–1311.

Kiwull-Schöne, H., Wiemann, M., Frede, S., Bingmann, D., and P. Kiwull, 2003, Tentative role of the Na^+/H^+ exchanger type 3 in central chemosensitivity of respiration. *Adv. Exp. Med. Biol.* **536**:415–421.

Nattie, E., Li, A., Meyerand, E., and J. F. Dunn, 2002, Ventral medulla pHi measured in vivo by ^{31}P NMR is not regulated during hypercapnia in anesthetized rat. *Respir. Physiol. & Neurobiol.* **130**:139–149.

Platzer, C., Ode-Kahn, S., Reinke, P., Döcke, W.-D., Ewert, R., and H.-D.Volk, 1994, Quantitative PCR analysis of cytokine transcription patterns in peripheral mononuclear cells after anti-CD3 rejection therapy using two novel multispecific competitor fragments. *Transplantation* **58**:264–268.

Ritucci, N. A., Chambers-Kersh, L., Dean, J. B., and R. W. Putnam, 1998, Intracellular pH regulation in neurons from chemosensitive and nonchemosensitive areas of the medulla. *Am. J. Physiol.* **275**:R1152–R1163.

Schultheis, P. J., Clarke, L. L., Meneton, P., Miller, M. L., Soleimani, M., Gawenis, L. R., Riddle, T. M., Duffy, J. J., Doetschman, T., Wang, T., Giebisch, G., Aronson, P. S., Lorenz, J. N., and G. E. Schull, 1998, Renal and intestinal absorptive defects in mice lacking the NHE3 Na^+/H^+ exchanger. *Nat.Genet.* **19**:282–285.

Tepel, M., Sanner, B.M., van der Giet, M., and W. Zidek, 2000, Increased sodium-proton antiporter activity in patients with obstructive sleep apnoea. *J. Sleep Res.* **9**:285–291.

Wiemann, M. and D. Bingmann, 2001, Ventrolateral neurons of medullary organotypic cultures: intracellular pH regulation and bioelectric activity. *Respir. Physiol.* **129**:57–70.

Wiemann, M., Baker, R. E., Bonnet, U., and D. Bingmann, 1998, CO_2-sensitive medullary neurons: activation by intracellular acidification. *NeuroReport* **9**:167 170.

Wiemann, M., Schwark, J.-R., Bonnet, U., Jansen, H.W., Grinstein, S., Baker, R. E., Lang, H.-J., Wirth, K., and D. Bingmann, 1999, Selective inhibition of the Na^+/H^+ exchanger type 3 activates CO_2/H^+-sensitive medullary neurones. *Pflügers Arch.* **438**:255–262.

Functional Connection From the Surface Chemosensitive Region to the Respiratory Neuronal Network in the Rat Medulla

Yasumasa Okada, Zibin Chen, Wuhan Jiang, Shun-ichi Kuwana, and Frederic L. Eldridge

1. Introduction

Although the ventral medullary surface (VMS) has previously been assumed to be the main site for central respiratory chemoreception in mammals, it has recently been argued that chemosensitive sites are widespread in the lower brainstem, including the deep medullary regions[1–3] (also see the reviews[4–6]). However, *c-fos* immunocytological studies have shown that CO_2-activated cells are primarily located in the superficial ventral medulla.[7–9] We have also recently suggested that the small cells surrounding fine vessels in the most superficial layer of the ventral medulla are the CO_2 chemoreceptor cells.[10] We also assume that there must be a neural connection from the VMS to the respiratory rhythm and pattern generating neuronal network, e.g., to the ventral respiratory group (VRG) region of the medulla. Here we analyze the respiratory output responses to local electrical or chemical stimulation at various sites in the VMS to establish a functional connection from the VMS to the respiratory neuronal network.

Yasumasa Okada • Department of Medicine, Keio University Tsukigase Rehabilitation Center, Tsukigase 380-2, Izu-City, Shizuoka-ken 410-3215 Japan. **Zibin Chen** • Department of Biochemical and Analytical Pharmacology, GlaxoSmithKline, Research Triangle Park, North Carolina 27709 USA. **Wuhan Jiang** • Lineberger Cancer Center, University of North Carolina, Chapel Hill, North Carolina 27599 USA. **Shun-ichi Kuwana** • Department of Physiology, Teikyo University, Tokyo 173-8605 Japan. **Frederic L. Eldridge** • Department of Cell and Molecular Physiology, University of North Carolina, Chapel Hill, North Carolina 27599 USA.

Post-Genomic Perspectives in Modeling and Control of Breathing, edited by Jean Champagnat, Monique Denavit-Saubié, Gilles Fortin, Arthur S. Foutz, Muriel Thoby-Brisson. Kluwer Academic/Plenum Publishers, 2004.

2. Methods

2.1. Preparations

Adult Sprague-Dawley rats (n = 12) were anesthetized first with diethyl ether inhalation, and then placed in a supine position on a stereotaxic holder. The femoral vein and femoral artery were catheterized for infusion and blood pressure measurement, respectively. A mixture of urethane (500 mg/kg) and chloralose (10 mg/kg) was administered intravenously, and the trachea had been cannulated. The animal was vagotomized, peripheral chemodenervated, paralyzed with gallamine triethiodide (3 mg/kg, intravenous injection), and artificially ventilated with 100% O_2. End-tidal PCO_2 ($P_{ET}CO_2$) was continuously monitored with an infrared CO_2 analyzer. The ventral medulla was exposed as previously described.[11] The phrenic nerve was exposed in the neck, immersed in a paraffin pool, and placed on a bipolar silver electrode. The phrenic neurogram was half-wave rectified, integrated and recorded.

The surgical procedure for the isolated brainstem-spinal cord preparation has been described elsewhere.[12,13] In brief, Sprague-Dawley neonatal rats (1–4 day-old; n = 16) were used. The rat was deeply anesthetized with diethyl ether, and the brainstem with cervical spinal cord was quickly isolated in a dissecting chamber filled with oxygenated mock cerebrospinal fluid (mock CSF). The pons and cerebellum were ablated. The preparation was transferred to a recording chamber and fixed with miniature pins on a silicon rubber base with the ventral side up. The preparation was then superfused at 26°C with oxygenated mock CSF, which was equilibrated with a gas mixture (2% CO_2 in O_2; mock CSF pH = 7.8). With this "alkaline" superfusate, the tissue pH of the superficial (<400 m) medullary layer was maintained in the physiological range.[14] The composition of the mock CSF was (in mM): 125 NaCl, 4 KCl, 2 $CaCl_2$, 1 $MgSO_4$, 0.5 NaH_2PO_4, 26 $NaHCO_3$ and 30 glucose. Central respiratory output was recorded from the C4 ventral roots with a glass suction electrode and integrated. The respiratory frequency was counted, based on the C4 burst frequency.

2.2. Electrical Stimulation

Electrical stimulation experiments were conducted only in *in vivo* preparations. The $P_{ET}CO_2$ was adjusted to between 36 and 38 mmHg. A metal microelectrode (tip diameter 1 m) mounted on a micromanipulator was inserted into the ventral medullary tissue in various regions, and the microelectrode tip was placed in the most superficial layer (at a depth of 50 m from the surface). Stimulation trains (intensity 2–10 V, duration 0.5 ms, frequency 25–50 Hz) were delivered for 60–90 sec through the microelectrode, and the phrenic responses to electrical stimulation were analyzed. Each response was judged as strongly positive or weakly positive, when either respiratory frequency or the amplitude of integrated phrenic neurogram increased more than 10% and 5%, respectively. At some positive response sites, we compared the responses induced by stimulation at various depths (at 50, 500, 1000 and 2000 m deep from the surface).

2.3. Chemical Stimulation

Chemical stimulation experiments were conducted in both *in vivo* and *in vitro* preparations. In the *in vivo* experiments, the animal was hyperventilated and $P_{ET}CO_2$ was maintained near the apneic threshold (about 28 mmHg) so that even weak chemical stimulation could evoke a respiratory response. Mock CSF equilibrated with 100% CO_2 was microinjected through a glass micropipette (tip diameter 3–4 m) with a pneumatic injector (Picospritzer-II, General Valve, Fairfield, NJ) at various sites in the superficial layer (at a depth of 50 m from the surface) of the ventral medulla. We controlled the microinjected volume (1–2 nl) by observing the change of the meniscus of the intra-pipette solution. In the *in vitro* experiments, chemical stimulation was conducted as in the *in vivo* experiments by microinjecting mock CSF (volume 3–4 nl) equilibrated with 100% CO_2 into the superficial ventral medulla. Larger volumes (10–50 nl) were injected when testing the dose-response relationship. The response to chemical stimulation was judged as positive in both *in vivo* and *in vitro* experiments when either respiratory frequency or the amplitude of integrated phrenic activity increased more than 10%, or when respiratory activity appeared in apneic preparations.

3. Results

3.1. Electrical Stimulation of VMS Regions

Increases in phrenic activity were elicited when we stimulated various sites in the superficial midline, parapypramidal and ventrolateral medullary regions (Fig. 1). The response pattern was characterized by a gradual increase in the amplitude of the integrated phrenic activity during electrical stimulation and a gradual recovery to the control level

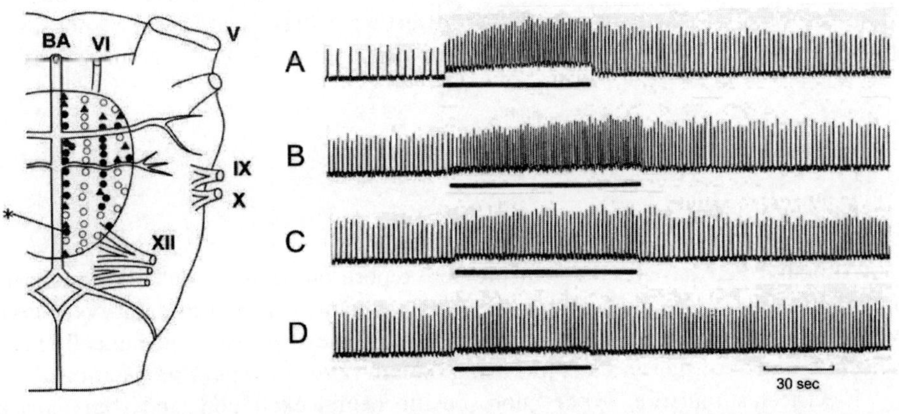

Figure 1. Responses of phrenic activity to electrical stimulation of the VMS in the *in vivo* adult rat. At the midline medullary region shown as *, the effect of stimulation at various depths from the surface was analyzed, and responses **A**, **B**, **C** and **D** correspond to depths of 50, 500, 1000 and 2000 m from the surface, respectively. The responses were stronger when stimulation was conducted at more superficial sites. Closed circles: sites with strongly positive responses. Triangles: sites with weakly positive responses. Open circles: sites with negative responses. Horizontal Bars: duration of electrical stimulation.

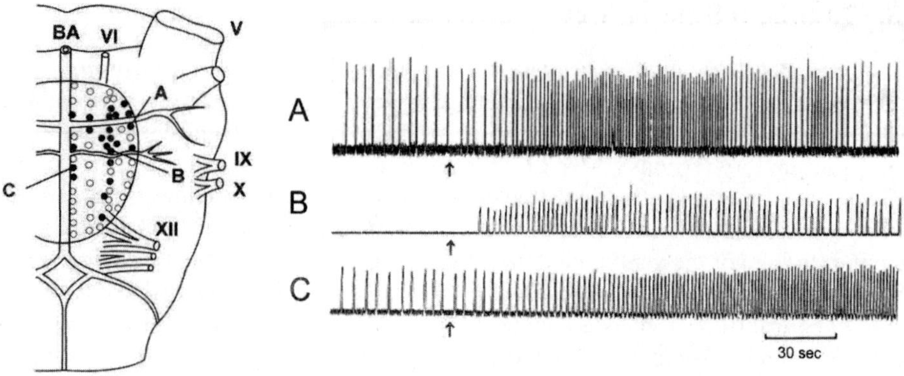

Figure 2. Responses of phrenic activity to microinjection of CO_2 enriched mock CSF into the VMS in the *in vivo* rat. The rat was hyperventilated and maintained near the apneic threshold. In trial **B**, the rat was apneic before the microinjection. Closed circles: sites with positive responses. Open circles: sites with negative responses. Arrows: injection. Injection sites **A**, **B** and **C** correspond to the recordings **A**, **B** and **C**, respectively.

after withdrawal of the stimulus, similar to the after-discharge elicited by carotid sinus stimulation.[15] The responses were stronger when we stimulated the more superficial sites (Fig. 1).

3.2. Chemical Stimulation of VMS Regions

When we chemically stimulated the VMS regions by microinjection of CO_2-enriched mock CSF at various ventral medullary sites in *in vivo* and *in vitro* rats, we found that respiratory output increased when the superficial midline, parapyramidal and ventrolateral regions (especially at the rostral medullary level) were stimulated. These responses were commonly observed in both *in vivo* and *in vitro* preparations (Figs. 2 and 3).

4. Discussion

4.1. Critique of Methods

Although there have been several previous reports on electrical[16–18] and chemical[19,20] stimulation of the VMS, we have for the first time compared the ventilatory responses to electrical and chemical stimulation of the VMS sites. We also have systematically mapped the positive response sites in the VMS using both *in vivo* and *in vitro* preparations.

Electrical stimulation evokes non-specific neural excitation, and stimulates both the cell body and nerve fibers passing around the tip of the stimulating microelectrode. Therefore, we cannot conclude from our electrical stimulation data either that positive response sites are chemosensitive or that there are chemoreceptor cell bodies at positive response sites. However, electrical stimulation has an obvious advantage over other chemical stimulation techniques; the stimulant does not diffuse out and it is possible to define the stimulated site only at the tip of the stimulating electrode.[21]

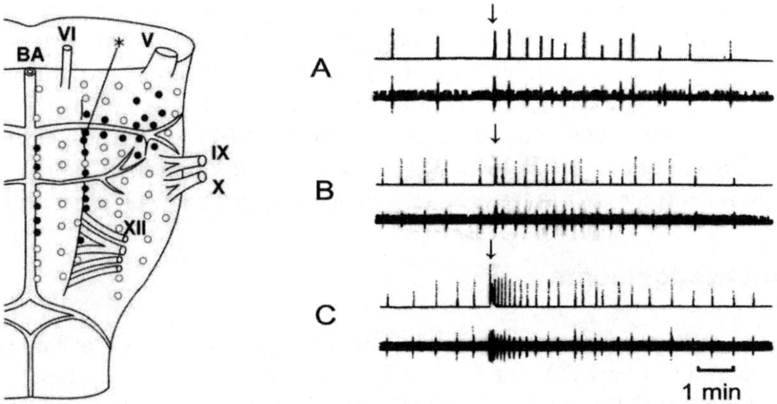

Figure 3. Responses of C4 activity to microinjection of CO_2-enriched mock CSF into the VMS in the *in vitro* preparation. Closed circles: positive response sites with an increase in respiratory frequency. Open circles: negative response sites. Arrows: injection. At the site shown by *, responses to chemical stimulation with various injection volumes were compared to confirm the dose-response relationship. Injection volumes were at **A** 4 nl, **B** 10 nl and C 50 nl.

Chemical stimulation with CO_2 has the advantage that CO_2 is the most physiological stimulant when stimulating respiratory chemoreceptors. However, in the *in vivo* animal experiments, injected CO_2 will rapidly diffuse to the surrounding tissue and injected CO_2 may be partially transported via blood circulation to discrete regions. Therefore, we cannot rule out the possibility that the injected CO_2 has stimulated regions at a distance from the injected site. Indeed, in one trial of the stimulation experiments of the midline regions, we have observed a delayed response (Fig. 2C). This delayed increase of respiratory activity might be induced by stimulation of distant chemosensitive regions by transported CO_2. Therefore, we have conducted experiments using an *in vitro* brainstem-spinal cord preparation to rule out possible CO_2 transportation through blood circulation. The results obtained from these *in vivo* and *in vitro* experiments were in close agreement, including the positive responses in the midline region.

4.2. Significance of Electrical and Chemical Stimulation of VMS

In the present study, we electrically and chemically stimulated the VMS. We observed that stimulation of the midline, parapyramidal and ventrolateral regions induced respiratory augmentation and that sites with positive responses to chemical stimulation were located preferentially in the rostral medulla, while sites that were responsive to electrical stimulation were distributed more evenly throughout the rostro-caudal axis. These results confirm the presence of a functional connection from the VMS to the respiratory neuronal network and also suggest that CO_2 chemoreceptors predominate in the rostral medulla rather than in the caudal medulla. This idea is compatible with our recent suggestion that CO_2 chemoreceptor cells are located in the midline, parapyramidal and ventrolateral medullary regions of the rostral VMS.[10]

There have been several anatomical reports that identify a neuronal connection between VMS and the VRG region using tract-tracing techniques.[22,23] It also has been reported that respiratory neurons that were stained intracellularly showed their dendritic projection to the VMS.[24] Kawai et al.[25] observed that dendrites of CO_2-excitable neurons reached the VMS, but those of CO_2-inhibitory or CO_2-insensitive neurons did not. The results in the present study are compatible with these previous morphological studies and reinforce the idea that a neural connection exists from the VMS to the respiratory neuronal network. Further studies will be necessary to identify the connection from superficially located chemoreceptor cells to the deep respiratory neuronal network at the cellular level.

5. Acknowledgements

This study was supported by a Grant from the Japanese Ministry of Education, Science, Sports and Culture (YO), the Specific Diseases Grant from the Japanese Ministry of Health and Welfare (YO) and the USPHS MERIT Award Grant HL-17689 (FLE).

References

1. E. L. Coates, A. Li, and E. E. Nattie, Widespread sites of brain stem ventilatory chemoreceptors, *J. Appl. Physiol.* **75,** 5–14 (1993).
2. E. E. Nattie, and A. Li, Central chemoreception in the region of the ventral respiratory group in the rat, *J. Appl. Physiol.* **81,** 1987–1995 (1996).
3. C. Solomon, N. H. Edelman, and M. H. O'Neal 3rd., CO_2/H^+ chemoreception in the cat pre-Bötzinger complex *in vivo*, *J. Appl. Physiol.* **88,** 1996–2007 (2000).
4. H. H. Loeschcke, Central chemosensitivity and the reaction theory, *J. Physiol. (Lond.)* **332,** 1–24 (1982).
5. D. E. Millhorn, and F. L. Eldridge, Role of ventrolateral medulla in regulation of respiratory and cardiovascular systems, *J. Appl. Physiol.* **61,** 1249–1263 (1986).
6. Y. Okada, Z. Chen Z, and S. Kuwana, Cytoarchitecture of central chemoreceptors in the mammalian ventral medulla, *Respir. Physiol.* **129,** 13–23 (2001).
7. M. Sato, J. W. Severinghaus, and A. I. Basbaum, Medullary CO_2 chemoreceptor neuron identification by *c-fos* immunocytochemistry, *J. Appl. Physiol.* **73,** 96–100 (1992).
8. N. Larnicol, F. Wallois, P. Berquin, F. Gros, and D. Rose, *C-fos*-like immunoreactivity in the cat's neuraxis following moderate hypoxia or hypercapnia, *J. Physiol. (Paris)* **88,** 81–88 (1994).
9. L. J. Teppema, J. G. Veening, A. Kranenburg, A. Dahan, A. Berkenbosch, and C. Olievier, Expression of *c-fos* in the rat brainstem after exposure to hypoxia and to normoxic and hyperoxic hypercapnia. *J. Comp. Neurol.* **388,** 169–190 (1997).
10. Y. Okada, Z. Chen, W. Jiang, S. Kuwana, and F. L. Eldridge, Anatomical arrangement of hypercapnia-activated cells in the superficial ventral medulla of rats, *J. Appl. Physiol.* **93,** 427–439 (2002).
11. Z. Chen, J. Hedner, and T. Hedner, Substance P-induced respiratory excitation is blunted by delta-receptor specific opioids in the rat medulla oblongata, *Acta Physiol. Scand.* **157,** 165–173 (1996).
12. Y. Okada, A. Kawai, K. Mückenhoff, and P. Scheid, Role of the pons in hypoxic respiratory depression in the neonatal rat, *Respir. Physiol.* **111,** 55–63 (1998).
13. S. Kuwana, Y. Okada, and T. Natsui, Effects of extracellular calcium and magnesium on central respiratory control in the brainstem-spinal cord of neonatal rat, *Brain Res.* **786,** 194–204 (1998).
14. Y. Okada, K. Mückenhoff, G. Holtermann, H. Acker, and P. Scheid, Depth profiles of pH and PO_2 in the isolated brain stem-spinal cord of the neonatal rat, *Respir. Physiol.* **93,** 315–326 (1993).
15. P. G. Wagner, and F. L. Eldridge, Development of short-term potentiation of respiration, *Respir. Physiol.* **83,** 129–139 (1991).
16. H. H. Loeschcke, J. de Lattre, M. E. Schläfke, and C. O. Trouth, Effects on respiration and circulation of electrically stimulating the ventral surface of the medulla oblongata, *Respir. Physiol.* **10,** 184–197 (1970).
17. J. L. Malcolm, I. H. Sarelius, and J. D. Sinclair, The respiratory role of the ventral surface of the medulla studied in the anaesthetized rat, *J. Physiol. (Lond.)* **307,** 503–515 (1980).

18. C. O. Trouth, H. H. Loeschcke, and J. Berndt, Topography of the respiratory responses to electrical stimulation in the medulla oblongata, *Pflügers Arch.* **339,** 153–170 (1973).
19. M. E. Schlaefke, W. R. See, and H. H. Loeschcke, Ventilatory response to alterations of H^+ ion concentration in small areas of the ventral medullary surface, *Respir. Physiol.* **10,** 198–212 (1970).
20. F. G. Issa, and J. E. Remmers, Identification of a subsurface area in the ventral medulla sensitive to local changes in PCO_2, *J. Appl. Physiol.* **72,** 439–446 (1992).
21. C. C. McIntyre, and W. M. Grill, Finite element analysis of the current-density and electric field generated by metal microelectrodes, *Ann. Biomed. Eng.* **29,** 227–235 (2001).
22. C. A. Connelly, H. H. Ellenberger, and J. L. Feldman, Are there serotonergic projections from raphe and retrotrapezoid nuclei to the ventral respiratory group in the rat?, *Neurosci. Lett.* **105,** 34–40 (1989).
23. H. H. Ellenberger, and J. L. Feldman, Origins of excitatory drive within the respiratory network: anatomical localization, *Neuroreport* **5,** 1933–1936 (1994).
24. L. Grelot, A. L. Bianchi, S. Iscoe, and J. E. Remmers, Expiratory neurones of the rostral medulla: anatomical and functional correlates, *Neurosci. Lett.* **89,** 140–145 (1988).
25. A. Kawai, D. Ballantyne, K. Mückenhoff, and P. Scheid, Chemosensitive medullary neurones in the brainstem-spinal cord preparation of the neonatal rat, *J. Physiol. (Lond.)* **492,** 277–292 (1996).

Chemosensory Inputs and Neural Remodeling in Carotid Body and Brainstem Catecholaminergic Cells

Christophe Soulage, Olivier Pascual, Jean-Christophe Roux, Monique Denavit-Saubié and Jean-Marc Pequignot

1. Introduction

Exposure to hypoxia elicits an immediate increase in ventilation in order to face the tissue oxygen deficit. The acute response to hypoxia develops gradually over several days despite a constant level of isocapnic hypoxia, before reaching a steady state level which has been termed ventilatory acclimatization to hypoxia (VAH). The functional acclimatization to hypoxia reveals a striking plasticity of the chemoreflex, which takes place within the first days of exposure and can be prolonged for weeks, months or years. There is clearcut evidence that the peripheral arterial chemoreceptors play a major role in initiating the ventilatory acclimatization to hypoxia. However, this does not preclude a role for central structures involved in the translation of chemosensory inputs and modulating the integration of carotid chemo-afferent inputs. Early and recent studies have shown that the ventilatory plasticity induced by sustained hypoxia is associated with changes in the morphology and phenotype of the carotid chemoreceptors, increases in neurotransmitter biosynthesis and release, modulation of neuroreceptor expression in the carotid body and increased firing rate of the carotid chemo-afferent neurons. More recent studies demonstrated that the neuroplasticity also takes place during long-term hypoxia in restricted areas of the central nervous system, which have been involved in respiratory and sympathetic responses to hypoxia. This short review is devoted to the neurochemical plasticity induced by sustained hypoxia in the carotid body and in brainstem structures involved in translation of the peripheral chemosensory inputs, and their possible role in triggering or modulating ventilatory acclimatization to hypoxia.

Christophe Soulage, Olivier Pascual, Jean-Marc Pequignot • Laboratoire de Physiologie Integrative, Cellulaire et Moléculaire, UMR CNRS 5123, Université Claude Bernard Lyon I, 69 622 VILLEURBANNE cedex. **Olivier Pascual, Monique Denavit-Saubié** • UPR CNRS 2216 NGI, Institut de Neurobiologie Alfred Fessard, 91198 Gif/Yvette, France. **Jean-Christophe Roux** • Neonatal Unit, Karolinska hospital, Q2:07 171 78 Stockholm Sweden.

Post-Genomic Perspectives in Modeling and Control of Breathing, edited by Jean Champagnat, Monique Denavit-Saubié, Gilles Fortin, Arthur S. Foutz, Muriel Thoby-Brisson. Kluwer Academic/Plenum Publishers, 2004.

2. The Carotid Body During Acclimatization to Hypoxia

The VAH is almost exclusively dependent on the carotid bodies as demonstrated by two studies carried out on sheep that showed that VAH is not achieved after chemodenervation (Smith et al., 1986), but can be induced by perfusion of the carotid bodies with hypoxic normocapnic blood (Busch et al., 1985). The fall of arterial oxygen tension in hypoxia induces a cascade of events leading to the associated anatomical, neurochemical and functional changes within the carotid bodies (Figure 1). Major features of chronically hypoxemic carotid bodies are a marked hypertrophy and an enhanced content, turnover and synthesis of catecholamines (dopamine and norepinephrine) that are among the most important neuromodulators stored in the carotid body. In adult male rats living at high altitude (La Paz, Bolivia, 3600 m), the content of norepinephrine and dopamine were respectively 24 and 43-fold higher in the carotid bodies compared to age-matched rats at sea level, while the activity of tyrosine hydroxylase, the rate-limiting enzyme in catecholamine synthesis, was almost 6-fold higher in high altitude native rats (Joseph et al., 2000). The component of carotid body hypertrophy in chronic hypoxia is an increased number and volume of chemosensitive cells, increased number of fibroblasts located in

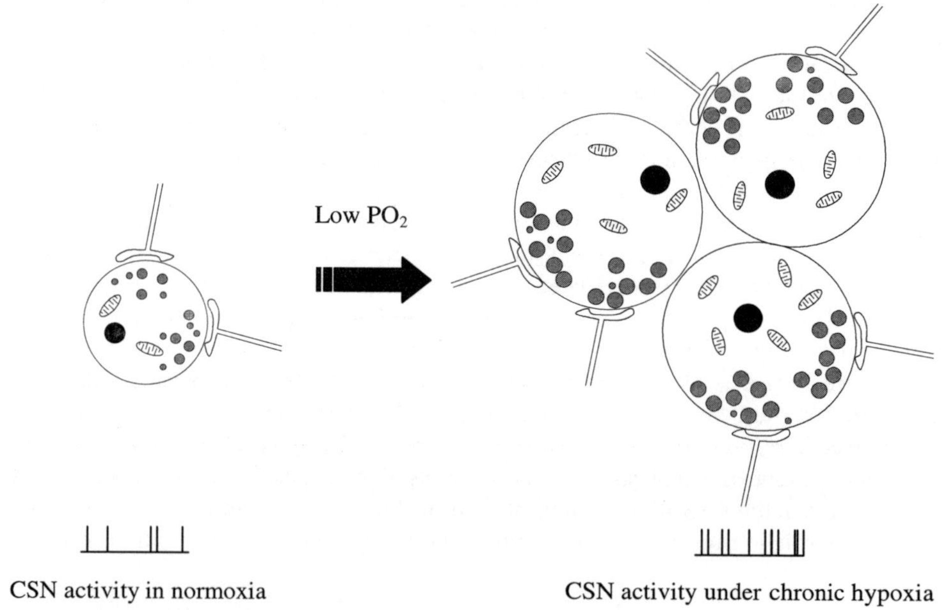

CB glomus cells in normoxia	CB glomus cells in chronic hypoxia
No dividing cells	Hyperplasia & hypertrophia
Low stores and turnover of catecholamines	High stores and turnover of catecholamines
Small and large vesicles	Large vesicles

Figure 1. Schematic drawing illustrating the plasticity of carotid body under chronic hypoxia. CB, carotid body, CSN, carotid sinus nerve.

the conjunctive walls surrounding the carotid body and increase vascularity caused both by vasodilatation and ingrowth of new blood vessels (Bee & Pallot, 1995; Pequignot & Hellström, 1983). Additionally glomic cells are submitted to ultrastructural changes in chronic hypoxia, which include an increased volume density of mitochondria and enlargement of dense-core vesicles that store dopamine and norepinephrine. There is also a gradual transformation of the morphological appearance of glomic cells, that shifts from a bimodal pattern (with both small- and large- vesicles cells) to an unimodal pattern exhibiting mainly large vesicles cells (Pequignot et al., 1984) and a global phenotypic transformation characterized by a very large increase in the proportion of noradrenergic cells (Verna et al., 1993).

Among the numerous neuroactive factors synthesized by carotid body glomic cells, the role of dopamine has been particularly studied during chronic hypoxia. Dopamine is found at high concentrations and has been recognized as a potent inhibitory neuromodulator of carotid body chemotransduction both under acute hypoxia or following VAH (Gonzalez et al., 1994). A proper acclimatization to hypoxia is very important and determines the overall ability to cope with hypoxia. Our own observations in rats inbred at 3600 m above sea level (Bolivian Institute for Altitude Biology—IBBA, La Paz, Bolivia) revealed that following surgical chemodenervation, the hematocrit rises from a baseline level around 50% to almost 70% within a few weeks. These hematological changes were accompanied by increasing mortality. The process of ventilatory control under acute or chronic hypoxic stimulation is gender-dependent (Joseph et al., 2000), with males having a blunted respiratory response under the chronic hypoxic conditions in La Paz. Gonadal steroids are critically involved in this gender discrepancy. Recent findings from our laboratory showed that this effect is dependent on an action of ovarian steroids that stimulate breathing by reducing the inhibitory dopaminergic drive in the carotid body (Joseph et al., 2002).

3. Neurochemical Acclimatization to Hypoxia in the Central Nervous System

In long-term hypoxia, the central translation of peripheral chemoreceptor inputs may be modulated by the plasticity of brainstem cardiorespiratory structures located in the nucleus tractus solitarii (NTS) and the ventrolateral medulla (VLM). In the rat, the afferent chemosensory fibres arising from the carotid body are contained in the carotid sinus nerve (CSN) and project primarily into the NTS within the dorsal medulla caudal to the obex. Chemosensitive neurons of caudal NTS project onto the VLM. The solitary complex and the VLM contain premotor neurons of the dorsal and ventral respiratory groups, respectively. The NTS and VLM also contain the catecholaminergic cell groups, A2C2 and A1C1, respectively, which are adjacent to, or intermingled with the respiratory neurons (Ellenberger et al., 1990; Pilowsky et al., 1990). The respiratory neurons do not synthetize catecholamines (Pilowsky et al., 1990). However, they possess adrenergic receptors (Champagnat et al., 1979) and receive close appositions from TH-immunoreactive neurons (Sun et al., 1994). Carotid body stimulation induces neuronal activation in brainstem catecholaminergic areas. A2 neurons in the caudal NTS are activated by long-term hypoxia, as shown by increases in noradrenaline turnover (Soulier et al., 1992), tyrosine hydroxylase (TH) activity (Soulier et al., 1995) and TH mRNA level (Dumas et al., 1996). The changes in the amount of

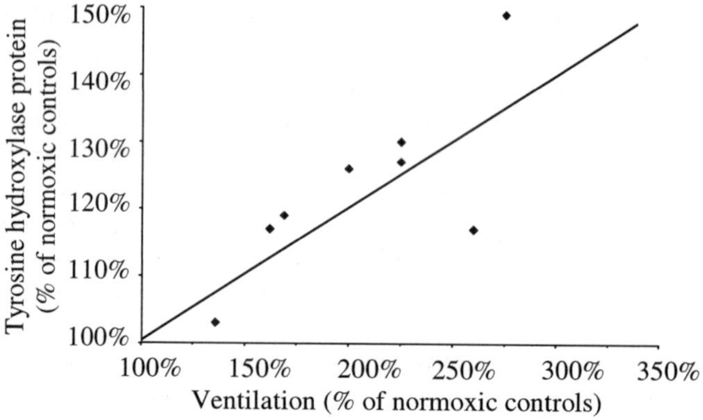

Figure 2. Relationship between the content of tyrosine hydroxylase protein in the caudal subset of A2C2 cell group and ventilation after 14 days of hypoxia. The two parameters measured in the same animal were significantly correlated ($r = 0.76$, $P < 0.01$). Modified from Schmitt et al. (1994).

TH protein correlate with changes in ventilatory output after acclimatization to hypoxia (Schmitt et al., 1994, see Fig. 2).

Carotid-chemodenervated rats can develop ventilatory responses to acute and long-term hypoxia and these functional responses were associated with upregulation of TH in brainstem catecholaminergic neurons (Dumas et al., 1996). These functional and molecular approaches showed that chronically CSN-transected rats can respond to hypoxia, suggesting that central O_2-chemosensitive mechanisms may develop independently of carotid chemosensory inputs to the brainstem (Roux et al., 2000). The brainstem plasticity of TH gene expression during long-term hypoxia in CSN-transected rats seems to be due to properties of the central neurons rather than to increases in synaptic activity due to peripheral influences. Therefore, neurons in the NTS appear not only as the site of integration of chemo-afferent inputs but also as the target for direct modulatory effects of hypoxia. The mechanism by which hypoxia stimulates TH gene expression in the brainstem involves the interaction of transcriptional factors with *cis*-acting regulatory elements within the proximal promoter of the TH gene. The most widespread molecular mechanism for hypoxia-dependent regulation is transcriptional induction via binding of the transcription factor HIF-1α (hypoxia inducible factor 1). In cultured PC12 cells physiological regulation of TH gene transcription by hypoxia has been shown to require activation of HIFs (Schnell et al., 2003). By using *in situ* hybridization of the mRNA encoding HIF-1α and immunocytochemical detection of the HIF-1α protein, we localized the HIF-1α expression in restricted areas of the brainstem of rats exposed to tolerable hypoxia (10% O2). HIF-1α is primarily and selectively expressed under tolerable hypoxia in medullary cardiorespiratory regions, where it could participate in the enhanced metabolic activity associated with the physiological responses to hypoxia (Pascual et al., 2001). The glia surrounding neurons of dorsomedian and ventrolateral cardiorespiratory areas was labelled after 1 hour of hypoxia. This finding suggests that glial cells are the early responsive element of O_2-sensing and then translate the stimulus to adjacent neurons located in cardiorespiratory areas. In these regions a subset of A2C2 and A1C1 catecholaminergic neurons colocalized TH and HIF-1 protein, suggesting that HIF-1 may participate in the control of hypoxia-induced expression of TH in cardiorespiratory structures involved in integration of chemo-afferent inputs.

4. Conclusion

The available evidence suggests a pivotal role for plasticity of catecholaminergic cells in acclimatization to hypoxia. The changes in neuronal noradrenergic function might act as a fine tuning of the increasing ventilatory output in long-term hypoxic rats: a short-term low noradrenergic activity in medullary areas implicated in respiratory control would be a factor favouring the gradual increase in hypoxic ventilatory response, while the delayed high noradrenergic neuronal activity might limit the magnitude of VAH and contribute to the ventilatory steady-state.

5. Acknowledgements

This work was supported by CNRS (PICS 917), University Claude Bernard Lyon I, Région Rhone-Alpes, HFSP n°101/97, ACI-BDPI n°57, DRET n°0034077 and the Ministère de la Recherche.

References

Bee, D., Pallot, DJ., 1995, Acute hypoxic ventilation, carotid body cell division, and dopamine content during early hypoxia in rats. *J. Appl. Physiol.* **79**:1504–1511.

Busch, MA., Bisgard, GE., Forster, HV., 1985, Ventilatory acclimatization to hypoxia is not dependent on arterial hypoxemia. *J. Appl. Physiol.* **58**:1874–1880.

Champagnat, J., Denavit-Saubie, M., Henry, JL., Leviel, V., 1979, Catecholaminergic depressant effects on bulbar repiratory mechanisms. *Brain Res.* **160**:57–68.

Dumas, S., Pequignot, JM., Ghilini, G., Mallet, J., Denavit-Saubie, M., 1996, Plasticity of tyrosine hydroxylase gene expression in the rats nucleus tractus solitarius after ventilatory acclimatization to hypoxia. *Mol. Brain Res.* **40**:188–194.

Ellenberger, HH., Feldman, JL., Zhan, WZ., 1990, Subnuclear organization of the lateral tegmental field of the rat. II: Catecholamine neurons and ventral respiratory group. *J. Comp. Neurol.* **294**:212–222.

Gonzalez, C., Almaraz, L., Obeso, A., Rigual, R., 1994, Carotid body chemoreceptors: from natural stimuli to sensory discharges. *Physiol. Rev.* **74**:829–898.

Joseph, V., Soliz, J., Pequignot, J., Sempore, B., Cottet-Emard, JM., Dalmaz, Y., Favier, R., Spielvogel, H., Pequignot, JM., 2000, Gender differentiation of the chemoreflex during growth at high altitude: functional and neurochemical studies. *Am. J. Physiol. Regul. Integr. Comp. Physiol.* **278**:R806–R816.

Joseph, V., Soliz, J., Soria, R., Pequignot, J., Favier, R., Spielvogel, H., Pequignot, 2002, Dopaminergic metabolism in carotid bodies and high altitude acclimatization in female rats. *Am. J. Physiol. Regul. Integr. Comp. Physiol.* **282**:R765–R773.

Pequignot, JM., Hellström, S., Johansson, C., 1984, Intact and sympathectomized carotid bodies of long-term hypoxic rats: a morphometric ultrastructural study. *J. Neurocytol.* **13**:481–493.

Pequignot, JM., Hellstrom, S., 1983, Intact and sympathectomized carotid bodies of long-term hypoxic rats. A morphometric light microscopical study. *Virchows Arch. A. Pathol. Anat. Histopathol.* **400**:235–243.

Pascual, O., Denavit-Saubie, M., Dumas, S., Kietzmann, T., Ghilini, G., Mallet, J., Pequignot, JM., 2001, Selective cardiorespiratory and catecholaminergic areas express the hypoxia-inducible factor-1a (HIF-1) under *in vivo* hypoxia in rat brainstem. *Eur. J. Neurosci.* **14**:1981–1991.

Pilowsky, PM., Jiang, C., Lipsky, J., 1990, An intracellular study of respiratory neurons in the rostral ventrolateral medulla of the rat and their relationship to catecholamine-containing neurons. *J. Comp. Neurol.* **301**:604–617.

Roux, JC., Pequignot, JM., Dumas, S., Pascual, O., Ghilini, G., Pequignot, J., Mallet, J., Denavit-Saubie, M., 2000, O2-sensing after carotid chemodenervation: hypoxic ventilatory responsiveness and upregulation of tyrosine hydroxylase mRNA in brainstem catecholaminergic cells. *Eur. J. Neurosci.* **12**:3181–3190.

Schmitt, P., Soulier, V., Pequignot, JM., Pujol, JF., Denavit-Saubie, M., 1994, Ventilatory acclimatization to chronic hypoxia: relationship to noradrenaline metabolism in the rat solitary complex. *J. Physiol. (Lond.)* **477**:331–337.

Schnell, PO., Ignacak, ML., Bauer, AL., Striet, JB., Paulding, WR., Czyzyk-Krzeska, MF., 2003, Regulation of tyrosine hydroxylase promoter activity by von Hippel-Lindau tumor suppressor protein and hypoxia-inducible transcription factors. *J. Neurochem.* **85**:483–491.

Smith, CA., Bisgard, GE., Nielsen, AM., Daristotle, L., Kressin, NA., Forster, HV., Dempsey, JA., 1986, Carotid bodies are required for ventilatory acclimatization to chronic hypoxia. *J. Appl. Physiol.* **60**:1003–1010.

Soulier, V., Cottet-Emard, JM., Pequignot, J., Hanchin, F., Peyrin, L., Pequignot, JM., 1992, Differential effects of long-term hypoxia on norepinephrine turnover in brain stem cell groups. *J. Appl. Physiol.* **73(5)**:1810–4.

Soulier, V., Dalmaz, Y., Cottet-Emard, JM., Kitahama, K., Pequignot, JM., 1995, Delayed increase of tyrosine hydroxylation in the rat A2 medullary neurons upon long-term hypoxia. *Brain Res.* **674(2)**:188–95.

Sun, QJ., Pilowsky, P., Minson, J., Arnolda, L., Chalmers, J., Llewellyn-Smith, IJ., 1994, Close appositions between tyrosine hydroxylase immunoreactive boutons and respiratory neurons in the rat ventrolateral medulla. *J. Comp. Neurol.* **340**:1–10.

Verna, A., Schamel, A., Pequignot, JM., 1993, Long term hypoxia increases the number of norepinephrine-containing glomus cells in the rat carotid body: a correlative immunohistochemical and biochemical study. *J. Auton. Nerv. Syst.* **44**:171–177.s

Role of Fe^{2+} in Oxygen Sensing in the Carotid Body

S. Lahiri, A. Roy, J. Li, S.M. Baby, A. Mokashi and C. Di Giulio

1. Introduction

Prolyl hydroxylase is an enzyme which oxidatively modifies the proline residue of hypoxia inducible factor (HIF-1α) in the presence of oxygen in an apparently irreversible reaction. This reaction also requires labile Fe^{2+}, 2-oxogluterate (2-OG) and ascorbic acid. Hypoxia retards this reaction and HIF-1α is accumulated. Similarly, Fe^{2+} chelation mimics hypoxia-like effect. Thus, the enzyme stands at the gateway between hypoxia and normoxia. The hydroxylated HIF-1α undergoes proteasomal degradation during normoxia whereas HIF-1α, accumulated during hypoxia, binds with β-subunits to form HIF-1 which is then transcripted to various genes in the nucleus.

It is remarkable that during normoxia the enzymes is most active and that it is retarded during hypoxia. This event provides a lynchpin for hypoxia response. Previous to this discovery, hypoxic response was thought to start from non-events situation during normoxia. Fe^{2+} chelation during normoxia prevents HIF-1α hydroxylation and proteasomal degradation so that oxygen is not utilized, and it mimics hypoxia. All these effects of prolyl hydroxylase have been worked out using non–excitable cells (Bunns and Poyton, 1996; Maxwell and Ratcliff, 2003; Wang and Semenza, 1993) as shown in Fig. 1.

There are no reports known to us on prolyl hydroxylase effects on excitable cells. In excitable cells, hypoxic effects are expected to be followed by the well known cascade of events: cell membrane K^+currents are suppressed which is followed by cell depolarisation, Ca^{2+} influx, $[Ca^{2+}]_i$ rise, neurotransmitters release and chemosensory response. All these events will take place within less than a few seconds (Lahiri, 2000). The same cascade of events will take place upon iron chelation and the events will be reversed by $FeSO_4$ administration (Daudu et al., 2002). A second event also begins almost simultaneously and grows for up to 10 and more min (Jewell et al., 2002) and that HIF-1α is accumulated as a result of hypoxia or of Fe^{2+} chelation (Table 1).

S. Lahiri, A. Roy, J. Li, S.M. Baby, and A. Mokashi • Dept. of Physiology, University of Pennsylvania Medical Center, Philadelphia, Pennsylvania, USA. **C. Di Giulio** • Dept. of Biomedical Sciences, G. d'Annunzio University, Chieti, Italy.

Post-Genomic Perspectives in Modeling and Control of Breathing, edited by Jean Champagnat, Monique Denavit-Saubié, Gilles Fortin, Arthur S. Foutz, Muriel Thoby-Brisson. Kluwer Academic/Plenum Publishers, 2004.

Table 1. Prolyl Hydroxylase Inhibition by Fe^{2+} Chelation.

Instantaneous effects	Delayed effects
K^+current-suppression →	HIF-1α accumulation →
Cell depolarization → Ca^{2+}gate opens →	Transcription →
Ca^{2+} influx → $[Ca^{2+}]_i$ rise →	Genetic expression
Neurotransmitters release →	
Chemosensory discharge	

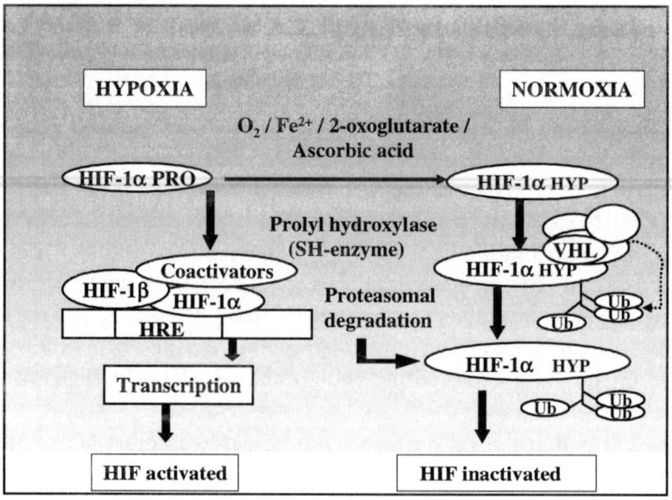

Figure 1. Schematic of prolyl hydroxylase effects on non-excitable cells. HIF-1α hydroxylated in normoxia in the presence of Fe^{2+}, 2-0G and ascorbic acid. The hydroxylated HIF-1α is degradated and hypoxia stabilized the HIF-1α. The same will happen in excitable cells.

Thus, the prediction that prolyl hydroxylase inhibition will result in an instantaneous excitation of glomus cell and a delayed accumulation of HIF-1α were tested according to the above scheme.

2. Methods

These phenomena were documented in the rat glomus cells separated and cultured over a period of hours to days. Freshly isolated glomus cells were used for electrophysiological (Lahiri et al., 2001) and for immunofluorescence studies of HIF-1α (Baby et al., 2004).

3. Result

3.1. Electrophysiological Effects of Labile Iron Chelation in Glomus Cells

3.1.1. K^+ current suppression: K^+-currents were suppressed as shown by voltage clamp experiments.

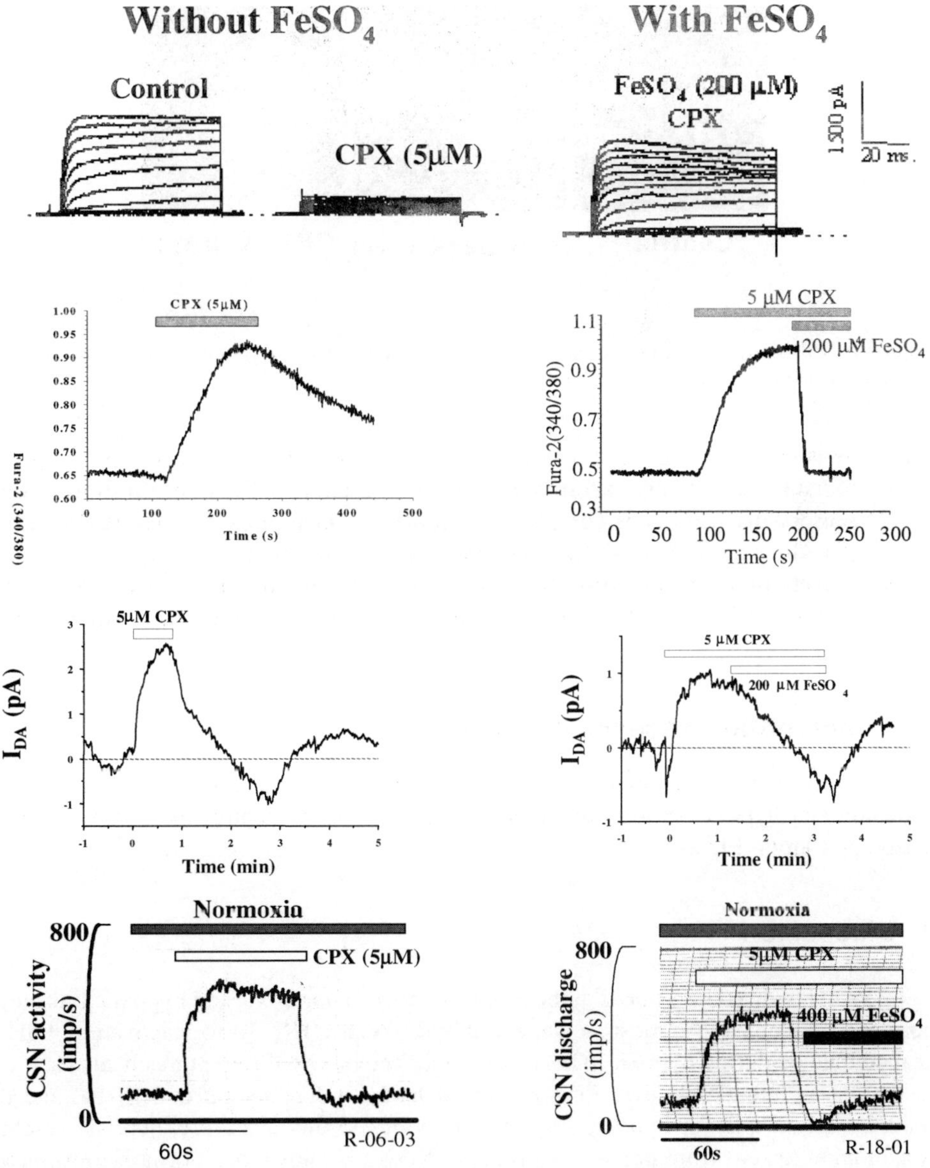

Figure 2. Top to Bottom: K^+Current suppressed. $[Ca^{2+}]_i$ increased. Neurotransmitter release increased and chemosensory discharge is also stimulated by CPX (left-hand column). All the effects by CPX were reversed by $FeSO_4$ (right-hand column).

3.1.2. Cell depolarization: Cells were depolarised by current clamp experiments (not shown).

3.1.3. $[Ca^{2+}]_i$ increase: As a result of iron chelation $[Ca^{2+}]_i$ increased as shown by increased Fura-2 fluorescence measurements (Fig. 2).

3.1.4. Neurotransmitter release: Neurotransmitter release also increased as a result of iron chelation (Fig. 2).

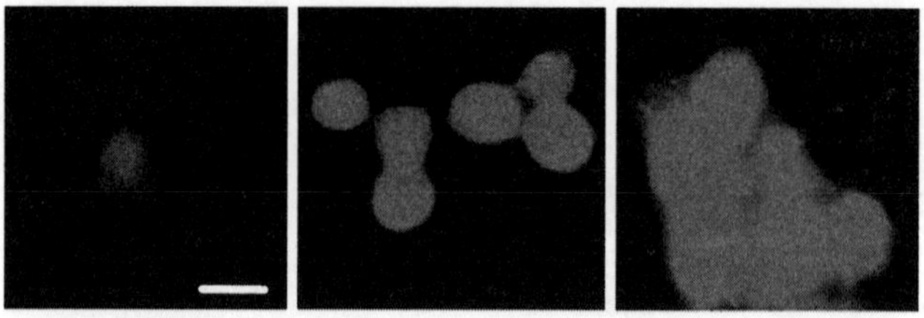

Figure 3. HIF-1α fluorescence increased by 5.0 μM CPX in 10 min. This is comparable to the effects of hypoxia (8 torr) as opposed to the effects of normoxia.

3.1.5. Sensory discharge: Sensory discharge increased (Fig. 2). This stimulation was found to be mutually exclusive to hypoxic stimulation. The hypoxia stimulation of sensory discharge was completely inhibited by ciclopirox olamine (CPX) effects (Daudu et al., 2002). All the effects were dose dependent.

3.1.6. Effects of $FeSO_4$on iron chelators: All the foregoing effects were immediately reversed (Fig. 2), showing that the iron chelation effects were due to specifically for iron.

3.2. Immunofluorescence of HIF-1α

It was shown that immunofluorescence increased in the cells as a result of Fe^{2+} chelation and hypoxia (8 torr) treatment as opposed to the control measurement over a period of 10 min (Fig. 3).

4. Discussion

Prolyl hydroxylase stands at the gateway between normoxia and hypoxia. It is during normoxia, the enzyme is most active, and it hydroxylates HIF-1α to inactivation. HIF-1α is accumulated during hypoxia. This is accomplished by two different mechanisms.

This is remarkable that normoxic event leads to degradation of HIF-1α, and that provides a lynchpin for hypoxic response. Previous to this discovery, hypoxic response was thought to start from non-event situation. Now it is known that normoxic response is retarded during hypoxia. As a result, HIF-1α is accumulated during hypoxia.

These results were known from experiments on non-excitable cells. Our results showed that it also happens in excitable glomus cells (Roy et al., 2004).

In addition, what we have shown here that glomus cells were also excited instantaneously by labile iron chelation just as hypoxia does. This is a new finding, and can be considered at the frontier in oxygen sensing.

Glomus cells require cell permeant iron chelator to inactivate the labile iron. Thus, desferrioxamine which is not cell permeant did not work for glomus cell (Daudu et al., 2003; Ren et al., 2000).

Labile iron in the cell functions as a Fenton reagent and participates in the generation of reactive oxygen species (ROS). Recently, we have shown (Roy et al., 2003) that chelation of glomus cell labile iron by CPX (5 μM) catalytically decreases (approx. 10% of control) the level of ROS measured indirectly by the CM-H_2DCFDA fluorescence. If ROS is involved, its contribution to the oxygen chemoreception was only minimal.

5. Acknowledgements

Supported by HL-R37-43414-14 and HL-R01-50180-15 and N00014-01-1-0948.

References

Baby SM, Roy A, Mokashi A and Lahiri S. Effects of hypoxia and intracellular iron chelation on hypoxia inducible factor-1α and -1ß in the rat carotid body and glomus cells. *Histochemistry and Cell Biology* 120: 343–352, 2003.

Bunn HF and Poyton RO. Oxygen sensing and molecular adaptation to hypoxia. *Physiol Rev* 76: 839–855, 1996.

Daudu PA, Roy A, Rozanov C, Mokashi A and Lahiri S. Extracellular and intracellular free iron and the carotid body responses. *Respir Physiol & Neurobiol* 130: 21–31, 2002.

Jewell UR, Kvietikova, Scheid A, Bauer C, Wenger RH and Gassman M. Induction of HIF1α in response to hypoxia is instantaneous. *FASEB J* 15: 1312–1314, 2001.

Maxwell PH and Ratcliffe PJ. Regulation of HIF-1α by oxygen. In *Oxygen Sensing: Responses and Adaption to Hypoxia.* Marcell Dekker, Inc. Vol. 475 Eds. Lahiri S, Semenza GL and Prabhakar NR. pp. 47–65, 2003.

Lahiri S. Historical perspectives of cellular oxygen sensing and responses to hypoxia. *J Appl Physiol* 88: 1467–1473, 2000.

Lahiri S, Rozanov C, Roy A, Storey B and Buerk DG. Regulation of oxygen sensing in peripheral arterial chemoreceptors. *IJBCB* 33: 755–774, 2001.

Ren X, Dorrington KL, Maxwell PH and Robbins PA. Effects of desferrioxamine on serum erythropoietin and ventrilatory sensitivity to hypoxia in humans. *J Appl Physiol* 89: 680–686, 2000.

Roy A, Buerk DG, Li J, Baby SM and Lahiri S. Intracellular iron chelation in carotid body chemotransduction. In *Oxygen and Cell Symposium*, Berlin, September 6–9, 2003.

Roy A, Li J, Baby SM, Mokashi A, Buerk DG, Lahiri S. Effects of iron-chelators on ion-channels and HIF-1α expression in the carotid body. *Respir Physiol Neurobiol* (in press).

Wang GL and Semenza GL. Characterization of hypoxia-inducible factor 1 and regulation of DNA binding activity by hypoxia. *J Biol Chem* 268: 21513–21518, 1993.

Added in proof

Not only do mitochondrial inhibitors mimic the hypoxic effects, but they also abolish oxygen sensitivity (Mulligan E et al., 1981; Wyatt and Buckler, 2004). Also, mitochondrial respiratory chain is required for HIF-1α accumulation (Agani et al., 2002). Accordingly mitochondrial inhibitors also abolish HIF-1α accumulation and therefore oxygen sensitivity will be lost. This is consistent with the thesis presented here.

Agani FH, pichiule P, Chavez JC and LaManna JC. Inhibitors of mitochondrial complex I attenuates the accumulation of hypoxia-inducible factor-1 during hypoxia in HEP3B cells. *Comp. Biochem. Physiol.* 132: 107–109, 2002.

Mulligan E, Lahiri S, Storey BT. Carotid body O_2 chemoreception and mitochondrial oxidative phosphorylation. *J Appl. Physiol.* 250: H202–207, 1981.

Wyatt CN and Buckler KJ. The effect of mitochondrial inhibitors on membrane currents in isolated neonatal rat carotid body type I cells. *J Physiol.* 556: 175–191, 2004.

Ventilatory Responsiveness to CO_2 Above & Below Eupnea: Relative Importance of Peripheral Chemoreception

Curtis A. Smith, Bruno J. Chenuel, Hideaki Nakayama, and Jerome A. Dempsey

1. Introduction

Sleep apnea is a highly prevalent problem occurring in the general working population at rates of 2–3% in children, 3–7% in middle-aged adults and 10–15% in the healthy elderly (>65 years old).[1,2] While anatomical dimensions or mechanical properties of the upper airway are an important risk factor for sleep apnea, neural control over the magnitude and stability of respiratory motor output to both the upper airway and chest wall pump muscles has also emerged as a major contributor to all types of sleep apnea.[3–7] Chemoreflexes are the most important determinant of respiratory drive during sleep. Given the rapidity with which hypopneas/apneas develop in a typical central and/or "mixed" apnea episode, our recent work has addressed the general hypothesis that carotid chemoreceptors have a dominant role in mediating ventilatory responses to transient increases and decreases in CO_2 as commonly occurs in sleep-disordered breathing.

2. Methods

We used awake or sleeping, chronically instrumented, trained dogs. Pressure support ventilation (PSV) was used to increase V_T and decrease $PaCO_2$ to allow determination of the hypocapnic apneic threshold.[8,9] Steady-state increases in F_ICO_2 were used to test hypercapnic responses. Carotid body (CB) denervation and a vascularly-isolated, intact CB perfusion preparation were used to address the relative contributions of peripheral

---

Curtis A. Smith, Bruno J. Chenuel, Hideaki Nakayama, and Jerome A. Dempsey • The John Rankin Laboratory of Pulmonary Medicine, Department of Population Health Sciences, University of Wisconsin School of Medicine, 504 North Walnut Street, Madison, WI, USA 53726-2368.

Post-Genomic Perspectives in Modeling and Control of Breathing, edited by Jean Champagnat, Monique Denavit-Saubié, Gilles Fortin, Arthur S. Foutz, Muriel Thoby-Brisson. Kluwer Academic/Plenum Publishers, 2004.

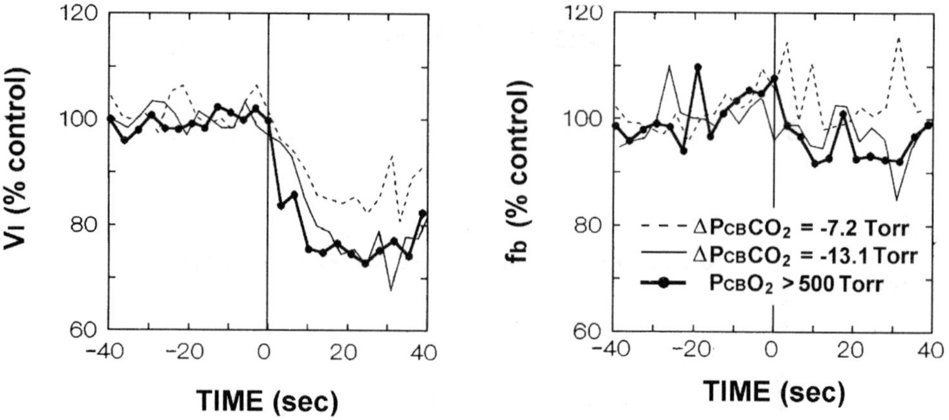

Figure 2. Mean data from 7 dogs in NREM sleep when the carotid bodies were maintained hypocapnic (~7 and 13 Torr < eupnea) or hyperoxic ($P_{CB}O_2$ > 500 Torr). Note the decrease in ventilation proportional to CB hypocapnia and the lack of effect on fb. (From ref. 10, with permission.)

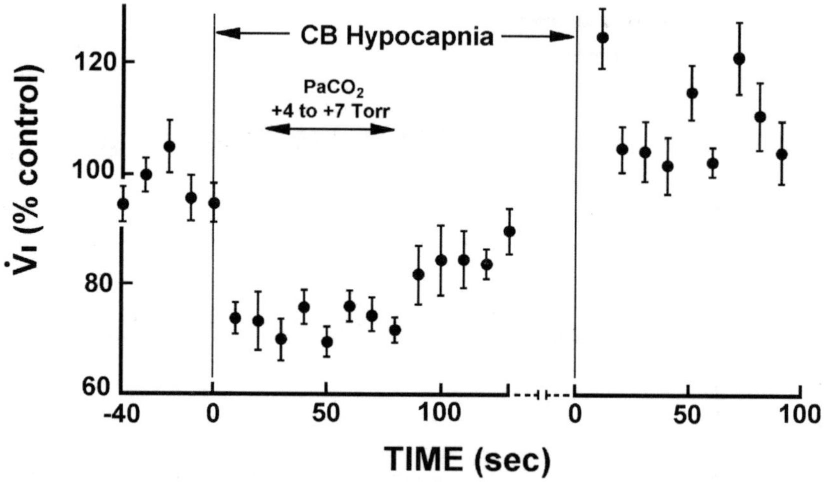

Figure 3. Mean ventilatory data in 10-s intervals for all trials (n = 14) in which blood gases were obtained during prolonged CB perfusions with normoxic and hypocapnic blood (CB PCO_2 range: –12 to –18 mmHg < eupnea) in NREM sleep. Perfusion was started at first vertical line. At right, after break, second vertical line indicates cessation of CB perfusion. The amount of CO_2 retention for the 20 to 80 second interval of prolonged CB hypocapnia is shown above the short double-headed arrow. (From ref. 10, with permission.)

3.4. Relative Contribution of Central vs. Peripheral Chemoreceptors to Hypercapnia

Preliminary data using extracorporeal perfusion to holed the isolated CB normocapnic showed: a) 40% of the steady-state ventilatory response to systemic hypercapnia of >30 seconds duration was attributable to CBs. b) All of the ventilatory response during the on-transient phase (<30 seconds) of hypercapnia was due to carotid chemoreceptors. These data suggest that CBs are responsible for initiating most types of ventilatory overshoots during sleep.

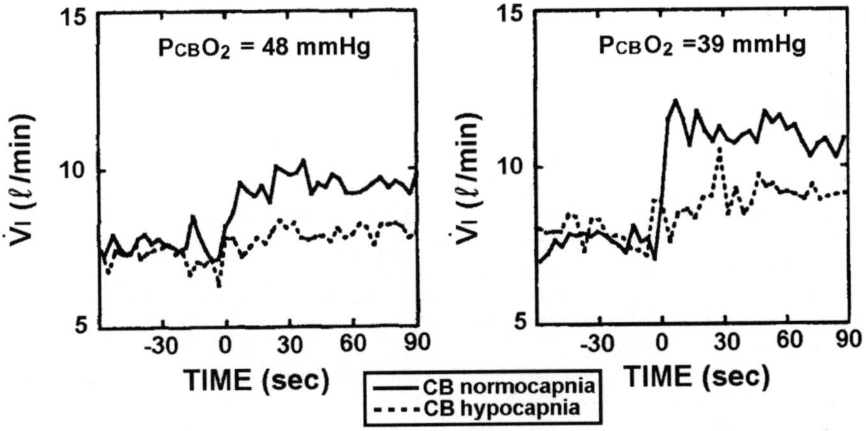

Figure 4. Mean ventilatory responses to mild and severe CB hypoxia against a background of CB normocapnia or hypocapnia in the canine isolated and perfused CB model. CB hypoxia began at time = 0. In both conditions, systemic arterial PCO_2 was allowed to fall as ventilation increased. Note that most, but not all, of the ventilatory response to CB hypoxia was prevented. (From ref. 11, with permission.)

4. Discussion

In summary, we have shown that: A) CB chemoreceptors are required for the apneas normally caused by transient hyperventilation or "ventilatory overshoot". B) Periodic breathing in sleep requires CB chemoreceptors. C) The hypocapnia-induced apneic threshold is 2-fold more sensitive when the CBs are intact. D) CB chemoreceptor hypocapnia alone causes progressive, linear reductions in ventilation below eupnea in wakefulness and NREM sleep. E) CB chemoreceptor hypocapnia causes persistent hypoventilation even in the face of brain respiratory acidosis. F) CB chemoreceptor hypocapnia alone cannot cause apnea. G) CB chemoreceptor hypocapnia alone contributes to much of the inhibitory feedback effects of hypocapnia on ventilation during hypoxia. H) Preliminary evidence to date from the canine isolated and perfused CB model shows that the entire transient (<35 seconds) ventilatory response to hypercapnia was attributable to the CB chemoreceptors.

Taken together, the foregoing is relevant to ventilatory chemoreflex control in sleep apnea because it demonstrates that the CB chemoreceptors' dynamic responsiveness to CO_2 above and below eupnea accounts for most of the ventilatory overshoots and undershoots experienced in sleep-disordered breathing both in normoxia and hypoxia. Further, since transient CB chemoreceptor hypocapnia alone is insufficient to initiate apnea, we propose that an interaction between CB hypocapnia and lung stretch in response to increased V_T during the ventilatory overshoot is required to initiate apnea.

5. Acknowledgements

This research was supported in part by NIH/NHLBI. We wish to thank K.S. Henderson for her many contributions to these studies.

References

1. T. Young, M. Palta, J. Dempsey, J. Skatrud, S. Weber, and S. Badr. The occurrence of sleep-disordered breathing among middle-aged adults. *N Engl J Med* 328: 1230–1235 (1993).
2. T. Young, P. E. Peppard, and D. J. Gottlieb. Epidemiology of obstructive sleep apnea: a population health perspective. *Am J Respir Crit Care Med* 165: 1217–1239 (2002).
3. M. Younes, M. Ostrowski, W. Thompson, C. Leslie, and W. Shewchuk. Chemical control stability in patients with obstructive sleep apnea. *Am J Respir Crit Care Med* 163: 1181–1190 (2001).
4. G. Warner, J. B. Skatrud, and J. A. Dempsey. Effect of hypoxia-induced periodic breathing on upper airway obstruction during sleep. *J Appl Physiol* 62: 2201–2211 (1987).
5. J. A. Dempsey, J. B. Skatrud, A. J. Jacques, S. J. Ewanowski, B. T. Woodson, P. R. Hanson, and B. Goodman. Anatomic determinants of sleep-disordered breathing across the spectrum of clinical and nonclinical male subjects. *Chest* 122: 1–13 (2002).
6. M. S. Badr, F. Toiber, J. B. Skatrud, and J. Dempsey. Pharyngeal narrowing/occlusion during central sleep apnea. *J Appl Physiol* 78: 1806–1815 (1995).
7. M. Younes. Contributions of upper airway mechanics and control mechanisms to severity of obstructive apnea. *Am J Respir Crit Care Med* 168: 645–658 (2003).
8. H. Nakayama, C. A. Smith, J. R. Rodman, J. B. Skatrud, and J. A. Dempsey. Effect of ventilatory drive on CO_2 sensitivity below eupnea during sleep. *Am J Respir Crit Care Med* 165: 1251–1259 (2002).
9. H. Nakayama, C. A. Smith, J. R. Rodman, J. B. Skatrud, and J. A. Dempsey. Carotid body denervation eliminates apnea in response to transient hypocapnia. *J. Appl. Physiol.* 94: 155–164 (2003).
10. C. A. Smith, K. W. Saupe, K. S. Henderson, and J. A. Dempsey. Ventilatory effects of specific carotid body hypocapnia in dogs during wakefulness and sleep. *J Appl Physiol* 79: 689–699 (1995).
11. C. A. Smith, C. A. Harms, K. S. Henderson, and J. A. Dempsey. Ventilatory effects of specific carotid body hypocapnia and hypoxia in awake dogs. *J Appl Physiol* 82: 791–798 (1997).
12. H. Ogawa, A. Mizusawa, Y. Kikuchi, W. Hida, H. Miki, and K. Shirato. Nitric oxide as a retrograde messenger in the nucleus tractus solitarii of rats during hypoxia. *J. Physiol.* 486: 495–504 (1995).
13. A. Liu, J. Kim, J. Cinotte, P. Homolka, and M. T. T. Wong-Riley. Carotid body denervation effect on cytochrome oxidase activity in pre-Botzinger complex of developing rats. *J. Appl. Physiol.* 94: 1115–1121 (2003).
14. A. Serra, D. Brozoski, N. Hedin, R. Franciosi, and H. V. Forster. Mortality after carotid body denervation in rats. *J Appl Physiol* 91: 1298–1306 (2001).

Carotid Body Tumors in Humans Caused by a Mutation in the Gene for Succinate Dehydrogenase D (SDHD)

Albert Dahan, Peter E.M. Taschner, Jeroen C. Jansen, Andel van der Mey, Luc J. Teppema and Cees J. Cornelisse

1. Introduction

Tumors of the carotid bodies (CB) are commonly associated with chronic tissue hypoxia from altitude, cyanotic heart disease and chronic pulmonary disease.[1–5] Here, we describe a hereditary form of carotid body tumors, which is not related to exposure to chronic hypoxia but is related to a missense mutation in the gene that encodes for succinate dehydrogenase D (SDHD). SDHD is a small part of cytochrome b588 of the mitochondrial respiratory chain complex II and an essential enzyme in the Krebs tricarboxylic-acid cycle.[6] These carotid body tumors are part of the hereditary paraganglioma type I (PGL1) syndrome.[7] The PGL1 syndrome is characterized by slowly growing tumors derived from paraganglia in the head and neck area (see Fig. 1) and (Fig. 2, color insert). Paraganglia are cell-clusters of neuro-ectodermal origin that have a close relationship with the autonomic nervous system and have the ability to synthesize catecholamines (*e.g.*, dopamine). The most common PGL tumor locations are the carotid bodies and the adrenal medulla. Other paraganglia which may be affected are: the vagal bodies at the nodose ganglion of the vagal nerve, the tympanic bodies at the promontory of the middle ear, the jugular bodies at the jugular foramen, the laryngeal bodies in the larynx, and the aortic bodies in the wall of the ascending aorta and aortic arch.

2. The Carotid Body

The CBs contain the peripheral chemoreceptors which are an essential part of the ventilatory control system regulating the chemical composition of the arterial blood.[8] They

Albert Dahan, Luc Teppema • Department of Anesthesiology. **Peter Taschner** • Department of Human Genetics. **Jeroen Jansen, Andel van der Mey** • Department of Otorhinolaryngology. **Cees Cornelisse** • Department of Pathology. Leiden University Medical Center, POBox 9600, 2300 RC Leiden, The Netherlands.

Post-Genomic Perspectives in Modeling and Control of Breathing, edited by Jean Champagnat, Monique Denavit-Saubié, Gilles Fortin, Arthur S. Foutz, Muriel Thoby-Brisson. Kluwer Academic/Plenum Publishers, 2004.

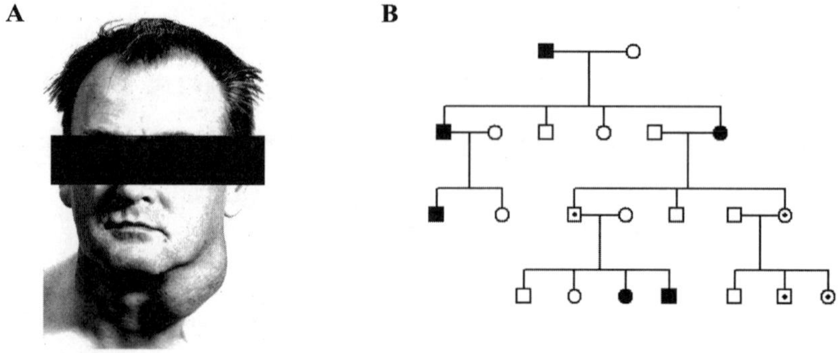

Figure 1. **A**. A male patient with a large carotid body and vagal body tumor. **B**. A pedigree of a family with PGL1. The filled symbols are patients with carotid body tumors; they received the mutated allele from their father. The dotted symbols are imprinted carrier; they received the mutated allele from their mother. Squares are males, circles females.

are strategically located at the bifurcation of the common carotid artery which supplies blood to the brain. The CBs respond to O_2, CO_2 and metabolic acidosis, causing brisk respiratory responses aimed at the supply of O_2 and removal of CO_2. The CBs further sense low glucose, temperature and osmolarity, and there are suggestions that they are involved in the regulation of airway resistance and brain blood flow.

The CBs contain type I and type II cells. The type I cells are involved in O_2-sensing; the type II cells are glia-like sustentacular cells. The cells are arranged in clusters formed by a central core of type I cells surrounded by a shell of type II cells. The full mechanisms of O_2-sensing is still poorly understood. At present it is thought that membrane ion channels are critically involved and that low O_2 inhibits K^+-currents through the CB type I cell membrane, which causes membrane depolarization and consequently the influx of calcium ions and the activation of a complex cascade of events within the type I cell. At the end of the cascade, the cell releases neurotransmitters (ATP, acetylcholine) which activate postsynaptic receptors located on afferent endings of the carotid sinus nerve (branch of the n. IX) that have their cell bodies in the petrosal ganglion with their axons terminating in the nucleus tractus solitarii. The oxygen sensitive K^+-channels may be modulated by other intracellular substances (*e.g.*, radical oxygen species, ATP) and/or organelles (*e.g.*, mitochondria, membrane bound heme containing protein complexes).[9]

3. Genetics, Inheritance and Occurrence of PGL1

Mitochondrial complex II of consists of four subunits (succinate dehydrogenase A, B, C and D, Fig. 3). The gene encoding for SDHD is associated with PGL1. This gene is located on the long arm of chromosome 11. In Dutch founder families a missense mutation causing $Asp^{92} \rightarrow$ Tyr in the SDHD gene product was observed. Similarly, mutations in SDHB and SDHC have also been associated with paragangliomas. In contrast, mutations in the SDHA gene cause Leigh-syndrome, a severe neurological disease.[10]

The inheritance of familiar PGL1-associated paraganglioma is characterized by an autosomal-dominant pattern with strong parent-of-origin effect resulting in generation-skipping upon maternal transmission of the disease allele (Fig. 1B). After van Baars *et al.*[11]

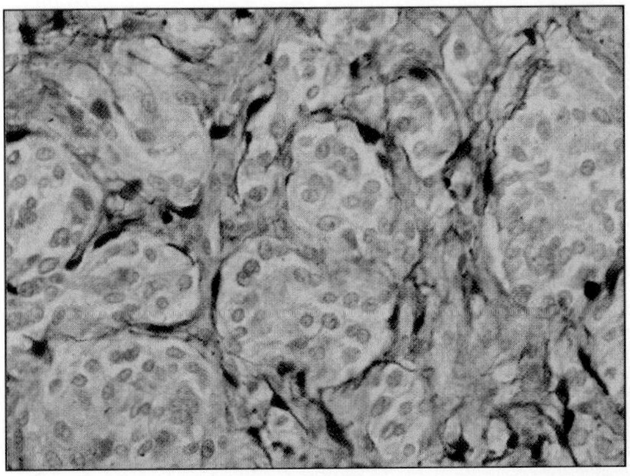

Figure 2. Histology of CB tumor showing typical 'Zellballen'-pattern. The globular clusters of type I cells are surrounded by elongated type II cells that immunohistochemically stain positive for the S-100 marker protein.

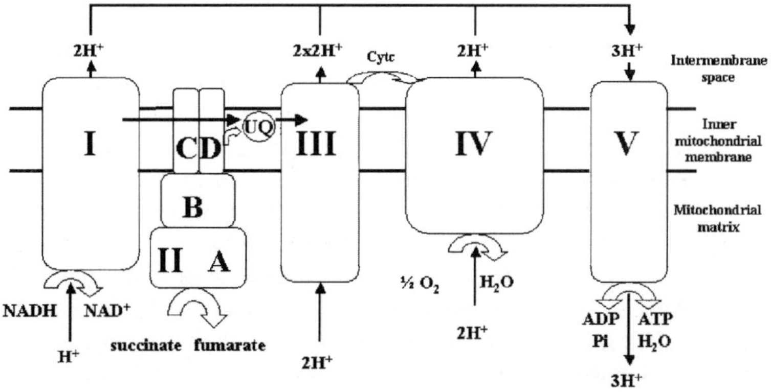

Figure 3. The respiratory chain in the mitochondrial membrane. Complex II (succinate dehydrogynase) is part of the oxidative pathway and oxidizes succinate to fumarate (part of the Krebs tricarboxylic-acid cycle). Complex II transfers an electron from succinate to co-enzyme Q (UQ), that transports the electron to the next complex. The proton gradient in the inter-membranal space is used for the production of ATP.

noticed a preponderance for the male sex in a single large Dutch family, van der Mey *et al.*[12] demonstrated exclusive paternal transmission of the disease in 15 extensive pedigrees and ascribed this phenomenon to genomic imprinting.[13] Genomic imprinting refers to a mono-allelic, reversible parent-of-origin dependent expression of mammalian genes due to methylation of genetic control elements. In a subsequent linkage study of these families, Heutink *et al.*[14] found linkage to a locus on 11q22-23, called PGL1. Mariman *et al.*[15] identified a second, locus PGL2 on 11q13 in a Dutch paraganglioma family which also showed exclusively paternal transmission of the disease. Genomic imprinting, as an epigenetic mechanism for parental influence on gene expression, is considered to play an important role in fetal development, in particular by influencing the transfer of nutrients to the fetus from the mother.[13] In the human genome, a major cluster of imprinted genes, including IGF2 and H19, is located at the chromosome 11p15.5 region but both SDHD and the still unknown gene at the PGL2 locus are located outside this imprinted region on chromosome 11. Sofar, there is no evidence that SDHD is an imprinted gene by itself since bi-allelic expression has been found in a variety of human tissues.[6] Moreover, loss of the wild-type maternal allele is frequently observed in SDHD-linked tumors,[6,16,17] suggesting SDHD expression in normal paraganglia. Interestingly, paragangliomas associated with two other sub-units of the SDH complex, SDHB and SDHC, located on chromosome 1p36 and 1q21, respectively, are transmitted in an autosomal dominant fashion without parent-of origin-effect.[18,19] It is still unclear which mechanism is responsible for the exclusive paternal transmission of SDHD and PGL2-associated paraganglioma although for SDHD, mono-allelic expression as predicted by a classical imprinting model, seems to be unlikely.

4. Link Between Hypoxia and Carotid Body Tumors

Unlike tumor suppressor genes like TP53, APC,Rb1, BRCA1,2 etc., the SDHD protein, as well as the two other complex II subuits SDHB and SDHC, has no known function in growth and cell cycle control, DNA-damage repair signaling or apoptosis. How functional inactivation of these three SDH subunits can initiate tumorigenesis is an intriguing question. In view of the role of the carotid body in hypoxia sensing, and the reported higher incidence of carotid body tumors in high-altitude dwellers, several authors have discussed the possibility that complex II dysfunction could generate a constitutive hypoxic signal, for example by production of reactive oxygen species (ROS) activating a mitogenic pathway.[10,20,21] In line with this hypothesis, Gimenez-Roqueplo *et al.*[22] found increased levels of HIF-1α, HIF-2α/EPAS1, VEGF and VEGFR1 in SDHD-associated phaeochromocytoma. The connection with oxygen-sensing was further explored by Astrom *et al.*[21] by studying genotype-phenotype correlations in relation to altitude and population genetics of PGL1. By comparing the data on the Dutch PGL1 families that live at almost sea-level to those in USA that live at approximately 260 m altitude, they found that subjects who were diagnosed with single tumors at first clinical evaluation lived at lower altitudes than those with multiple tumors. Germline-founder mutations in sporadic patients were also more frequently detected in those living at low altitude. In contrast, the incidence of phaeochromocytomas was increased in persons living at higher altitude. On the basis of these results, the investigators concluded that altitude is a phenotypic modifier in PGL1-associated paraganglioma. They postulate that the low altitudes and thus higher atmospheric pressure in The Netherlands reduces the penetrance and relax the selection pressure on SDHD mutations. This would account for the occurrence of three different Dutch founder mutations. They propose that SDHD mutations result in a defect in O_2-sensing inducing proliferation which is exacerbated at moderately higher altitudes. Furthermore, they found that SDHD nonsense/splicing mutations developed symptoms 8.5 years earlier than missense mutation carriers. They suggest that the earlier clinical manifestation of tumors in carriers of nonsense/splicing mutations might be an effect of the total dissolution of complex II due to the absence of a functional SDHD protein.

A somewhat different view is presented by Eng *et al.*[10] In addition to the the likelihood that generation of ROS may activate a mitogenic pathway, they point to the possibility that mitochondrial dysfunction also may have an inhibitory effect on apoptosis which could equally contribute to tumorigenesis. Of further interest is the finding that germline mutations in fumarate hydratase, the next step in the Krebs tricarboxylic-acid cycle to SDH, are associated with a different tumor spectrum including uterine fibroid, cutaneous leiomyomas and type 2 papillary-renal cell carcinomas.[23] It is still an open question if these two non-overlapping hereditary tumor syndromes, associated with dysfunction of two closely interacting mitochondrial enzymes share a common tumorigenic pathway or reflect a hitherto unknown complex connection between disturbed energy metabolism and tumorigenesis.

5. Blunted Hypoxic Responses as Early Phenotypical Expression of PGL1?

Taken into account all of the above it is of interest to speculate whether indeed there is a defect in the mechanism of oxygen-sensing in type I CB cells of PGL1 patients. In a first approach we tested the ventilatory response to hypoxic pulses (SpO_2 ~85%; duration of hypoxia 3 min) and acute hypercapnia (against a background of normoxia and hypoxia) in a male carrier of the PGL1 mutation who was still without CB tumor (as tested by MRI). His hypoxic sensitivity averaged to 0.24 L/min per % desaturation. Although this contrasts sharply with the mean hypoxic sensitivity of a healthy, sex and aged matched control population (1.2 $L.min^{-1}.\%^{-1}$; $n = 50$), the value of our patient was at the low range of values observed within this population (0.2 to 2.0 $L.min^{-1}.\%^{-1}$). Interestingly, this same subject lacked the characteristic synergistic O_2-CO_2 interaction at the carotid body since he was unable to increase the slope of the ventilatory CO_2 response from normoxia to hypoxia. It is this latter observation which indicates a fundamental defect in the mechanism of oxygen-sensing at the CB of this subject, possibly related to a dyfunction of SDHD. Evidently further studies are needed, for example aimed at the question whether the blunted hypoxic sensitivity of our patient has a similar origin as the blunted hypoxic ventilatory response observed in chronic hypoxia.[24]

6. Concluding Remarks

We believe that patients with the PGL1 syndrome may serve as an important 'model' to elucidate the mechanism of oxygen sensing. Furthermore, the study of this intriguing disease will increase our insight in the link between hypoxia and tumorigenesis and tumor metastasis.

References

1. L. Pacheco-Ojeda, D. Enrique, C. Rodriguez and N. Vivar. Carotid body tumors at high altitude: Quito, Equador, 1987. *World J. Surg.* **12**, 856–860 (1988).
2. M.J. Nissenblatt. Cyanotic heart disease: "low altitude" risk for carotid body tumor? *Johns Hopkins Med. J.* **142**, 18–21 (1978).
3. J.H. Hirsch, F.C. Killien and R.H. Troupin. Bilateral carotid body tumors and cyanotic heart disease. *Am. J. Radiol.* **134**, 1073–1075 (1980).
4. H. Gruber and R. Metson. Carotid body paraganglioma regression with relief of hypoxemia. *Ann. Int. Med.* **92**, 800–802 (1980).
5. J. Herget, C.A. Brown, S. Kutilek, G.R. Barer and F. Palacek. Enlargement of carotid bodies in rats with lung emphysema or silicosis. *Bull. europ. Physiopath. resp.* **18**, 75–79 (1982).
6. B.E. Baysal, R. Ferrell, J. Willet-Brozik, E. Lawrence, D. Myssiorek, A. Bosch, A. van der Mey, P. Taschner, W. Rubinstein, E. Myers, C. Richard, C.J. Cornelisse, P. Devilee and B. Devlin. Mutations in SDHD, a mitochondrial complex II gene, in hereditary paraganglioma. *Science* **297**, 848–851 (2000).
7. J.C. Jansen. *Paragangliomas of the head and neck: Clinical implications of growth rate and genetics* (PhD thesis, Leiden University, Leiden, 2001).
8. A. Dahan, J. DeGoede, A. Berkenbosch and I. Olievier. The influence of oxygen on the ventilatory response to carbon dioxide in man. *J. Physiol. (Lond.)* **428**, 485–499 (1990).
9. L.J. Teppema, D. Nieuwenhuijs, E. Sarton, R. Romberg. C.N. Olievier, D.S. Ward and A. Dahan. Antioxidants prevent depression of the acute hypoxic ventilatory response by subanaesthetic halothane in men. *J. Physiol. (Lond.)* **544**, 931–938 (2002).
10. C. Eng, M. Kiuru, M.J. Fernandez and L.A. Aaltonen. A role for mitochondrial enzymes in inherited neoplasia and beyond. *Nat. Rev. Cancer* **3**, 193–202 (2003).
11. F.M. van Baars, C.W. Cremers, P. vanden Broek and J.E. Veldman. Familiar non-chromaffinic paragangliomas (glomus tumors). Clinical and genetic aspects. *Acta Otolaryngol.* **91**, 589–593 (1981).
12. A. van der Mey, P. Maaswinkel-Mooy, C. Cornelisse, P. Schmidt and J. vande Kamp. Genomic imprinting in hereditary glomus tumours: evidence for new genetic theory. *Lancet* **2.8675**, 1291–1294 (1989).
13. W. Reik and J. Walter. Genomic imprinting: parental influence on the genome. *Nat. Rev. Genet.* **2**, 21–32 (2001).

14. P. Heutink, A.G. van der Mey, L.A. Sandkuijl, A.P. van Gils, A. Bardoel, G.J. Breedveld, M. van Vliet, G.J. van Ommen, C.J. Cornelisse and Oostra BA. A gene subject to genomic imprinting and responsible for hereditary paragangliomas maps to chromosome 11q23-qter. *Hum. Mol. Genet.* **1**, 7–10 (1992).
15. E.C. Mariman, S.E. van Beersum, C.W. Cremers, P.M. Struycken and H.H. Ropers. Fine mapping of a putatively imprinted gene for familial non-chromaffin paragangliomas to chromosome 11q13.1: evidence for genetic heterogeneity. *Hum. Genet.* **95,** 56–62 (1995).
16. E.M. van Schothorst, M. Beekman, P. Torremans, N.J. Kuipers-Dijkshoorn, H.W. Wessels, A.F. Bardoel, A.G. van der Mey, M.J. van der Vijver, G.J. van Ommen, P. Devilee and C.J. Cornelisse. Paragangliomas of the head and neck region show complete loss of heterozygosity at 11q22-q23 in chief cells and the flow-sorted DNA aneuploid fraction. *Hum. Pathol.* **29**, 1045–49 (1998).
17. P.E. Taschner, J. Jansen, B.E. Baysal, A. Bosch, E.H. Rosenberg, A.H. Brocker-Vriends, A.G. vander Mey, G. van Ommen, C.J. Cornelisse and P. Devilee. Nearly all hereditary paragangliomas in the Netherlands are caused by two founder mutations in the SDHD gene. *Gen. Chrom. Canc.* **31**, 274–281 (2001).
18. D. Astuti, F. Latif, A. Dallol, P.L. Dahia, F. Douglas, E. George, F. Skoldberg, E.S. Husebye, C. Eng and E.R. Maher. Gene mutations in the succinate dehydrogenase subunit SDHB cause susceptibility to familial pheochromocytoma and to familial paraganglioma. *Am .J. Hum Genet.* 69(1): 49–54, 2001.
19. S. Niemann and U. Muller. Mutations in SDHC cause autosomal dominant paraganglioma, type 3. *Nat. Genet.* **26**, 268–270 (2000).
20. J. Arias-Stella and J. Valcarcel. The human carotid body at high altitudes. *Pathol. Microbiol. (Basel)* **39**, 292–297 (1973).
21. K. Astrom, J. Cohen, J. Willett, C. Aston and B. Baysal. Altitude is a phenotypic modifier in hereditary paraganglioma type 1: evidence for an oxygen-sensing defect. *Hum. Genet.* **113**, 228–237 (2003).
22. A.P. Gimenez-Roqueplo and the COMETE Network. Mutations in the SDHB gene are associated with extra-adrenal and/or malignant phaeochromocytomas. *Cancer Res.* **63**, 5615–5621 (2003).
23. I.P. Tomlinson and the Multiple Leiomyoma Consortium. Germline mutations in FH predispose to dominantly inherited uterine fibroids, skin leiomyomata and papillary renal cell cancer. *Nat. Genet.* **30**, 406–410 (2002).
24. N.H. Edelman, S. Lahiri, L. Braudo, N.S. Cherniack and A.P. Fishman. The blunted ventilatory response to hypoxia in cyanotic congenital heart disease. *N. Eng. J. Med.* **282**, 405–411 (1970).

A SIDS-Like Phenotype is Associated With Reduced Respiratory Chemoresponses in PACAP Deficient Neonatal Mice

Kevin J. Cummings, Jonathan D. Pendlebury, Frank R. Jirik, Nancy M. Sherwood and Richard J.A. Wilson

1. Introduction

Pituitary adenylate cyclase-activating polypeptide (PACAP) is an abundant neuropeptide within the CNS[1]. Mice deficient in PACAP or the PACAP-preferring receptor (PAC1), though normal at birth, are more susceptible than littermates to death during the neonatal period[2,3], with death sometimes occurring suddenly[4]. The reason why PACAP signaling increases the chance of surviving the neonatal period is unknown, but previous studies have indicated it is important for a proper physiological response to hypothermia[5].

Here we summarize our previously published data suggesting that the higher neonatal mortality in PACAP-null mice is principally the result of defective respiratory control[6]. Ventilation in PACAP deficient animals was evaluated using whole-body plethysmography and electrocardiography (ECG). We found that PACAP deficiency leads to a reduction in ventilation with blunted responses to both hypoxia and hypercapnia. In addition, under anaesthetic-induced hypothermia, PACAP-null animals suffered from prolonged apnea that preceded atrio-ventricular block. Based on these findings, as well as other characteristics of the PACAP knockout phenotype[4], we speculate that mutations within critical regions of genes coding PACAP or PACAP-signaling components might predispose human infants to Sudden Infant Death.

Kevin J. Cummings, Jonathan D. Pendlebury and Richard J.A. Wilson • Faculty of Medicine, Department of Physiology and Biophysics **Frank R. Jirik** • Department of Biochemistry and Molecular Biology, University of Calgary, 3330 Hospital Dr. NW, Calgary AB, Canada, T2N 4N1. **Nancy M. Sherwood** • Department of Biology, University of Victoria, Victoria BC, Canada, PO Box 3020, STN CSC, V8W 3N5.

Post-Genomic Perspectives in Modeling and Control of Breathing, edited by Jean Champagnat, Monique Denavit-Saubié, Gilles Fortin, Arthur S. Foutz, Muriel Thoby-Brisson. Kluwer Academic/Plenum Publishers, 2004.

2. Methods

2.1. Whole-Body Plethysmography

Respiration was assessed in non-anesthetized PACAP$^{+/+}$, $^{+/-}$ and $^{-/-}$ post-natal day 4 (P4) pups using continuous-flow, unrestrained, whole-body plethysmography with an ambient temperature within the thermoneutral range (33° ± 0.5°C). Pups were subjected to either hypoxia (10% O_2/balance N_2) or hypercapnia (8% CO_2/balance air) using the following protocol: 15 minute calm-down period, 5 minute baseline (room air) period, 5-minute hypoxia or hypercapnia treatment, and 15 minute washout (room air) period. Data from minutes 3–5 of baseline, minutes 2–3 and 4–5 of treatment and minutes 2–3 and 14–15 of washout were analyzed. Breathing parameters studied were rate (breaths · min^{-1}), tidal volume (V_T) and minute ventilation (V_E). We integrated the area under the inspiratory pressure curve as to obtain an index of tidal volume (V_T). Experiments and analysis were completed before genotyping animals, ensuring both were performed blind. A one-way, between-subject ANOVA was used for statistical analysis of baseline data. Two-factor, repeated-measures ANOVA was used to assess the effect of treatment, genotype and genotype-treatments interaction.

2.2. Electrocardiography (ECG)

Given that PACAP$^{-/-}$ mortality increases with reduced ambient temperature, we assessed the consequence of PACAP deficiency on heart function and breathing during anesthesia-induced hypothermia. Surface ECG tracings were recorded from 10 sets of paired PACAP$^{+/+}$ and PACAP$^{-/-}$ littermates on P7-14. After 10 min with rectal temperature (T_b) clamped at a normothermic level, hypothermia was induced with isoflurane. With T_b clamped at 30°C, and recordings were continued for an additional 10 minutes. Breaths and heart rate (beats · min^{-1}) were counted over minutes 6 of normothermia and hypothermia. Breathing data were normalized (reciprocal transform). 3 of 10 PACAP$^{-/-}$ animals died upon application of anesthetic. In addition, another PACAP$^{-/-}$ animal had no recordable breathing during minute 6 of hypothermia. Data from the 6 pairs of animals remaining were analyzed for significant differences using two-way repeated-measures ANOVA.

3. Results

3.1. Plethysmography: Effects of Genotype on Baseline Respiration in Air

Genotype did not affect breathing rate ($P = 0.19$; Figure 1). However, −/− mice had a smaller tidal volume ($P < 0.01$) than +/+ littermates resulting in a ~25% reduction in V_E ($P < 0.01$).

3.2. Plethysmography: Respiratory Response to Hypoxia and the Effect of Genotype

Genotype had significant influences on hypoxia-mediated changes in rate ($P < 0.001$): PACAP$^{-/-}$ neonates having blunted responses in comparison to PACAP$^{+/+}$ littermates

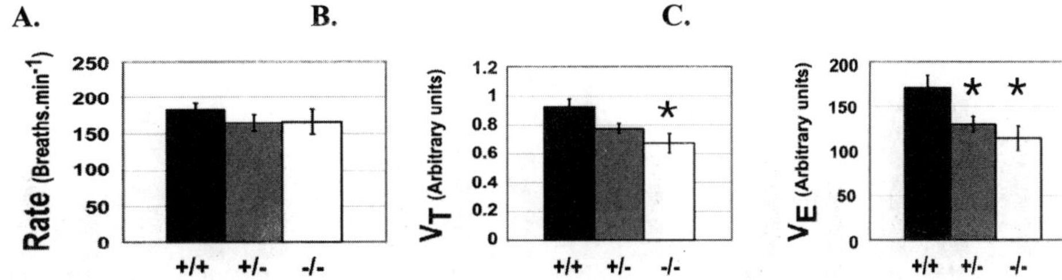

Figure 1. PACAP genotype and baseline breathing. *A*. Breathing rate. *B*. Index of tidal volume (V_T). *C*. Index of minute ventilation (V_E). In this and subsequent figures, mean values were obtained for two-minute windows and bars represent S.E.M. Asterisk (*) indicates a significant difference compared to wildtype.

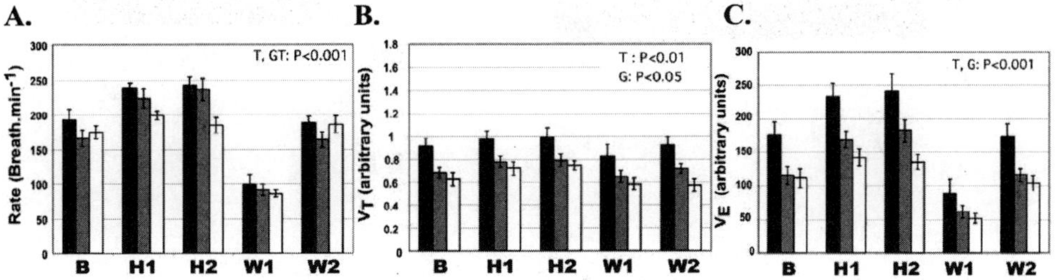

Figure 2. PACAP genotype and response to hypoxia (10% O_2 bal. N_2). *A*, *B*, *C* as in Fig 1. *Baseline* (B), *Early and late Hypoxic Response* (H1 and H2, respectively), *Wash1* (W1) (1 min. after return to air) and *Wash2* (W2) *T*, *G*, *GT*: $P < 0.05$ for treatments, genotypes, and interaction between treatment and genotype, respectively.

(Figure 2). However, genotype did not affect the hypoxia-mediated increase in V_T ($P = 0.54$) nor the overall ventilatory response to hypoxia (i.e. V_E, $P = 0.57$). The genotype-hypoxia interaction effect on rate is counteracted by the genotype effect alone on tidal volume.

3.3. ECG: Respiratory Response to Hypercapnia and the Effect of Genotype

Hypercapnia significantly increased breathing rate, V_T and V_E (Figure 3). Genotype had no effect on the rate response to hypercapnia ($P = 0.51$). However, the increase in V_T with hypercapnia was blunted in PACAP$^{-/-}$ neonates compared with that of PACAP$^{+/+}$ littermates ($P < 0.005$). This contributed to a reduction in the V_E response to hypercapnia ($P < 0.001$), from an increase of 104% (PACAP$^{+/+}$) to an increase of 59% (PACAP$^{-/-}$).

3.4. Effects of Conditions that Selectively Exacerbate PACAP$^{-/-}$ Neonatal Mortality

Figure 4 shows the ECG from one of the pairs of animals during hypothermia. Immediately after the time at which T_b reached 30°C the heart rate is comparable between the genotypes, the respiratory rate of the PACAP-null mouse is much less than its wild-type littermate (compare Figure 4A2 and 4B2). After 10 minutes of hypothermia, the breathing and heartbeat of the wild-type mice remained regular with frequencies only slightly less than

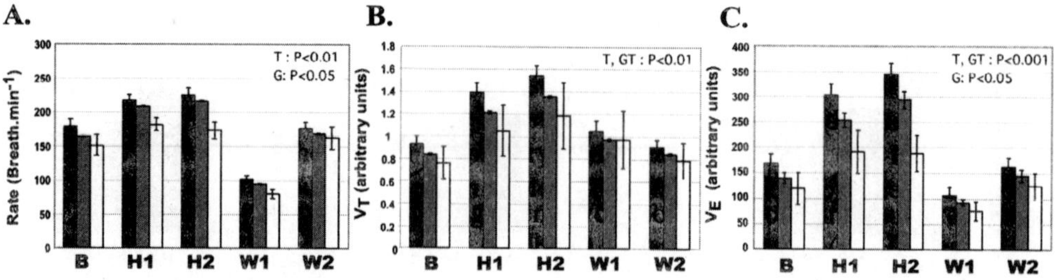

Figure 3. PACAP genotype and response to hypercapnia (8% CO_2 bal. air). *See Fig 2 legend for details.*

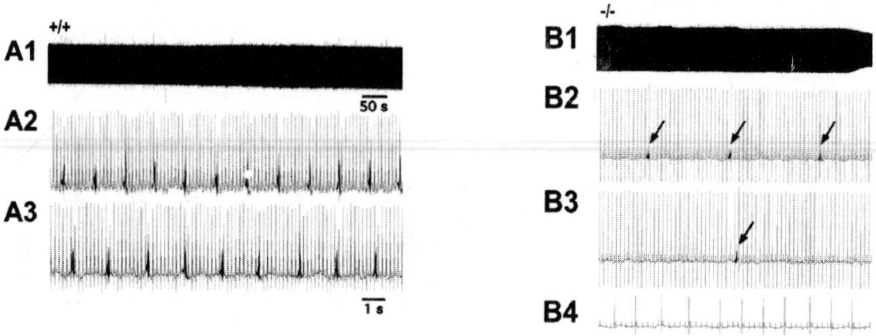

Figure 4. ECGs during steady-state hypothermia of anesthetized P4 neonates. *A1-3*: PACAP$^{+/+}$. *B1-4*: PACAP$^{-/-}$. *A1, B1*: ECG of first 10 min of hypothermia. *A2, B2*: Expanded trace of first 15 s of hypothermia. *B3*: After 7 min 30 s of hypothermia. *A3, B4*: After 9 min, 45 s of hypothermia. Arrows in *B2* & *B3* indicate noise in ECG, indicative of ventilatory efforts. Note: respiratory rate fell more rapidly following the onset of hypothermia in PACAP$^{-/-}$ neonates (*B2*) than in PACAP$^{+/+}$ littermates (*A2*).

those observed soon after the onset of hypothermia (Figure 4A3). In contrast, in PACAP-null mice the frequency of breathing continues to decline as the duration of hypothermia increases (Figure 4B3). Despite these dramatic effects on the respiratory rhythm, the heart rate remained relatively stable. Only towards the end of the 10 minutes of hypothermia, when breathing in PACAP knockout animals is vastly reduced or absent, is cardiac function seriously compromised. Note the declining R-wave amplitude (Figure 4B1) and an uncoupling of atrial and ventricular firing (AV-block) (Figure 4B4). In total, 3 of 6 PACAP$^{-/-}$ neonates succumbed by the end of the hypothermic treatment with AV-block. These three mice had long duration apneas preceding AV-block, suggesting that chronic hypoxemia led to cardiac arrest. A 4th mouse had no detectable breathing with an absence of AV-block. In contrast, none of the PACAP$^{+/+}$ neonates died, and ECG tracings remained relatively normal, even after 10 minutes of hypothermia (*e.g.*, Figure 4A1).

4. Discussion

A central finding of this study is that a PACAP deficiency compromises respiration in neonatal mice. The respiratory phenotype of the PACAP-null mice precedes the vulnerable period in post-natal development when they have a higher mortality. In addition to studying the effect of PACAP on respiration, we examined whether the respiratory phenotype contributes to increased neonatal mortality. During anaesthetic-induced hypothermia, we found that the PACAP-deficient neonates have prolonged apneas that can lead to AV-block and death. Thus, the present data suggest that a respiratory phenotype, including susceptibility to apnea, may account for the higher mortality of PACAP-null mice when housed at slightly lower ambient temperatures.

4.1. Direct Role for PACAP in Respiratory Control?

Our data suggest that PACAP influences respiration, both during baseline conditions and in response to hypoxic and hypercapnic challenges. The influence of PACAP may be direct, through neuromodulation of the respiratory circuit, or it may be indirect through augmentation of metabolism. However, differences in metabolic rate are unlikely to explain the differential effects of genotype on the overall ventilatory response to CO_2, since the ratio of ventilation to metabolic rate is not affected by hypercapnia[7]. An alternate hypothesis is that PACAP has a role in respiratory chemosensitivity, acting directly on elements of the respiratory circuits to modulate their response to respiratory challenges. We note that PACAP is widely distributed within the CNS and is located in brainstem regions implicated in mediating the response to peripheral chemoreceptor activation, the generation of respiratory rhythm and central chemosensitivity (e.g., nucleus of the solitary tract, ventro-lateral medulla, nucleus ambiguous, dorsal vagal nucleus and hypothalamus)[8].

4.2. PACAP Mutations as a Genetic Background for SIDS?

Our understanding of the genetic component of SIDS remains in its infancy. Recent advances suggest that there may be multiple genes in which congenital abnormalities predispose to SIDS. For example, a mutation in a gene encoding a cardiac sodium channel, which almost never occurs in the adult population, is present in only 2% of SIDS victims[9]. Phenotypic analysis of transgenic and knockout mice provides us with a possible strategy to uncover other gene mutations that may play a role in the development of SIDS. But which phenotypes should we look for?

Abnormalities in chemosensitive brainstem regions are common in SIDS victims[10] and some studies have indicated that near-miss SIDS infants have blunted arousal responsiveness to hypoxia and hypercapnia[11,12]. Kelly et al. (1982)[13] found a higher frequency of apneic periods in newborn sibs of SIDS victims than in normal infants. Further, mortality resulting from a failure to arouse during suffocation when neonates are put to sleep on their stomachs may explain the success of the back to sleep campaign. Another phenotype that may help identify genes important in SIDS is a critical developmental period. In SIDS, the highest rate of mortality occurs well into neonatal life (two and a half months after birth). This distinguishes SIDS from a host of other congenital diseases where death is most likely at, or soon after birth. Finally, a potential SIDS phenotype is likely to be more susceptible to environmental stressors: hyperthermia and exposure to cigarette smoke are important risk factors implicated in human SIDS[14].

Interestingly, PACAP-deficient mice respond poorly to environmental stress, i.e., hypothermia (the effect of hyperthermia has not been determined). While death at birth is a phenotype common to several strains of knockout mice (including a number with breathing defects)[15,16,17], there is no difference in mortality of PACAP+/+ and −/− littermates until after 4 days postpartum. Finally, as the results herein describe, PACAP-null mice have abnormal responses to hypoxia and hypercapnia.

We also note that a study by Boles et al. (1998)[18], examining the livers from 418 SIDS cases, found that 23% showed at least one abnormality with respect to elevated fatty acids (8%), glucose depletion (14%) and steatosis of the liver (9%). Similar abnormalities have been documented in the PACAP-knockout mice[4]. In addition, after the first post-natal week, some PACAP knockout mice display abnormal growth[4], a documented risk factor for SIDS[19,20]. Given the above data, we speculate that mutations in PACAP, or components of the PACAP-signaling pathway, may be present in some cases of SIDS.

5. Acknowledgments

Salary and operating support for this project was provided by grants from the Canadian Heart & Stroke Foundation/Canadian Stroke Network/CIHR/AstraZeneca Partnership Program, CIHR and NSERC (RJAW).

References

1. N.M. Sherwood, S.L. Krueckl, and J.E. McRory, The origin and function of the pituitary adenylate cyclase-activating polypeptide (PACAP)/glucagon superfamily, *Endocr Rev* **21** (6) 619–70 (2000).
2. H. Hashimoto, N. Shintani, K. Tanaka, W. Mori, M. Hirose, T. Matsuda, M. Sakaue, J. Miyazaki, H. Niwa, F. Tashiro, K. Yamamoto, K. Koga, S. Tomimoto, A. Kunugi, S. Suetake, and A. Baba, Altered psychomotor behaviors in mice lacking pituitary adenylate cyclase-activating polypeptide (PACAP), *Proc Natl Acad Sci USA* **98** (23) 13355–60 (2001).
3. C. Hamelink, O. Tjurmina, R. Damadzic, W.S. Young, E. Weihe, H.W. Lee, and L.E. Eiden, Pituitary adenylate cyclase-activating polypeptide is a sympathoadrenal neurotransmitter involved in catecholamine regulation and glucohomeostasis, *Proc Natl Acad Sci USA* **99**(1) 461–6 (2002).
4. S.L. Gray, K.J. Cummings, F.R. Jirik, and N.M. Sherwood, Targeted disruption of the pituitary adenylate cyclase-activating polypeptide gene results in early postnatal death associated with dysfunction of lipid and carbohydrate metabolism, *Mol Endocrinol* **15**(10) 1739–47 (2001).
5. S.L. Gray, N. Yamaguchi, P. Vencova, and N.M. Sherwood, Temperature-sensitive phenotype in mice lacking pituitary adenylate cyclase-activating polypeptide, *Endocrinology* **143**(10) 3946–54 (2002).
6. K.J. Cummings, J.D. Pendlebury, N.M. Sherwood and R.J.A. Wilson, Sudden neonatal death in PACAP deficient mice is associated with reduced respiratory chemoresponse and susceptibility to apnea, *J. Physiol: in press.*
7. J.P. Mortola and C. Lanthier, The ventilatory and metabolic response to hypercapnia in newborn mammalian species, *Respir Physiol* **103**(3) 263–70 (1996).
8. J. Hannibal, Pituitary adenylate cyclase-activating peptide in the rat central nervous system: an immunohistochemical and in situ hybridization study, *J. Comp Neurol* **453**(4) 389–417 (2002).
9. M.J. Ackerman, B.L. Siu, W.Q. Sturner, D.J. Tester, C.R. Valdivia, J.C. Makielski, and J.A. Towbin, Postmortem molecular analysis of SCN5A defects in sudden infant death syndrome, *Jama* **286**(18) 2264–9 (2001).
10. J.J. Filiano, Arcuate nucleus hypoplasia in sudden infant death syndrome: a review, *Biol Neonate* **65**(3–4) 156–9 (1994).
11. C.E. Hunt, Abnormal hypercarbic and hypoxic sleep arousal responses in Near-Miss SIDS infants, *PediatrRes* **15**(11) 1462–4 (1981).

12. K. McCulloch, R.T. Brouillette, A.J. Guzzetta, and C.E. Hunt, Arousal responses in near-miss sudden infant death syndrome and in normal infants, *J Pediatr* **101**(6) 911–7 (1982).
13. D.H. Kelly, J. Twanmoh, and D.C. Shannon, Incidence of apnea in siblings of sudden infant death syndrome victims studied at home, *Pediatrics* **70**(1) 128–31 (1982).
14. W.G. Guntheroth and P.S. Spiers, The triple risk hypotheses in sudden infant death syndrome, *Pediatrics* **110**(5) e64 (2002).
15. F. Guillemot, L.C. Lo, J.E. Johnson, A. Auerbach, D.J. Anderson, and A.L. Joyner, Mammalian achaete-scute homolog 1 is required for the early development of olfactory and autonomic neurons, *Cell* **75**(3) 463–76 (1993).
16. E. Hummler, P. Barker, J. Gatzy, F. Beermann, C. Verdumo, A. Schmidt, R. Boucher, and B.C. Rossier, Early death due to defective neonatal lung liquid clearance in alpha-ENaC-deficient mice, *Nat Genet* **12**(3) 325–8 (1996).
17. K. Tokieda, J.A. Whitsett, J.C. Clark, T.E. Weaver, K. Ikeda, K.B. McConnell, A.H. Jobe, M. Ikegami, and H.S. Iwamoto, Pulmonary dysfunction in neonatal SP-B-deficient mice, *Am J Physiol* **273**(4 Pt 1) L875–82 (1997).
18. R.G. Boles, E.A. Buck, M.G. Blitzer, M.S. Platt, T.M. Cowan, S.K. Martin, H. Yoon, J.A. Madsen, M. Reyes-Mugica, and P. Rinaldo, Retrospective biochemical screening of fatty acid oxidation disorders in postmortem livers of 418 cases of sudden death in the first year of life, *J Pediatr* **132**(6) 924–33 (1998).
19. N. Oyen, R. Skjaerven, R.E. Little, and A.J. Wilcox, Fetal growth retardation in sudden infant death syndrome (SIDS) babies and their siblings, *Am J Epidemiol* **142**(1) 84–90 (1995).
20. E.A. Mitchell, P.G. Tuohy, J.M. Brunt, J.M. Thompson, M.S. Clements, A.W. Stewart, R.P. Ford, and B.J. Taylor, Risk factors for sudden infant death syndrome following the prevention campaign in New Zealand: a prospective study, *Pediatrics* **100**(5) 835–40 (1997).

Selective Alteration of the Ventilatory Response to Hypoxia Results from Mutation in the Myelin Proteolipid Protein Gene

Martha J. Miller, Musa A. Haxhiu, Cindy D. Kangas, Paraskevi Georgiadis, Tatyana I. Gudz, and Wendy B. Macklin

1. Introduction

Central nervous system myelin is formed from the lipid-rich, highly specialized oligodendrocyte plasma membrane, which ensheaths axons, leaving spaces only at the nodes of Ranvier. Within the myelin sheath are many myelin-associated lipophilic proteins, the most abundant of which are myelin proteolipid protein (PLP) and myelin basic protein (MBP). These two proteins constitute 60–90% of myelin protein in mammalian species.[1]

The gene for mammalian PLP is 17 kb in size, and is located on the X-chromosome. Myelin PLP is believed to be a tetraspan membrane protein, bearing a structural resemblance to a number of ion channels and receptors.[2] Remarkably, the sequence of the PLP protein is almost 100% conserved in mammals. Postnatal expression of PLP protein within the central nervous system correlates with the development of myelination,[3] which proceeds from a caudal to rostral direction and is complete within roughly the first two weeks of postnatal life in the rodent hindbrain.

The cellular functions of the proteolipid protein gene are of considerable interest, for mutations in this gene are associated with a human dysmyelinating disorder, Pelizaeus-Merzbacher disease (PMD).[4] In its classic form, the initial onset of the disease occurs in the first three months of life, with slowly progressing psychomotor retardation, ataxia, and death in the second decade of life.[5] The connatal type of PMD shows a more rapid postnatal onset and is fatal in infancy or childhood. The complex pathophysiology which leads to death in this form of PMD is poorly understood.

Martha J. Miller, P. Georgiadis • Dept. Pediatrics, Case Western Reserve University, Cleveland, Ohio, 44106. **M.A. Haxhiu** • Dept. Physiol and Biophys, Howard University, Washington, D.C., 20059. **T.I. Gudz, C.D. Kangas, W.B. Macklin** • Dept. Neurosci., Lerner Research Inst., Cleveland Clinic Foundation, Cleve., Ohio, 44195.
Post-Genomic Perspectives in Modeling and Control of Breathing, edited by Jean Champagnat, Monique Denavit-Saubié, Gilles Fortin, Arthur S. Foutz, Muriel Thoby-Brisson. Kluwer Academic/Plenum Publishers, 2004.

Spontaneous mutations in the PLP gene have been found in rats, mice, dogs and rabbits which exhibit many of the characteristics of the human disease PMD, and thus are good experimental models.[6] Of particular interest, the majority of animals with mutations in this gene exhibit severe CNS hypomyelination and early death.

Our group has used the MD rat as a model for the pathophysiology of *plp* gene mutations. In this animal an A-T conversion in the *plp* gene on the X-chromosome results in a single amino acid change in the second transmembrane segment of PLP.[7] After P10, symptoms of tremor, ataxia, and seizures develop accompanied by severe central nervous system dysmyelination and there is early death at postnatal age 21–24d.

We have focused our studies on the hypothesis that *plp* mutation could produce fatal alteration of autonomic control of breathing at a critical period in development. To test this hypothesis, we compared resting ventilation and the ventilatory responses to hypoxia and hypercapnia in the first three weeks of life in the affected male MD rat, using normal male littermates for comparison). These studies were accompanied by detailed histologic study of myelination, expression of PLP protein, and neurotransmitter receptors at the critical age at which death occurs in the MD rat.[8]

2. Respiratory Output and Responses to Hypoxia and Hypercapnia in the MD Rat

Barometric plethysmography was used to measure baseline ventilation and the ventilatory responses to hypoxia (8% 02, bal N2, for 5 min) and hypercapnia (10% CO2, 30% 02, bal N2 for 10 min) at postnatal days 14, 18 and 21–24 in 18 affected MD pups and 19 normal male littermates. At all ages studied, there was no difference in respiratory rate, tidal volume, and minute ventilation between affected males and normal male littermates nor did the MD pups exhibit apnea. When the ventilatory responses to hypoxia and hypercapnia were compared, a striking increase in late hypoxic ventilatory depression (rolloff) occurred during hypoxia in the MD rats at P21 (Fig. 1). In contrast, the response to hypercapnia did not differ between MD and normal males at any age. Approximately 70% of affected MD males died in response to a single hypoxic exposure. The striking alteration in late hypoxic ventilatory depression suggested that the *plp* gene mutation could have selectively altered central neural pathways in the brainstem which control breathing.

3. Histopathology of the Brainstem in the MD Rat

Based on our physiologic studies, we hypothesized that severe late hypoxic ventilatory depression in the MD rat could be caused by one or more of the following processes: altered myelination, altered cytoarchitecture; or altered balance of neurotransmitters and their receptors. We focused our investigations on the caudal brainstem at the level of the area postrema, at which central processing of sensory input from the carotid bodies occurs.

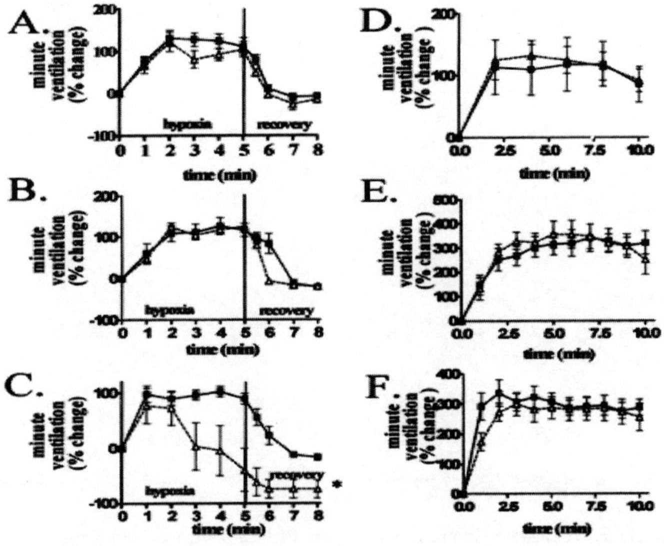

*P < 0.001

Figure 1. Ventilatory response to inhalation of 8% O_2 (balance N2) for 5 min. was normal at P14 and P18 in the MD rat (open triangle) as compared to the normal littermates (closed squares) (A, B). At P21–24, MD rats exhibited severe depression of ventilation in response to 8% O_2 in comparison to wild type males (C). At all ages, the ventilatory response to inhalation of 10% CO_2 of MD and normal rats did not differ. (P14:D, P18: E, P21–24:F). Copyright 2003 by the Society for Neuroscience.[8]

3.1. Myelination and Expression of PLP in Neurons in the Caudal Medulla of the MD rat

As a marker for myelination, PLP protein immunoreactivity was compared in MD and wild type littermates. As expected, there was a striking decrease in immunoreactive fibers in the caudal brainstem of the MD rat, as compared to the normal male littermate (Fig. 2 A,B). Unexpectedly, we observed immunostaining for PLP in large cells resembling neurons in the hypoglossal nucleus. These cells were definitively identified as neurons by double labeling with antibody against PLP and the neuron-specific antibody neuron specific nuclear antigen (NeuN).[8]

To confirm that the *plp* gene can be expressed in neurons in the caudal brainstem during development, we analyzed this region of the brain in the transgenic *plp*-EGFP mouse developed in the laboratory of Dr. W. Macklin.[9] This transgenic mouse expresses EGFP under the control of the *plp* promoter. By this means, neurons expressing the *plp* promoter were identified in the caudal brainstem of the normal mouse at postnatal day 12.

3.2. Glutamatergic NMDA Receptor Expression in the Caudal Medulla of the MD Rat

Integration of afferent sensory input from the carotid bodies within the nucleus tractus solitarius requires glutamatergic NMDA receptors on postsynaptic neurons This input is

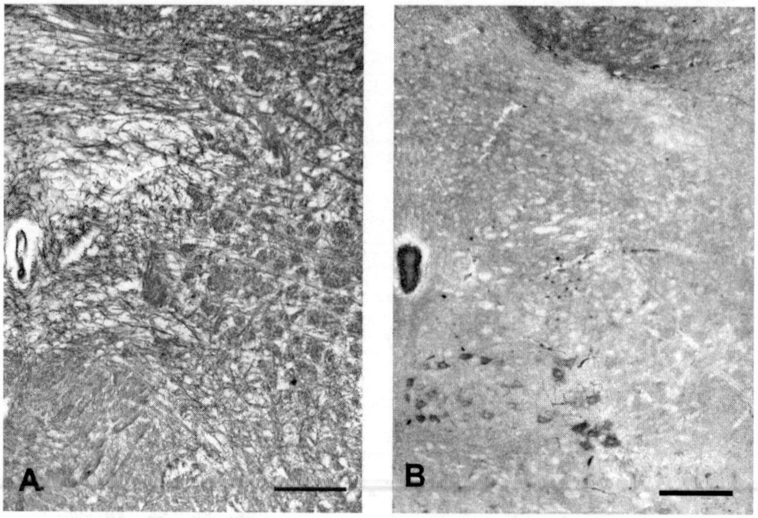

Figure 2. In the caudal brainstem, immunostaining for PLP was greatly diminished in the MD rat (B) as compared to the wild type (A). Furthermore, neurons immunoreactive for PLP were present in the hypoglossal nucleus (B). A, B, Bar = 100. Copyright 2003 by the Society for Neuroscience.[8]

critical for the normal increase in ventilation in response to hypoxia.[11, 12] When we compared the expression of the NMDAR1 receptor subunit in the caudal brainstem of the MD and normal rats at P21, we found a striking decrease in the expression of this critical receptor subunit (Fig. 3).

In order to determine whether the decrease in immunoreactivity for NMDAR1 was correlated with a loss of this specific receptor protein, we performed immunoblot analysis for NMDAR1, the GABAAβ2 and γ subunit proteins as well as PLP and MBP proteins in whole brainstem samples from P21 MD and WT rats (Fig. 3). We found that NMDAR1 receptor protein was significantly diminished in the brainstem of the MD rat, along with PLP and MBP. The decrease in PLP and MBP proteins has been previously described in the CNS of animals with mutations in the *plp* gene, however, the downregulation of the NMDA receptor is an entirely new finding.

3.3. Severe Hypoxic Ventilatory Depression is a Characteristic of Animals with Early-lethal Mutations in the *plp* Gene

A number of spontaneous *plp* mutations with an early-lethal phenotype have also been found in the mouse. These include the myelin-synthesis deficient mouse (*msd*) and the jimpy (Jp) mouse.[6] These rodents suffer an early death between P18 and P21. The site of mutation in the *plp* gene in these animals differs from the mutation in the MD rat. The *jp* mouse has a frameshift mutation which leads to a deletion of a large part of the carboxy-terminus of the PLP protein. The MSD mouse has a single base-pair missense mutation in exon 6 of the *plp* gene. We hypothesized that the shared early lethal phenotype in these rodents could be due in part to the same development of extreme sensitivity to hypoxia, as shown in the MD rat.

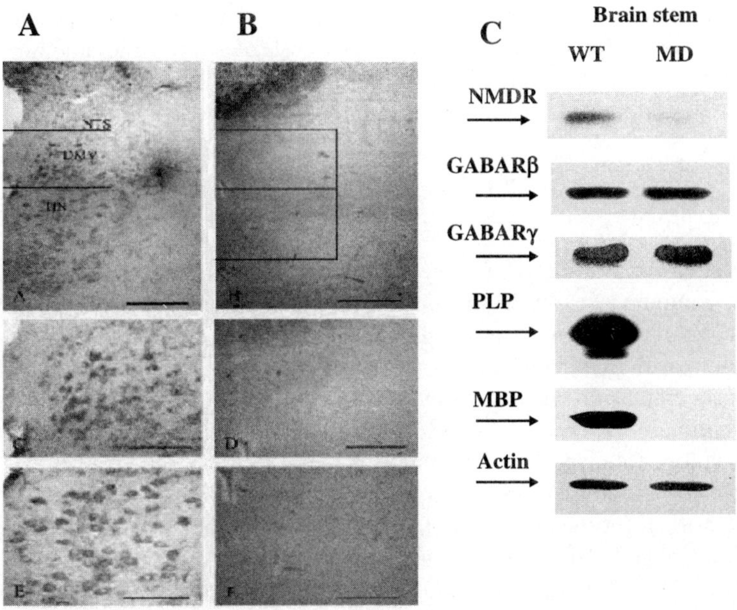

Figure 3. In the caudal brainstem NMDAR1 reactive neurons were observed in the nucleus tractus solitarius, hypoglossal nucleus, and dorsal motor nucleus of the vagus (A). In the MD rat, neurons immunoreactive for NMDAR1 were strikingly diminished in these areas (B). Boxed areas in A and B are enlarged in C-F. Upper box: C, D; lower box E, F. In (C), immunoblot analysis shows downregulation of NMDAR1, PLP and MBP (myelin basic protein). Copyright 2003 by the Society for Neuroscience.[8]

In order to test the hypothesis that dysmyelination alone could not lead to altered hypoxic ventilatory depression, we also evaluated the ventilatory response to hypoxia in the shiverer (*shi*) mouse, which carries a dysmyelinating mutation in the *mbp* gene, which does not result in an early-lethal phenotype.

The ventilatory response to hypoxia was compared in the *msd*, *jp*, and *shi* mice at P21, P18, and P30, respectively. Both the *msd* and *jp* mice exhibited severe hypoxic ventilatory depression when compared to their normal littermates (Fig. 4). In contrast, the *shi* mouse

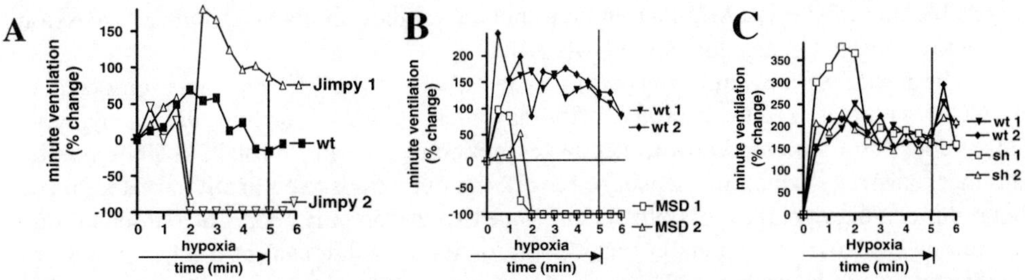

Figure 4. Comparison of the respiratory response to inhalation of 8% 02 (5 min) in A. jimpy (jp)(P18), B.myelin synthesis deficient (msd) (P21), and C. shiverer (sh) mice (P60-80), and respective age-matched controls (N = 2 mutant and 2 controls, respectively, in each study). A. After 2 min inhalation of 8% 02, jp mice exhibited severe ventilatory depression. One died, the other recovered after a severe apnea. B. Both msd mice had severe depression of breathing after .5 min of hypoxia, and died C. In contrast, sh mice exhibited a normal response.

did not exhibit altered ventilatory response to hypoxia. Thus, we conclude that *plp* gene mutations involving different sites in the molecule may lead to the same pathologic alteration in control of breathing during hypoxia. Furthermore, dysmyelination as characteristic of the Sh mouse does not alter the ventilatory response to hypoxia (Fig. 4,C).

4. Discussion

These studies have shown, for the first time, that early-lethal mutations in the *plp* gene share a common phenotype: increased hypoxic ventilatory depression at the critical age at which death occurs in these mutants. Furthermore, neither the ventilatory response to hypercapnia nor rhythm pattern generation were altered, indicating that the effect of *plp* mutations is specific for the neural pathways that control breathing during hypoxia.

The ventilatory response to hypoxia in a 21d rat consists of an initial increase in minute ventilation, followed by a sustained increase above baseline. In the MD rat, the initial increase in ventilation is preserved, but is followed by an exaggerated rolloff. NMDA receptors are critical for the normal ventilatory response to hypoxia in the adult rat[10,11,12] and we speculate that the almost complete absence of the NMDAR1 receptor subunit in the medulla in the P21 MD rat may account for the inability to sustain ventilation in response to continued hypoxia.

The mechanism by which NMDA receptors are downregulated in the MD rat is unknown. Myelination of the CNS proceeds in a caudal to rostral direction in the rat, and is complete by P21 in the brainstem. It is conceivable that dysmyelination in the MD rat could alter developmental expression of the NMDA receptor. However, we failed to find alteration of the hypoxic ventilatory response in the shiverer mouse, which has dysmyelination due to mutation in the *mbp* gene.[15] Thus, it appears that dysmyelination alone cannot account for the pathophysiology of the MD rat.

Our finding that the *plp* promoter and MD-PLP protein can be expressed in neurons in the caudal medulla raises the possibility that the gradual accumulation of mutant MD-PLP protein within some neurons can alter expression of NMDAR. The nonmyelin functions of the *plp* gene in neurons are currently unknown. It has been shown that MD-PLP accumulates within the endoplasmic reticulum of the oligodendrocyte, and is poorly transported to the cell surface, resulting in reduced myelination and cell death.[14] Furthermore, the MD *plp* gene downregulates its own expression, as well as the expression of the *mbp* gene Thus, it is possible that misfolded MD-PLP in neuronal cell bodies similarly can alter expression of specific neuronal genes, including NMDAR1.

The results of our study suggest that the effects of the MD *plp* mutation occur in the CNS, within a specific neuronal network which controls breathing during hypoxia. This is consistent with the known selective expression of the PLP protein in CNS myelin. The mechanism by which the sensory network for hypoxia is specifically targeted in *plp* mutations is unknown. It is possible that *plp* gene expression only occurs in a limited group of neurons which are functionally linked. Alternatively, withdrawal of trophic support of neighboring oligodendrocytes, which are dying due to MD *plp* expression, could lead to altered neuronal function.

In summary, we have shown that, in addition to dysmyelination throughout the CNS, the MD *plp* mutation causes pathologic alteration of the ventilatory response to hypoxia, as well as loss of NMDAR1 within the brainstem. The finding that all early lethal mutations of the *plp* gene exhibit a similar pathophysiologic alteration of ventilation during hypoxia may be relevant to understanding the most severe forms of Pelizaeus-Merzbacher disease. It is conceivable that expression of the mutant PLP protein in the human disease could also lead to a severe disturbance of respiratory homeostatic mechanisms, and contribute to early death. Finally, these studies suggest that the *plp* mutants may be excellent models in which to investigate the mechanisms of ventilatory responses to hypoxia in the developing CNS.

5. Acknowledgments

This work was supported by NIH grant NS25304, NS39407, HL 62527, HL 56527, The American Heart Association, and a grant from the National Multiple Sclerosis Society.

References

1. Lees, M.B., Brostoff, S.W., 1984, Proteins of myelin: in Morell P (ed): *Myelin*, New York, Plenum Press, 1984, pp. 197–224.
2. Unwin, N., 1989, The structure of ion channels in membranes of excitable cells. *Neuron* 3: 3665–676.
3. Macklin, W.B., Oberfeld, E., Lees, M.B., 1983, Electroblot analysis of rat myelin proteolipid protein and basic protein during development. *Dev. Neurosci.* 6: 161–168.
4. Garbern, J., Cambi, F., Shy, M., and Kamholtz, J., 1999, The molecular pathogenesis of Pelizaeus-Merzbacher disease, *Arch. Neurol.* 56: 1210–1214.
5. Renier W.O., Gabreels F.J.M., Hustinx T.W.J., Jaspar H.H.J., Geleen J.A.G., Van Haelst U.J.G., Lommen E.J.P., Ter Haar, B.G.A., 1981, Connatal Pelizaeus-Merzbacher Disease with Congenital Stridor in Two Maternal Counsins. *Acta. Neuropathol.* (Berl) 54: 11–17.
6. Knapp, P.E., 1996, Proteolipid protein: is it more than just a structural component of Myelin? *Devel. Neurosci.* 18: 297–308.
7. Boison, D., and Stoffel, W., 1989, A point mutation in exon3 (A-C, Thr 75-Pro) of the myelin proteolipid protein causes dysmyelination and oligodendrocyte death. *EMBO J* 8: 3295–3302.
8. Miller, M.J., Haxhiu, M.A., Gudz, T.I., Kangas, C., and Macklin, W.B., 2003, Proteolipd protein gene mutation induces altered ventilatory response to hypoxia in the myelin deficient rat. *J. Neurosci.*, 23, 2265–2273.
9. Mallon, B.S., Shick, H.E., Kidd, G.J. and Macklin, W.B., 2002, Proteolipid promotor activity distinguishes two populations of NG2-positive cells throughout neonatal cortical development. *J. Neurosci.* 22(3): 876–885
10. Ang, R.C., Hoop, B. and Kazemi, H., 1992, Role of glutamate as the central neurotransmitter in the hypoxic ventilatory response, *J. Applied Physiol.* 72: 1480–1487.
11. Ohtake, P.J., Simakajornboon, N., Fehniger, M.D., Xue, Y. and Gozal, D., 2000, N-methyl-D-aspartate receptor expression in the nucleus tractus solitarii and maturation of hypoxic ventilatory response in the rat, *Am. J. Resp Crit. Care Med.*, 162: 1140–1147.
12. Ohtake, P.J., Torres, J.E., Gozal, Y.M., Graff, G.R. and Gozal, D., 1998, NMDA receptors mediate peripheral chemoreceptor afferent input in the conscious rat, *J. Appl. Physiol.* 84: 853–861
13. Mullen, R.J., Buck, C.R. and Smith, A., 1992, NeuN, a neuronal specific nuclear protein in vertebrates, *Devel* 116: 201–211.
14. Gow, A., Friedrich, V.L. and Lazzarini, R.A., 1994, Many natural occurring mutations of myelin proteolipid protein impair its intracellular transport, *J. Neurosci. Res* 15: 383–394.
15. Roach, A., Takahashi, N., Pratcheve D., Ruddle, F., Hood, L., 1985, Chromosomal mapping of mouse myelin basic protein gene and structure and transcription of the partially deleted gene in shiverer mutant mice, *Cell* 42: 149–155.

3

From Neurons to Neural Assemblies: Brainstem Control of Rhythm Generation

Organization of Central Pathways Mediating the Hering-Breuer Reflex and Carotid Chemoreflex

Chi-Sang Poon

1. Introduction

How are respiratory reflex pathways organized centrally? Surprisingly little data are available in the literature that clearly address this basic question. It is generally assumed that afferents from carotid chemoreceptors, vagally-mediated slowly-adapting pulmonary stretch receptors (PSR) and other cardiopulmonary receptors modulate the respiratory rhythm through paucisynaptic central pathways that directly excite or inhibit the respiratory oscillator. The distinct response patterns specific to differing respiratory afferents suggest that the corresponding central pathways are not organized the same. However, precisely how varying central reflex pathways are organized to selectively modulate the respiratory rhythm is not clear.

Because neural organization is a complex issue that is not readily amenable to conventional experimental investigation, we used a computational modeling approach to address this fundamental problem. Model analysis revealed some basic neural architectures that are common to the mediation of the Hering-Breuer reflex (HBR) and carotid chemoreflex (CCR), including frequency and phase filtering of the corresponding afferent inputs via parallel conditioning pathways to the respiratory oscillator. In addition, our modeling results suggested certain differences in the central organization of HBR and CCR afferent conditioning pathways, which may account for their distinct response patterns.

2. Working Model of Hering-Breuer Reflex Modulation of T_E

One of the most well identified central respiratory afferent pathways is the reported excitatory projection of PSR afferents via nucleus tractus solitarius (NTS) "pump cells"

Chi-Sang Poon • Harvard-MIT Division of Health Sciences and Technology, Massachusetts Institute of Technology, Cambridge, MA, USA.

Post-Genomic Perspectives in Modeling and Control of Breathing, edited by Jean Champagnat, Monique Denavit-Saubié, Gilles Fortin, Arthur S. Foutz, Muriel Thoby-Brisson. Kluwer Academic/Plenum Publishers, 2004.

to ventrolateral medulla decrementing expiratory (E) neurons, which are critical for respiratory rhythmogenesis.[1] This excitatory paucisynaptic pathway linking PSR afferents to E-related neurons provides a plausible explanation of the HBR prolongation of expiratory duration (T_E). Further, by virtue of an implicit mutual inhibition of E neurons and inspiratory (I) neurons this afferent relay model may also account for the HBR shortening of T_I.

The afferent relay model of HBR suggests some general organization principles of central respiratory reflex pathways. Thus, respiratory afferents may selectively prolong (or shorten) one phase of the respiratory rhythm through paucisynaptic excitation (or inhibition) of respiratory neurons critical for that phase. In addition, through mutual inhibition of the E and I neurons such phase-specific afferent relay pathway may also exert an indirect and opposite effect on the other phase.

Although the afferent relay model provides a plausible explanation of the immediate HBR effects on T_E and T_I, it does not account for higher-order effects of HBR. As shown in Fig. 1, during electrical vagal stimulation in rat the initial HBR prolongation of T_E is gradually offset by a bi-exponential recovery.[2] The slow decay component is independent of NMDA receptors and is probably correlated with the reported synaptic accommodation of NTS neurons during afferent stimulation.[3] This NMDA receptor-independent decay effect conforms with response habituation (the simplest form of nonassociative learning).[4] The fast decay component during stimulation is mirrored by a similar rapid decay of the post-stimulation rebound decrease in T_E. Both fast decay components are abolished by NMDA receptor blockade and bilateral lesions of the pneumotaxic area in the rostrolateral pons, evidencing a biphasic short-term potentiation (STP) effect.[2,4] Such biphasic STP via pontine pathway suggests a novel form of nonassociative learning called response "desensitization,"

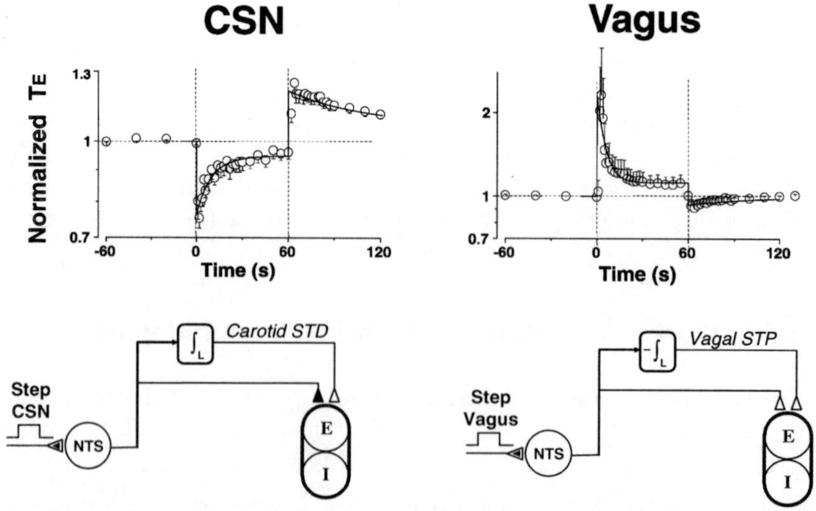

Figure 1. Differentiator models of HBR and CCR control of T_E. *Upper* panels: responses in T_E to low-intensity 1-min electrical stimulation of carotid sinus nerve (20 Hz, *left*) or vagus nerve (80 Hz, *right*) in anesthetized, paralyzed and ventilated rat (adapted from ref. 2). *Lower* panels: schematic of conditioning pathways that are consistent with corresponding direct reflex and indirect differentiator effects via ventrolateral/rostrolateral pons.

which is secondary to PSR activity but independent of the primary relay between NTS pump cells and decrementing E neurons.[4]

The observed HBR response habituation and desensitization suggest that PSR afferents are relayed centrally by at least two parallel pathways (Fig. 1): a direct, primary reflex pathway with paucisynaptic connection from NTS pump cells to E neurons as previously proposed and a secondary, indirect differentiator pathway via the pontine pneumotaxic center. (Here, a neural differentiator is simply a neural integrator that counteracts rather than augments the primary reflex). Both pathways are modulated by synaptic accommodation typical of most NTS relay neurons (including pump cells)[8] resulting in a differentiator-like habituation effect that attenuates the PSR input. In addition, pontine desensitization results in a secondary differentiator effect that attenuates neurotransmission in the pontine pathway, perhaps in a manner similar to that observed during direct stimulation of the pontine pneumotaxic center.[5] These differentiator effects provide dynamic conditioning of the PSR input analogous to high-pass filtering.[2]

3. Working Model of Carotid Chemoreflex Modulation of T_E

Similar differentiator-like effects are also evident in CCR. During carotid sinus nerve (CSN) stimulation.[2] or acute hypoxia.[6] both the initial reflex shortening of T_E and post-stimulation rebound increase in T_E decay exponentially, resulting in short-term depression (STD) of respiration (Fig. 1). Both decay effects are abolished by NMDA receptor blockade.[2,6] or bilateral lesions of a critical area in the ventrolateral pons.[6,7] Certain expiratory neurons in this area have also been found to exhibit similar complex response pattern following acute hypoxia, with a successive sequence of initial inhibition and subsequent sustained excitation coincident with the reflex shortening and subsequent STD prolongation of TE.[6] These findings suggest a model of CCR control of T_E (Fig. 1) in which respiratory STD is correlated to sensitization of a pontine pathway that promotes expiratory activity. The combination of reflex shortening and STD prolongation of T_E results in a differentiator-like effect that is equivalent to high-pass filtering of the carotid chemoreceptor input.

Because of the relatively low CSN stimulation frequency (20 Hz) a habituation effect correlated with synaptic accommodation (with short-term or long-term depression) in NTS is not discerned during the 1-min stimulation period but is evident over a longer period.[8] Thus, the conditioning pathways mediating CCR modulation of T_E appear to be organized in a similar fashion as those for HBR except the effects are opposite.

4. Working Model of Carotid Chemoreflex Modulation of T_I

In contrast to the control of T_E, the CCR and HBR control of T_I may involve very different neural architectures. Presumably, the HBR shortening of T_I could stem from similar relay pathways mediating the HBR prolongation of T_E, since E and I neuronal responses are inverted through mutual inhibition (Fig. 1). However, the simultaneous (non-inverted) shortening of T_E and T_I by CCR implies that the corresponding relay pathways are distinct and independent of E-I mutual inhibition.

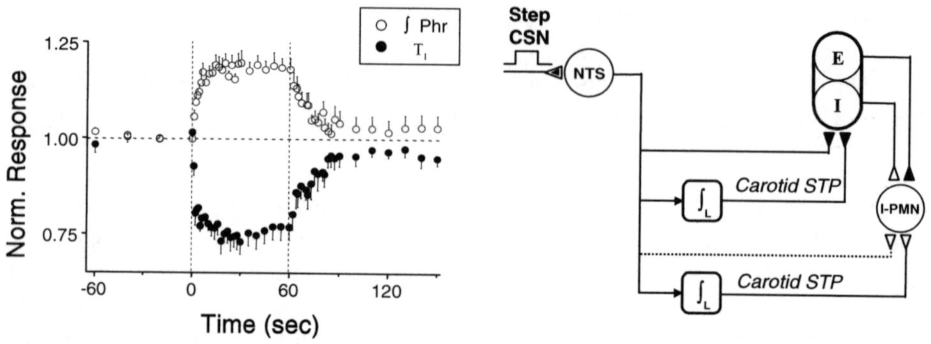

Figure 2. Integrator model of CCR control of phrenic activity and T_I. *Left*: responses in amplitude of integrated phrenic activity and T_I to low-intensity 1-min CSN stimulation (20 Hz) in anesthetized, paralyzed and ventilated rat (adapted from ref. 10). *Right*: schematic of corresponding conditioning pathways relayed by NTS. I-PMN, inspiratory premotor neuron.

Typically, acute hypoxia or electrical CSN stimulation may induce STP of inspiratory activity characterized by incrementing phrenic nerve discharge and post-stimulation afterdischarge.[9] Recently, it has been shown that this inspiratory STP is also expressed as gradual shortening of T_I during CSN stimulation and gradual recovery post-stimulation.[10] The carotid STP response in T_I and in amplitude of integrated phrenic activity (Phr) follow a similar integrator-like time course equivalent to low-pass filtering of carotid chemoreceptor input (Fig. 2). Importantly, the corresponding integrator time constants during response onset and offset are different than those of carotid STD response in T_E suggesting separate CCR control of T_I and T_E. The carotid STP response in T_I and Phr conform with response sensitization (another form of nonassociative learning) and both contribute to an augmentation of ventilatory output. Again, habituation of the CCR modulation of T_I could be discerned over a longer stimulation period.[8]

5. Carotid Chemoreflex is not a Primary Reflex

The above working models of CCR assume that carotid chemoreceptor afferents modulate T_E and T_I via separate primary reflex and secondary integrator-differentiator pathways. But are these "primary reflex pathways" for CCR genuine reflex pathways, i.e., comprising of paucisynaptic neuronal projections like those for HBR? To address this question we have developed an analytical procedure (*stroboscopic interferometric filtering technique*, SIFT).[8] that readily distinguishes a primary reflex pathway and secondary integrator-differentiator pathway based on their distinct responses to stroboscopic stimuli gated to either the E or I phase. Thus, a primary reflex pathway may respond rapidly whenever the stimulus comes on or off, whereas a secondary pathway with short-term memory generally responds more slowly but may persist post-stimulus.

In Fig. 3A, stroboscopic CSN stimuli are seen to elicit similar integrator- and differentiator-like responses in T_E, T_I and Phr as with a continuous-step CSN stimulus (Figs. 1 and 2). Moreover, the effects of the E- or I-gated stimuli on T_E and T_I are strikingly

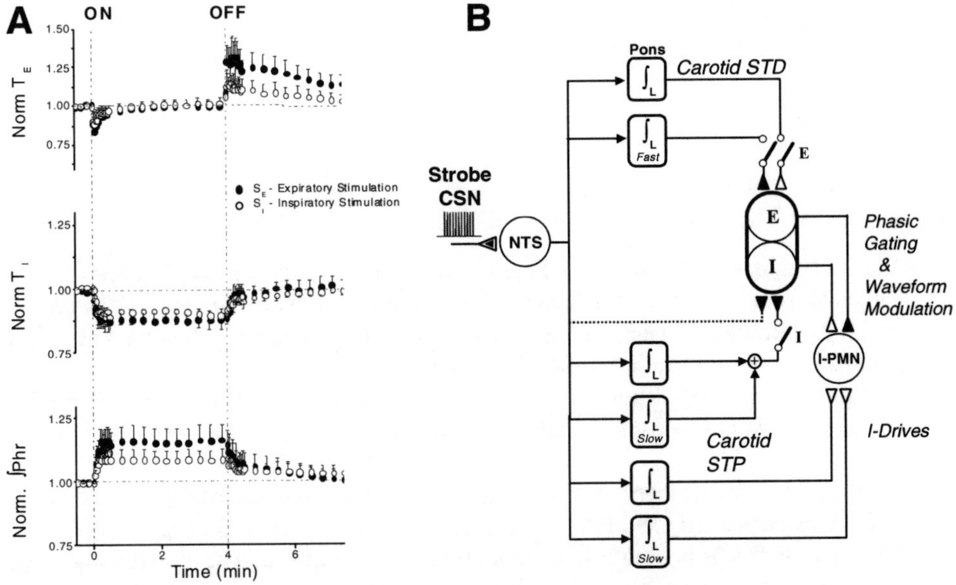

Figure 3. Comprehensive integrator-differentiator model of CCR revealed by stroboscopic CSN stimulation. **A**. Responses in TE, TI and Phr to low-intensity 4-min CSN stimulation (20 Hz) during either E or I phase in anesthetized, paralyzed and ventilated rat. **B**. Schematic of corresponding integrator-differentiator pathways which are relayed by NTS and are separately gated to E or I phase. Note that the "primary reflex pathways" in the CCR models of Figs. 1 and 2 are replaced by fast-responding integrator pathways. (Adapted from ref. 8)

similar, in contrast to the opposite effects that are expected of a primary reflex pathway (such as the E-promoting and I-shortening effects of HBR). These observations suggest that the "primary reflex pathways" depicted above for CCR are not paucisynaptic pathways but may be comprised of occult secondary integrators with response dynamics so fast that they are often mistaken for a primary reflex pathway except when activated with a stroboscopic stimulus (Fig. 3B).

6. Carotid Chemoreflex Pathways are Phase-Selective

A dilemma in the above models of CCR is: even if TE and TI are controlled by separate integrator-differentiator pathways, the corresponding effects may still interfere with one another through mutual inhibition of E and I neurons. Thus, it is conceivable that a conditioning pathway that serves to shorten TE would also lengthen TI and such crossover correlation may negate the resultant ventilatory effect.

However, analysis of the CCR data (Figs. 1–3) revealed that the TE and TI responses have distinct onset and offset dynamics, with no evidence of crossover correlation. This observation suggests the possibility that CCR conditioning pathways may be individually gated to either the E or I phase and are cutoff in the opposite phase by some intrinsic "phase filter" (as opposed to frequency filter) similar to a traffic light at a two-way intersection. Thus, the corresponding effects are phase-specific and do not interfere in the opposite phase.

Such phase filtering effect may also occur in the efferent pathway where inspiratory motor activity may be phasically gated by I and E neurons (Fig. 3B).

7. Summary

The above modeling results suggest several general and special principles of neural organization that may underlie the central mediation of HBR and CCR:

1. The immediate HBR effects on TE and (via mutual inhibition) on TI appear to be mediated by an identified paucisynaptic central pathway. The immediate CCR effects on TE, TI and Phr may be mediated by fast secondary integrator (or low-pass filtering) pathways instead of primary reflex pathways.
2. Additional secondary differentiator (or high-pass filtering) pathways via the rostro-lateral/ventrolateral pons may underlie the desensitization of HBR/CCR response in TE.
3. Additional secondary integrator pathways (loci unclear) may underlie the sensitization of CCR response in TI and Phr.
4. These parallel CCR integrator-differentiator pathways appear to be individually gated to a specific respiratory phase by corresponding E or I phase filters.
5. NTS synaptic depression may underlie response habituation of HBR and CCR.

8. Acknowledgments

The author's research was supported by National Institutes of Health grants HL60064 and HL67966.

References

1. Hayashi, F. Coles, S.K. & McCrimmon, D.R. Respiratory neurons mediating the Breuer-Hering reflex prolongation of expiration in rat. *J Neurosci* **16**, 6526–36 (1996).
2. Poon, C.-S. Young, D.L. & Siniaia, M.S. High-pass filtering of carotid-vagal influences on expiration in rat: role of N-methyl-D-aspartate receptors. *Neurosci. Lett.* **284**, 5–8 (2000).
3. Zhou, Z., Champagnat, J. & Poon, C.-S. Phasic and long-term depression in brainstem nucleus tractus solitarius neurons: differing roles of AMPA receptor desensitization. *J. Neurosci.* **17**, 5349–56 (1997).
4. Siniaia, M.S., Young, D.L. & Poon, C.-S. Habituation and desensitization of the Hering-Breuer reflex in rat. *J. Physiol. (Lond.)* **523**, 479–491 (2000).
5. Younes, M., Baker, J. & Remmers, J.E. Temporal changes in effectiveness of an inspiratory inhibitory electrical pontine stimulus. *J. Appl. Physiol.* **62**, 1502–1512 (1987).
6. Dick, T.E. & Coles, S.K. Ventrolateral pons mediates short-term depression of respiratory frequency after brief hypoxia. *Respir Physiol* **121**, 87–100 (2000).
7. Coles, S.K. & Dick, T.E. Neurones in the ventrolateral pons are required for post-hypoxic frequency decline in rats. *J Physiol* **497 (Pt 1)**, 79–94 (1996).
8. Young, D.L., Eldridge, F.L. & Poon, C.S. Integration-differentiation and gating of carotid afferent traffic that shapes the respiratory pattern. *J Appl Physiol* **94**, 1213–29. (2003).
9. Wagner, P.G. & Eldridge, F.L. Development of short-term potentiation of respiration. *Respir Physiol* **83**, 129–39 (1991).
10. Poon, C.-S. Siniaia, M.S. Young, D.L. & Eldridge, F.L. Short-term potentiation of carotid chemoreflex: An NMDAR-dependent neural integrator. *NeuroReport* **10**, 2261–65 (1999).

Converging Functional and Anatomical Evidence for Novel Brainstem Respiratory Compartments in the Rat

Donald R. McCrimmon, George F. Alheid, Minchun Jiang, Tara Calandriello and Anupama Topgi

1. Introduction

Central nervous system circuits controlling respiratory motoneurons have usually been divided into a dorsal medullary respiratory group—associated with portions of the nucleus of the solitary tract, a ventrolateral medullary respiratory group—mostly ventral to the motoneurons of nucleus ambiguus, and a pontine respiratory group—associated with portions of the parabrachial nuclear complex. When visualized via retrograde labeling of neurons from injections within the ventrolateral medulla (Fig. 1A), the labeled aggregates of respiratory-related neurons in the pons and medulla appear to form a practically continuous column of cells. These are concentrated in lateral brainstem areas stretching from the caudal mesencephalon, through the pons and medulla[1–4]. This pattern of labeling highlights the rostrocaudal columnar organization of respiratory neurons that integrates respiratory control across the entire extent of the rhombencephalon. A second organizational theme is represented by a rostrocaudal segmentation of respiratory neurons within this column into functional compartments (Fig. 1B).

Within the medulla, the ventral respiratory column (VRC) is generally subdivided into successive rostrocaudal compartments based principally on the phasic firing patterns of the constituent neurons (e.g. inspiratory in the rostral part of the ventral respiratory cell group [rVRG], or expiratory in the Bötzinger complex and caudal VRG [cVRG; Fig. 1B]).

Physiological identification remains the definitive means for distinguishing between respiratory compartments. However, with the growing availability of reliable antibodies to neural antigens and the development of novel technical methods for the histochemical demonstration of brainstem neurochemical markers, the ability to discriminate established,

---

Donald R. McCrimmon • Dept. Physiology—M211, Northwestern University Feinberg School of Medicine, 303 E. Chicago Ave. Chicago IL 60611-3008 USA. **George F. Alheid, Minchun Jiang, Tara Calandriello and Anupama Topgi** • Northwestern University Feinberg School of Medicine, Chicago IL 60611 USA.

Post-Genomic Perspectives in Modeling and Control of Breathing, edited by Jean Champagnat, Monique Denavit-Saubié, Gilles Fortin, Arthur S. Foutz, Muriel Thoby-Brisson. Kluwer Academic/Plenum Publishers, 2004.

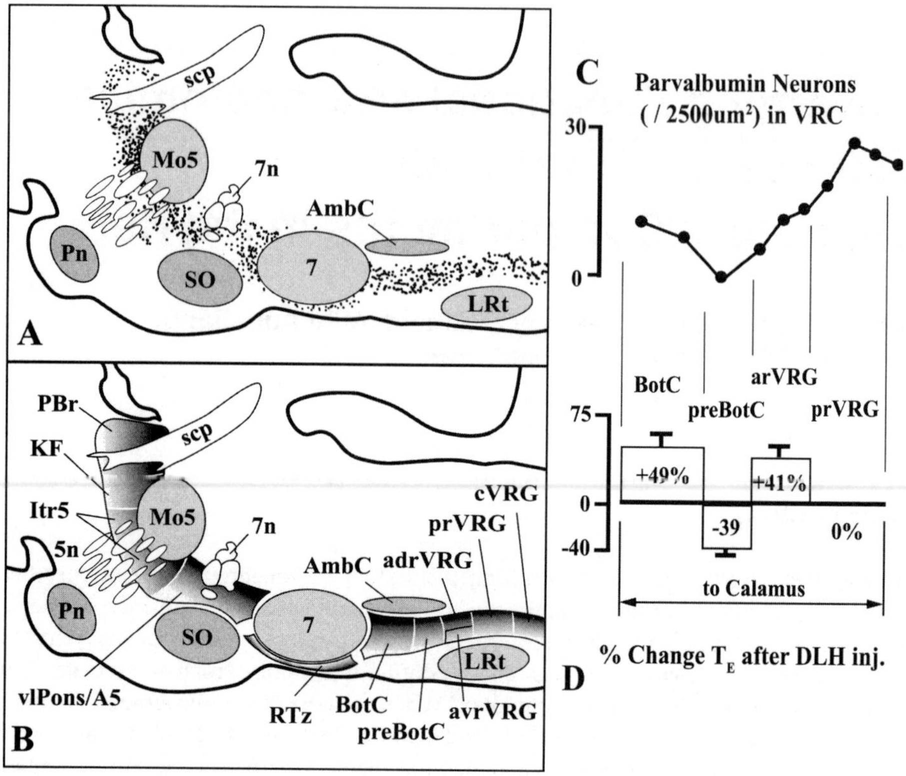

Figure 1. A. Retrograde labeling in the pontine respiratory column and in the ventral respiratory column of the rat brain after a FluoroGold injection in the region of the Bötzinger/preBötzinger complex. Neurons were mapped in multiple adjacent sections and projected onto a single sagittal plane.
B. Schematic diagram of pontomedullary respiratory regions corresponding to the retrograde labeling depicted in Figure 1A.
C. The rostrocaudal distribution of parvalbumin immunolabeled neurons in the VRC. Labeled neurons were counted in a 500 × 500 um^2 area in the VRC located just ventral to the compact part of nucleus ambiguus or at approximately the same dorsoventral level in VRC areas caudal to the compact part of the nucleus.
D. Changes in the expiratory time (TE) resulting from DL-homocysteate microinjections in successive rostrocaudal levels of the VRC.
Abbreviations: 5n, trigeminal nerve; 7, facial motor nucleus; adrVRG, anterodorsal part of the rostral ventral respiratory group; AmbC, nucleus ambiguous, compact part; arVRG anterior part of the rostral ventral respiratory group; avrVRG, anteroventral part of the rVRG; BotC, Bötzinger complex; cVRG, caudal VRG; Itr5, intertrigeminal nucleus; KF, Kölliker-Fuse nucleus; LRt, lateral reticular nucleus; Mo5, motor nucleus of the trigeminal nerve; PBr, parabrachial nucleus; Pn, basilar pontine nuclei; preBötC, preBötzinger complex; prVRG, posterior part of the rostral ventral respiratory group; PV, parvalbumin; RTz, retrotrapezoid nucleus; rVRG, rostral ventral respiratory group; scp, superior cerebellar peduncle; SO, superior olive; vlPons/A5, ventrolateral pons & A5 cell areas.

or even novel respiratory compartments is considerably enhanced. For example, the presence of neurokinin-1 (NK1) receptors in small propriobulbar neurons serve to distinguish the preBötzinger complex from the rostrally adjacent Bötzinger complex[5,6] and from a population of larger NK1 receptive bulbospinal neurons located in the anterior ventral portion of the rVRG[4,7–10]. It has also been argued that the dense somatostatin immunoreactivity of a subset of neurons within the preBötzinger complex is sufficient to distinguish the preBötzinger complex[11]. In the same vein, preproenkephalin appears to identify glutamatergic bulbospinal neurons in the rVRG[12]. In our experiments, we sought to provide a further neurochemical characterization of VRC neurons using antigenicity for parvalbumin, an additional neurochemical marker prevalent in VRC regions.

2. Parvalbumin Bulbospinal Respiratory Neurons

Parvalbumin is a calcium binding protein that fills the role of an endogenous neuronal calcium buffer and has a widespread but heterogeneous distribution within the central nervous system[13–15]. Antibodies to this protein are widely available and antigenicity is relatively high within neuronal cell bodies, making cell labeling relatively straightforward. Based on normal immunocytochemistry, it was suggested[16] that parvalbumin might be a marker for respiratory neurons in the VRC. This suggestion was applied in labeling VRC regions for a recent neurochemical atlas of the rat brainstem[15], however, direct identification of parvalbumin in respiratory neurons had not been demonstrated. Accordingly, we immunolabeled parvalbumin neurons in the brainstem of rats counterstained for the NK1 receptor which is prevalent in preBötzinger propriobulbar neurons[5,8,10], and also examined parvalbumin in VRC neurons counterstained by retrograde labeling from injections in either the phrenic nucleus or the VRC[4]. In parallel experiments[17], we mapped the cardiorespiratory response to stimulation of VRC neurons with nanoliter injections of an excitatory amino acid, DL-homocysteate acid (DLH). The results of these experiments ultimately provided data complementing our anatomical analysis.

These data indicate that the majority (66%) of bulbospinal VRC neurons projecting to the phrenic nucleus contain parvalbumin. A much lower percentage of neurons with axons that target the VRC (22%; i.e. propriobulbar projections) are positively stained for parvalbumin, and it is likely that some of the latter include bulbospinal neurons with local medullary collaterals[18].

Parvalbumin immunocytochemistry in the rat is particularly effective in identifying transitions between established VRC compartments (Fig. 2A). Since parvalbumin appears to be preferentially expressed in bulbospinal VRC neurons, it is consequently rarely found in VRC neuronal compartments populated mainly by propriobulbar neurons. For example, areas such the retrotrapezoid nucleus and preBötzinger complex largely exclude parvalbumin neurons.

The combination of parvalbumin immunolabeling, NK1 receptor immunoreactivity, and chemical microstimulation, suggests a novel subdivision of the rVRG into anterior and posterior parts (Fig. 1B-D). It has previously been indicated that NK1 receptor labeling is not restricted to the preBötzinger region of the VRC but also labels a subset of bulbospinal neurons[8–10]. However, the NK1-receptive bulbospinal neurons have a relatively restricted distribution. They are only found ventrally in the anterior half of the rVRG. In addition, NK1 receptors are not found on parvalbumin positive bulbospinal neurons. Consequently, within the anterior rVRG, parvalbumin neurons and NK1-receptive neurons represent two different populations of bulbospinal respiratory neurons. These appear in more or less separate anteroventral and anterodorsal portions of the rVRG, respectively (Fig. 1B).

A functional difference between anterior and posterior aspects of the rVRG is also consistent with the distribution of parvalbumin containing VRC neurons. The cell density of parvalbumin neurons within the VRC peaks in the posterior part of the rostral VRG

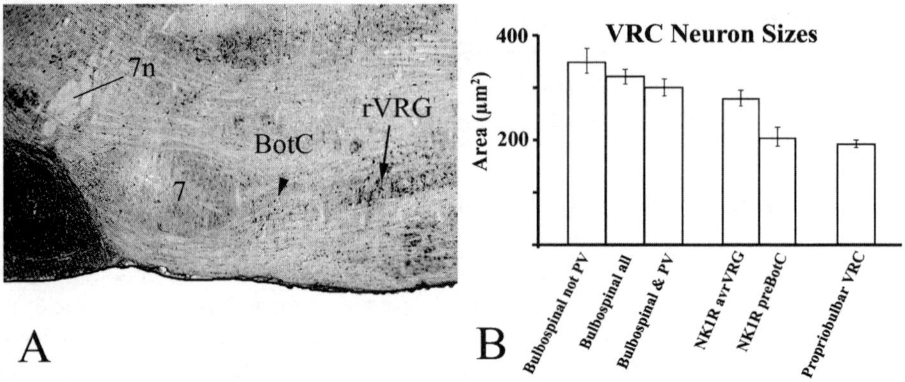

Figure 2. Parvalbumin immunoreactivity in the ventral respiratory column. **A**. Parasagittal section is shown through the level of the compact part of nucleus ambiguus. Notably, parvalbumin neurons are evident in the Bötzinger complex (BotC) and in a column beginning in the rostral VRG (rVRG) and extending caudally into the caudal VRG, but parvalbumin neurons are nearly entirely absent from the preBötzinger complex (preBotC). **B**. Average cross-sectional areas (±SE) of VRC neuronal populations. Fluorescent labeled neurons were measured in cleared and coverslipped wet-mounted sections to avoid shrinkage. Sizes of DAB labeled neurons were corrected for shrinkage by a factor derived from comparison of photos of individual cells before and after dehydration (see reference 4 for details). Abbreviations as in Fig. 1.

(Fig. 1C), a region that appears to be mainly populated by respiratory premotor bulbospinal neurons[19]. Very few propriobulbar (third order) respiratory neurons seem to be found in this region. Injecting an excitatory amino acid into the midst of physiologically recorded respiratory neurons in the anterior half of the rVRG (3 nl of 5 mM DLH) elicits bradypnea as does an injection into the Bötzinger complex[17]. Stimulation of neurons between these two regions, i.e., within the preBötzinger complex, resulted in tachypnea. However, no effect on respiratory rhythm was elicited by stimulation of the posterior half of the rostral VRG (Fig. 1D). Similar results were observed, in part, by Wang et al[20], i.e., bradypnea and tachypnea with stimulation in the Bötzinger and preBötzinger complexes respectively, with little or no effect on respiratory rhythm after stimulation in the posterior part of the rostral VRG. The inability of DLH stimulation to reset respiratory rhythm when injected into the posterior rVRG is concordant with the suggestion that this region consists mainly of bulbospinal rather than propriobulbar respiratory neurons[19], since it has previously been reported that electrical stimulation of descending bulbospinal axons, while momentarily disrupting breathing, does not result in resetting of the respiratory rhythm[21]. This suggests that the feedback from respiratory bulbospinal premotor neurons to the rhythm generating circuits is low.

Finally, the comparison of immunolabeled and retrogradely labeled neurons also indicates that the different populations of VRC neurons can be distinguished from one another on the basis of cell size[4] (corrected cross-sectional areas; Fig. 2B). Based on retrograde labeling alone, bulbospinal VRC neurons are slightly larger on average, than the VRC parvalbumin cell population. Parvalbumin neurons, on the other hand, are approximately the same size as the NK1 receptive neurons in the anteroventral rVRG, and on average, both of these cell types are significantly larger than propriobulbar neurons. This includes the NK1 receptor-positive propriobulbar neurons within the preBötzinger complex[4,7,10] (Fig. 2B). Thus even without direct double labeling, multiple VRC neuronal classes may be discriminated based on the somatic size of the cell population, their neurochemical content, and their compartmental location.

Acknowledgement

We gratefully acknowledge the support of NIH grants HL 72415 and HL 73474.

References

1. H.H. Ellenberger, J.L. Feldman, Brainstem connections of the rostral ventral respiratory group of the rat, *Brain Res.* **513**, 35–42 (1990).
2. H.H. Ellenberger, J.L. Feldman, Subnuclear organization of the lateral tegmental field of the rat. I: Nucleus ambiguus and ventral respiratory group, *J. Comp. Neurol.* **294**, 202–211 (1990).
3. P.A. Núñez-Abades, A.M. Morillo, R. Pásaro, Brainstem connections of the rat ventral respiratory subgroups: afferent projections, *J. Auton. Nerv. Syst.* **42**, 99–118 (1993).
4. G.F. Alheid, P.A. Gray, M.C. Jiang, J.L. Feldman, D.R. McCrimmon, Parvalbumin in respiratory neurons of the ventrolateral medulla of the adult rat, *J. Neurocytol.* **31**, 693–717 (2002).
5. P.A. Gray, J.C. Rekling, C.M. Bocchiaro, J.L. Feldman, Modulation of Respiratory Frequency by Peptidergic Input to Rhythmogenic Neurons in the PreBötzinger Complex. *Science* **286**, 1566–1568 (1999).
6. P.A. Gray, W.A. Janczewski, N. Mellen, D.R. McCrimmon, J.L. Feldman, Normal breathing requires preBötzinger complex neurokinin-1 receptor-expressing neurons, *Nature Neurosci.* **4**, 927–930 (2001).
7. Y.-Y. Liu, G. Ju, M.T.T. Wong-Riley, Distribution and colocalization of neurotransmitters and receptors in the pre-Bötzinger complex of rats, *J. Appl. Physiol.* **91**, 1387–1395 (2001).
8. H. Wang, R.L. Stornetta, D.L. Rosin, P.G. Guyenet, Neurokinin-1 receptor-immunoreactive neurons of the ventral respiratory group in the rat, *J. Comp. Neurol.* **434**, 128–148 (2001).
9. J.M. Makeham, A.K. Goodchild, P.M. Pilowsky, NK1 receptor and the ventral medulla of the rat: bulbospinal and catecholaminergic neurons, *NeuroReport* **12**, 3663–3667 (2001).
10. P.G. Guyenet, C.P. Sevigny, M.C. Weston, R.L. Stornetta, Neurokinin-1 receptor-expressing cells of the ventral respiratory group are functionally heterogeneous and predominantly glutamatergic, *J. Neurosci.* **22**, 3806–3816 (2002).
11. R.L. Stornetta, D.L. Rosin, H. Wang, C.P. Sevigny, M.C. Weston, P.G. Guyenet, A group of glutamatergic interneurons expressing high levels of both neurokinin-1 receptors and somatostatin identifies the region of the pre-Bötzinger complex, *J. Comp. Neurol.* **455**, 499–512 (2003).
12. R.L. Stornetta, C.P. Sevigny, P.G. Guyenet, Inspiratory augmenting bulbospinal neurons express both glutamatergic and enkephalinergic phenotypes, *J. Comp. Neurol.* **455**, 113–124 (2003).
13. M.R. Celio, Calbindin D-28k and parvalbumin in the rat nervous system, *Neurosci.* **35**, 375–475 (1990).
14. G. Paxinos, L. Kus, K.W.S. Ashwell, C. Watson, *Chemoarchitectonic Atlas of The Rat Forebrain.* (Academic Press, San Diego, 1999).
15. G. Paxinos, P. Carrive, H. Wang, P.-Y. Wang, *Chemoarchitectonic Atlas of the Rat Brainstem* (Academic Press, San Diego, 1999).
16. M. Cox, G.M. Halliday, Parvalbumin as an anatomical marker for discrete subregions of the ambiguus complex in the rat, *Neurosci. Lett.* **160**, 101–105 (1993).
17. A. Monnier, G.F. Alheid, D.R. McCrimmon, Defining ventral medullary respiratory compartments with a glutamate receptor agonist in the rat, *J. Physiol. (Lond.)* **548**, 859–874 (2003).
18. J. Lipski, X. Zhang, B. Kruszewska, R. Kanjhan, Morphological study of long axonal projections of ventral medullary inspiratory neurons in the rat, *Brain Res.* **640**, 171–184 (1994).
19. E.G. Dobbins, J.L. Feldman, Brainstem network controlling descending drive to phrenic motoneurons in rat, *J. Comp. Neurol.* **347**, 64–86 (1994).
20. H. Wang, T.P. Germanson, P.G. Guyenet, Depressor and tachypneic responses to chemical stimulation of the ventral respiratory group are reduced by ablation of neurokinin-1 receptor-expressing neurons, *J. Neurosci.* **22**, 3755–3764 (2002).
21. J.L. Feldman, D.R. McCrimmon, D.F. Speck, Effect of synchronous activation of medullary inspiratory bulbo-spinal neurones on phrenic nerve discharge in cat, *J. Physiol. (Lond.)* **347**, 241–254 (1984).

Eupneic Respiratory Rhythm in Awake Goats is Dependent on an Intact Pre-Bötzinger Complex

H.V. Forster, J.M. Wenninger, L.G. Pan, M.R. Hodges and R. Banzett

1. Introduction

Different views exist regarding sites within the brain responsible for eupneic respiratory rhythmogenesis. Studies by Lumsden in the 1920's[1, 2] indicated that the minimum neural substrate for respiratory rhythm was the medulla but an intact pons was necessary for eupneic rhythm. Recent studies by St. John et al.[3, 4] support this concept. Another concept is that the pre-Bötzinger Complex (PBC) is the site of rhythmogenesis, and in the neonatal rat spinal cord-brainstem preparation, the minimum neural substrate for respiratory rhythm is the PBC.[5] Another study on awake adult rats showed that greater than 80% destruction of PBC neurons expressing the neurokinin 1 receptor (NK1-R) resulted in an "ataxic" breathing pattern and attenuation of eupneic breathing and CO_2 sensitivity, leading to the conclusion that "normal breathing in an intact animal requires an intact PBC".[6] However, St. John et al[4] contend that an ataxic breathing pattern is evident with lesions at other medullary sites, and we[7] found that lesions in rostral medullary nuclei result in transient interruptions of eupneic breathing. In addition, pre-inspiratory neurons rostral to the PBC are capable of generating a respiratory rhythm.[8, 9]

Because of the different views summarized above, we designed studies to gain insight into the role of the PBC in the control of breathing in the awake state. We studied goats before and after bilateral injections into the PBC area of saporin conjugated to substance P (SAP-SP) (neurotoxic for NK1-R expressing neurons) and then two weeks later ibotenic acid (IA) (neurotoxic for glutamate receptor expressing neurons). A portion of the data have been published;[10, 11] thus herein, we summarize the previous reports and present and discuss findings not previously reported.

H.V. Forster, J.M. Wenninger, L.G. Pan, M.R. Hodges and R. Banzett • Department of Physiology, Medical College of Wisconsin, Zablocki VA and Department of Physical Therapy, Marquette University, Milwaukee, WI 53226. Physiology Program, Harvard School of Public Health & Dept of Medicine, Harvard Medical School, MA 02115.

Post-Genomic Perspectives in Modeling and Control of Breathing, edited by Jean Champagnat, Monique Denavit-Saubié, Gilles Fortin, Arthur S. Foutz, Muriel Thoby-Brisson. Kluwer Academic/Plenum Publishers, 2004.

2. Method

Goats were cared for in accordance with and protocols approved by the Medical College of Wisconsin Animal Care Committee. Eleven adult female and 2 castrated male goats (weight = 35–55 kg) were housed and studied in a temperature and light controlled chamber. Except for study periods, food and water were available.

In an initial surgery, the carotid arteries were elevated, and electrodes were implanted into respiratory muscles. Three weeks later, microtubules were chronically implanted bilaterally into the PBC area (just ventral to rostral nucleus ambiguous). In 4 goats, a tracheotomy was created 3 days before the IA injections.

To monitor inspiratory flow in airway intact goats, a pneumotach was attached to the inspired port of a breathing valve attached to a mask taped to the goat's snout while in tracheostomized goats the valve was attached to an inserted, cuffed endotracheal tube. Diaphragm, upper airway, and abdominal muscle activities were recorded. Arterial blood was sampled (arterial blood gas determination) and arterial blood pressure was measured from a chronically indwelling catheter. Rectal temperature was recorded. Three weeks after implanting the microtubules, SAP-SP (n = 10) or SP (n = 1) was injected bilaterally. Measurements were made for 30 minutes before, 4 to 5 hours after the injections, and daily up to 15 days thereafter. Then in 10 goats bilateral injections of IA were made. Seven goats were studied for only 3 to 5 hours (see results), but 3 were studied daily for 2 weeks. Two goats had only IA injections and they were studied for 2 weeks thereafter. Other daytime studies assessed CO_2 and hypoxic sensitivity and the exercise hyperpnea.

Upon termination of the goats (3 died spontaneously and 10 were euthanized with Beuthanasia), the brain was flushed and fixed by vascular perfusion with buffered saline prior to removal of the medulla which was cryoprotected and sectioned (25 m). One set of slides was stained with hematoxylin and eosin for identification of dead or dying neurons and a second set was prepared for NK1-R immunoreactivity.

3. Results

During the first hour after bilateral injection of SAP-SP (50 pM in 10 l) into the PBC area of awake goats, there was a 50 to 100% increase ($P < 0.05$) from control in the frequency of augmented breaths (figure 1b) and there was a 50% increase ($P < 0.05$) in the frequency of fractionated breaths (defined as a transient interruption of the normal respiratory rhythm by 3 or more rapid, small complete breaths (figure 1c). Another main effect within the first 5 hours after the injection was a discoordination of diaphragm, abdominal, and upper airway muscle activities (figure 1d). The frequency of fractionated breaths was 10 fold above control 10 to 15 days after the injection. Fractionated breaths were associated with stronger than normal activity of the abdominal muscles. Often there was co-activation of the diaphragm and abdominal muscles (figure 1d) resulting in no or minimal flow and often the stronger abdominal contraction produced active expiration. Consequently, when the muscles relaxed there was passively driven inspiratory flow. CO_2 sensitivity, eupneic $PaCO_2$, and the exercise hyperpnea were at control levels between 2 and 15 days after the SAP-SP injections. Smaller doses of SAP-SP (n = 2) or injection of SP only (n = 1) into the PBC did not alter breathing patterns.

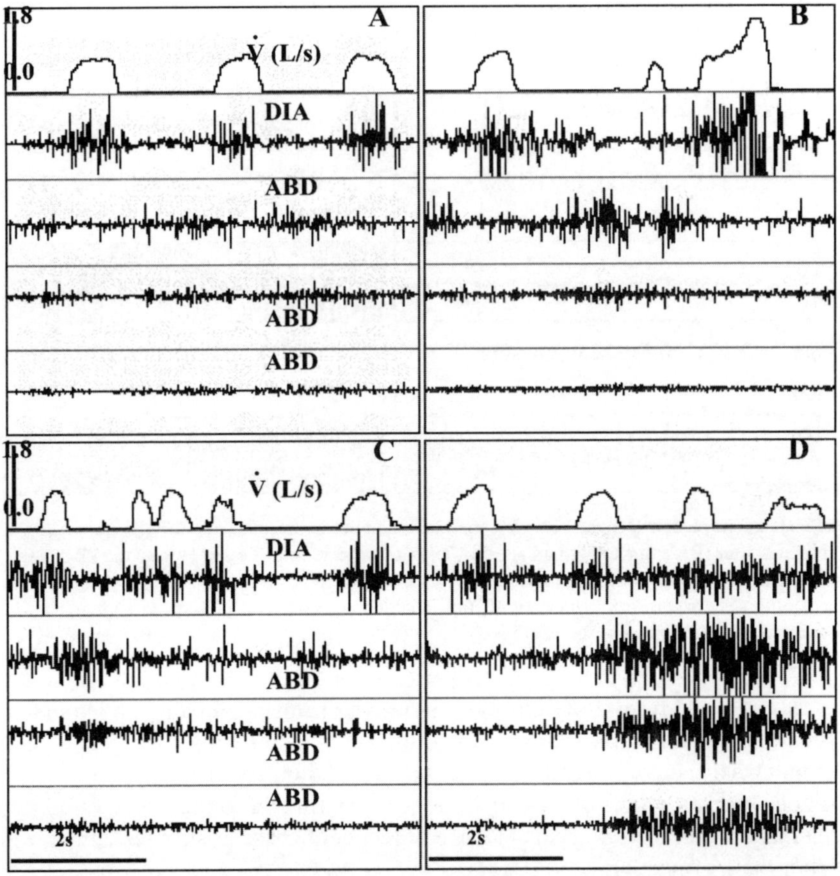

Figure 1. Inspiratory flow (V) and raw diaphragm and abdominal muscle activities in one awake goat before (A) and at 1.5 (B), 2 (C), and 3 (D) hours after bilateral injection of SAP-SP into the PBC area.

Within minutes after unilateral injection of IA into the PBC area, breathing and phasic respiratory muscle activity increased. (figure 2b) An hour later with an injection into the contralateral PBC area, breathing frequency and respiratory muscle activity were increased further, to a maximum (~150 breaths/min) about an hour later. In the 7 goats that previously had large bilateral injections of SAP-SP, periods of central apnea, hypoxemia and hypercapnia became evident. Eventually all breaths were gasp-like (decrementing diaphragm activity). In 3 airway intact goats, the frequency of gasps rapidly decreased and within minutes, respiratory arrest was followed by cardiac arrest. With onset of gasps in 4 tracheostomized animals, we began to mechanically ventilate the goats to avoid severe hypoxemia and hypercapnia. Over the next 4 hours respiratory muscle activity was minimal as long as $PaCO_2$ was at or below normal. (figure 2c) If mechanical ventilation was reduced resulting in hypercapnia, tonic diaphragm (figure 2d) and abdominal muscle activity increased. When the ventilator was briefly shut off, there was minimal tonic or phasic respiratory muscle activity until blood gasses changed or until the goat moved.

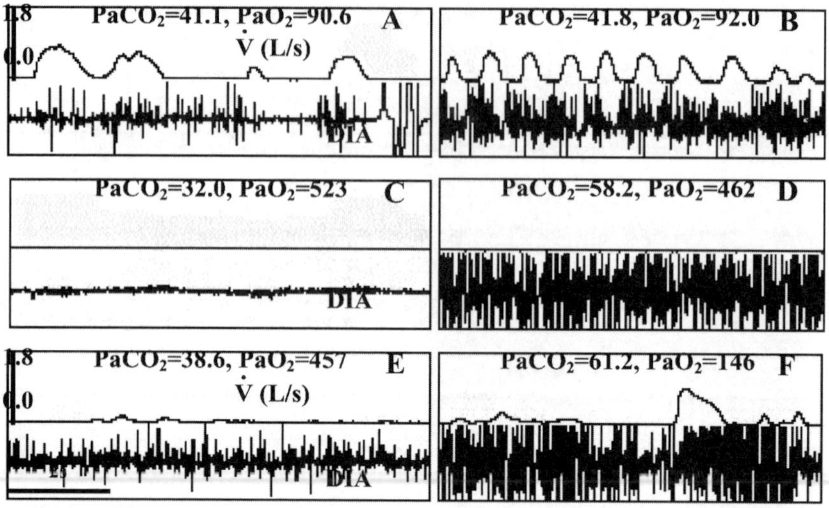

Figure 2. Inspiratory flow (V) and raw diaphragm and abdominal muscle activities of one awake goat before (A), 30 minutes after (B), and 4 to 5 hours after (C, D, E, F) bilateral injection of IA into the PBC area. The goat was mechanically ventilated at different rates during panels C and D. Panels E and F were obtained about 20 and 80 seconds after cessation of mechanical ventilation. The larger breath and high diaphragm activity in panel F were associated with movement of the entire body.

(figure 2e, f). Phasic respiratory muscle activity was primarily in all 3 abdominal muscles and synchronous with diaphragm activity, resulting in inspiratory flow 180 degrees out of phase with muscle activity, signifying a passive inspiration as described above.[11] Since any spontaneous breathing during these periods was insufficient to prevent severe hypoxemia and hypercapnia and there was no evidence of recovery, these goats were euthanized 5 to 6 hours after the IA injection.

In 3 goats that had no or only a small dose of SAP-SP previously injected into the PBC area, bilateral injection of IA into the PBC area initially had the same effect as described above. However, central apneas and gasping never occurred. Rather the tachypnea was sustained for up to 7 hours with subsequently a gradual restoration of breathing to normal. These goats recovered with no deficits in eupneic $PaCO_2$, CO_2 sensitivity or the exercise hyperpnea over the following 2 weeks. In 3 other goats, IA injections into nuclei outside the PBC had no acute or chronic effect on breathing.

These experiments were not designed to measure behaviour; but we did make informal observations. The interpretation of these changes must be tempered by possible involvement of non-respiratory motor pathways. When bilateral IA injections were made outside of the PBC area there were no behavioural responses. In contrast, after IA injections into the PBC area, there was a behavioural response, which varied between goats. These goats preferred not to or were unable to maintain a standing posture with the second injection. Once recumbent some goats maintained the normal sternal posture for all or part of the subsequent 2 to 5 hour period. The most prevalent response was a catatonic-like motor reaction characterized by: a) lying on one side and resisting attempts to change posture, b) over-all diminished motor initiative with long periods of inactivity broken by tonic-clonic whole body muscle contractions occurring always but not restricted to periods of hypoxemia and hypercapnia, and c) general restlessness and staring, bulging eyes interpreted as discomfort. In goats that fully recovered, the behavioural characteristics gradually decreased 4 to 5 hours after the injection and a standing posture was soon evident. In the 3 airway intact goats that died, there was no excessive reaction beyond that described above. In those goats that were mechanically ventilated, the intensity and frequency of whole body movements increased with time, which led to the decision of euthanasia.

4. Discussion

Others have shown that glutamate receptor agonists elicit a tachypnea when injected into the PBC area in anesthetized rats and cats.[12, 13] In addition, injections of SP into the same area elicits augmented breaths.[12] The same injection at sites immediately surrounding the PBC area do not elicit these responses. These data indicate the PBC area can be identified physiologically. Since in our awake goats, injections of SAP-SP elicited augmented breaths and the glutamate receptor agonist ibotenic acid induced a marked tachypnea only when injected into a site corresponding to the PBC in rats, we believe that we have identified the PBC in goats.

Injections of SAP-SP resulted in a 29% reduction in NK1-R expressing neurons in the area we designated the PBC.[11] The major respiratory effect was fractionated breathing and discoordination of respiratory muscle activity. Particularly evident was abnormally increased abdominal expiratory muscle activity associated with fractionated breathing and discoordination of muscle activity. These effects could be due to disruption of the normal respiratory rhythm and/or pattern generating mechanisms such as altered communication: a) between rhythm and pattern generation neurons, b) between the bilaterally located neurons, and/or c) between the PBC inspiratory and the more rostrally located expiratory rhythm generators. Whatever the mechanism, it is clear that it is specific to a reduction in NK1-R expressing neurons and only a moderate reduction in these neurons elicits the effects.

In the 7 goats that had reduced NK1-R neurons, the major finding with the subsequent IA injections was the eventual elimination of eupneic respiratory rhythm. In goats mechanically ventilated to sustain life, brief cessation of support resulted in no or minimal phasic diaphragm activity. Any phasic activity was associated with hypercapnia and usually whole body movements, (figure 2f) and it was in synchrony with phasic abdominal muscle activity that was greater than diaphragm activity.[11] We believe these findings support the concept of an expiratory rhythm generator, which usually is inhibited by a normally functioning inspiratory PBC rhythm generator.[8, 9] Interestingly, when goats were insufficiently mechanically ventilated, the ensuing hypercapnia (figure 2c, 2d) increased tonic respiratory muscle activity, suggesting that chemoreceptor input to respiratory neurons remained intact. Also of interest was that there did not appear to be any "volitional" breathing. These findings may indicate that cortical pathways for breathing utilize the PBC for rhythm.

Since these experiments contained no formal behavioural observations, inferences based on behaviour of the goats are only speculative; nonetheless they give us a possible insight heretofore impossible. We report them to encourage further thought and study. We observed signs interpreted as expressions of discomfort. There is no reason for these animals to have experienced pain as a result of the IA injections; furthermore, these expressions seemed to wax as ventilation decreased and wane as ventilation was restored by mechanical ventilation. We speculate that some of the behaviour interpreted as discomfort may have resulted from dyspnea, specifically 'air hunger', the sense of not getting enough air. Air hunger is increased by increasing $PaCO_2$ and is increased by decreased tidal volume even at constant $PaCO_2$.[14] Although it has never been well proven, the leading hypothesis to explain air hunger is projection of a copy of respiratory center motor activity to forebrain perceptual areas.[15,16,17] The present observations suggest that rhythmic respiratory center activity is not necessary for the genesis of air hunger. However observations in at least one goat show that signs of air hunger increased along with graded tonic activity of respiratory muscles. Is there an "amplitude" component of the respiratory center that survived the destruction of rhythmic activity, and that drives both air hunger and the amplitude of respiratory muscle contraction?

5. Acknowledgements

These studies were supported by NIH Grants 25739 and 46690 and by the Veterans Administration.

References

1. Lumsden, T. Observations on the respiratory centres. *J. Physiol Lond.* 57: 153–160, 1923.
2. Lumsden, T. The regulation of respiration I. *J Physiol. Lond.* 58: 81–91, 1923.
3. John, St. W.M. Medullary regions for neurogenesis of gasping: noeud vital or noeuds vitals? *J. Appl. Physiol.* 81(5): 1865–1877, 1996.
4. John, St. W.M., and J.F. Paton. Characterizations of eupnea, apneusis and gasping in a perfused rat preparation. *Respir Physiol.* Nov;123(3): 201–13, 2000.
5. Smith, J.C., H.H. Ellenberger, K. Ballanyi, D.W. Richter, and J.L. Feldman. Pre-Bötzinger Complex: A Brainstem Region That May Generate Respiratory Rhythm in Mammals. *Science*, 254: 726–729, 1991.
6. Gray, P.A., W.A. Janczewski, N. Mellen, D.R. McCrimmon, and J.L. Feldman. Normal breathing requires pre-Bötzinger complex neurokinin-1 receptor-expressing neurons. *Nature Neuroscience* (4)9: 927–930, 2001.
7. Feroah T.R., H.V. Forster, C.G. Fuentes, P. Martino, M. Hodges, J. Wenninger, L. Pan, and T. Rice. Perturbations in three medullary nuclei enhances fractionated breathing in awake goats. *J. Appl. Physiol.* 94: 1508–18, 2003.
8. Feldman, J.L., G.S. Mitchell, and E.E. Nattie. Breathing: Rhythmicity, plasticity chemosensitivity. *Annu. Rev. Neurosci.* 26: 239–66, 2003.
9. Janczewski, W.A., H. Onimaru, I. Homma, and J.L. Feldman. Opioid-resistant respiratory pathway from the preinspiratory neurons to abdominal muscles: in vivo and in vitro study in the newborn rat. *J Physiol.* Dec15; 545(Pt3): 1017–26, 2002.
10. Wenninger, J.M., H.V. Forster, L.G. Pan, L. Klum, T. Leekley, J. Bastastic, M.R. Hodges, S. Davis, and T.R. Feroah. Lesioning pre-Bötzinger complex-area NK1R-expressing neurons alters breathing of awake and asleep goats. *J. Appl. Physiol.* (Under Review).
11. Wenninger, J.M., H.V. Forster, L.G. Pan, L. Klum, T. Leekley, J. Bastastic, M.R. Hodges, S. Davis, and T.R. Feroah. Injections of two neurotoxins into the pre-Bötzinger Complex area eliminates eupneic breathing in awake goats. *J. Appl. Physiol.* (Under Review).
12. Solomon, I.C., N.H. Edelman, and J.A. Neubauer. Patterns of phasic motor output evoked by chemical stimulation of neurons located in the pre-Bötzinger Complex in vivo. *J. Neurophysiol.* 81: 1150–1161, 1999.
13. Monnier, A., G.F. Alheid, and D.R. McCrimmon. Defining ventral medullary respiratory compartments with a glutamate receptor agonist in the rat. *J Physiol.* Mar 14, 2003.

14. Banzett R.B., and R.W. Lansing. Respiratory sensations arising from chemoreceptors and pulmonary receptors: air hunger and lung volume. *Respiratory sensation (one ed)* edited by L. Adams and A. Grey. New York: Marcel Dekker. 155–180, 1996.
15. Banzett R.B., R.W. Lansing, M.B. Reed, L. Adams, and R. Brown. Air hunger arising from increased PCO2 in mechanically ventilated quadriplegics. *Respiration Physiology.* 76: 53–67, 1989.
16. Eldridge, F.L., and Z. Chen. Respiratory sensation a neurophysiological perspective. *Respiratory Sensation (1st ed.)* edited by L. Adams and H. Grey. New York: Marcel Dekker. 19–68, 1996.
17. Wright G.W., and B.V. Branscomb. The origin of the sensations of dyspnea? *Trans. Of Amer. Clin. And Clin. Assn.* 66: 116–125, 1954.

BDNF Preferentially Targets Membrane Properties of Rhythmically Active Neurons in the pre-Bötzinger Complex in Neonatal Mice

Muriel Thoby-Brisson, Sandra Autran, Gilles Fortin, and Jean Champagnat

1. Introduction

Neurotrophins are a class of factors known to be involved in the development, maintenance, plasticity and survival of the peripheral and central nervous systems[1]. Neurotrophins are active through high-affinity receptors having an intrinsic tyrosine-kinase activity[2,3]. One of these neurotrophins, the brain-derived neurotrophic factor (BDNF) has been shown to be involved in a wide range of developmental and plasticity mechanisms and its biological action depends mainly upon activation of the TrkB receptor sub-type and to a less extent to the p75 receptors. BDNF binding on TrkB receptors initiates a cascade of events of phosphorylation that activate a complex of signal proteins and the induction of several genes[4].

Mice carrying a targeted deletion in the gene encoding BDNF exhibit abnormal breathing pattern with a depressed and irregular ventilation[5,6]. This perturbed respiration could arise from developmental and/or neuromodulatory deficits. Respiration is generated by a neural network located in a sub-region of the medulla called pre-Bötzinger complex (PBC)[7,8]. This network can be isolated in vitro in a transverse medullary slice while remaining spontaneously rhythmically active. To address the possibility that BDNF can play an acute role in the respiratory neural network we examined expression of BDNF binding sites in PBC neurons as well as the effects of exogeneous BDNF application on neuronal properties. We found that PBC neurons expressed TrkB and p75 receptors and that BDNF affects membrane properties of rhythmically active PBC neurons specifically.

Muriel Thoby-Brisson, Sandra Autran, Gilles Fortin, and Jean Champagnat • Neurobiologie Génétique et Intégrative—Institut de Neurobiologie A. Fessard—CNRS, Gif sur Yvette, France. Supported by European grant QLG2/CT/2001-01467.

Post-Genomic Perspectives in Modeling and Control of Breathing, edited by Jean Champagnat, Monique Denavit-Saubié, Gilles Fortin, Arthur S. Foutz, Muriel Thoby-Brisson. Kluwer Academic/Plenum Publishers, 2004.

2. Methods

Transverse brainstem slices were obtained from 1- to 4-d-old OF1 mice as described previously[9]. A 400 μm thick slice containing the PBC at its rostral surface, was placed in a recording chamber maintained at 30°C and continuously perfused with oxygenated artificial CSF (composition in mM: 128 NaCl, 8 KCl, 1.5 $CaCl_2$, 1 $MgSO_4$, 24 NaHCO, 0.5 Na_2HPO_4, 30 glucose, pH = 7.4).

Recordings of population activity were performed using a suction electrode positionned at the surface of the slice on top of the PBC. Signals were amplified (AC amplifier Grass 7P511), filtered (widthband between 3Hz and 3KHz), integrated (Neurolog system, time constant 100 ms) and analyzed (Pclamp9, Digidata1320, Axon Instruments). Whole-cell patch clamp recordings were performed under visual control using an IR-DIC microscope. Neurons were selected according to their location (80–100 μm around the extracellular electrode and located ventral to the nucleus ambiguus). Rhythmic (R) and non-rhythmic (NR) neurons were discriminated according to their discharge pattern: rhythmic neurons were activated in phase with population activity, whereas non-rhythmic neurons were either silent or tonically active. Patch electrodes were filled with a solution composed of (in mM): 140 K-gluconic acid, 1 $CaCl_2$, 10 EGTA, 2 $MgCl_2$, 4 Na_2-ATP, 10 Hepes (pH 7.2).

Single-cell multiplex RT-PCR experiments were performed on identified (R or NR) PBC neurons (n = 42) as described previously[10,11]. At the end of the recording the cytoplasm was harvested into the recording pipette by application of a gentle negative pressure. The content of the pipette was then expelled into a test tube in which reverse transcription was performed. We simultaneously detected expression of BDNF, TrkB and p75 receptors mRNAs. Primers used (from 5′ to 3′, position 1 being the first base of the initiation codon): ATGTCTATGAGGGTTCGGCG (BDNF sense, position 370), GCGAGTTCCAGTGCCTTTTG (BDNF antisense, 606), ACTGTGAGAGGCAACCCCAA (TrkB sense, 916), ATCACCAGCAGGCAGAATCC (TrkB antisense, 1327), GGAGCCAACCAGACCGTGTG (p75 sense, 151), TCTGGGCACTCTTCACACACTG (p75 antisense, 410). These primers generated PCR fragments of 151, 344 and 165 bp for BDNF, TrkB and p75 receptor respectively. The cDNAs obtained after the first PCR were then amplified by 20 PCR cycles (94°C, 30 sec; 60°C, 30 sec; 72°C, 35 sec) using 2.5U of Taq polymerase and 10 pmol of each of the sets of primers cited above. Thirty five cycles of a second PCR were then performed using 2 μl of the first PCR products as template and the same primers as above. Ten microliters of each individual second step PCR reaction were then run on a 1.5% agarose gel using $\Phi \times 174$ cut by *Hae*III as molecular weight markers and stained with ethidium bromide.

3. Results

3.1. BDNF Modulates Rhythmic Activity in the PBC

On brainstem slices *in vitro* (n = 18), respiratory frequency was measured by extracellularly recording population activity from the PBC. BDNF bath application (100 ng/ml during 30 minutes, n = 12) induced a significant decrease in respiratory frequency from 0.21 ± 0.01 Hz to 0.16 ± 0.01 Hz (Fig. 1A, B; see ref 11). To test whether this effect involved

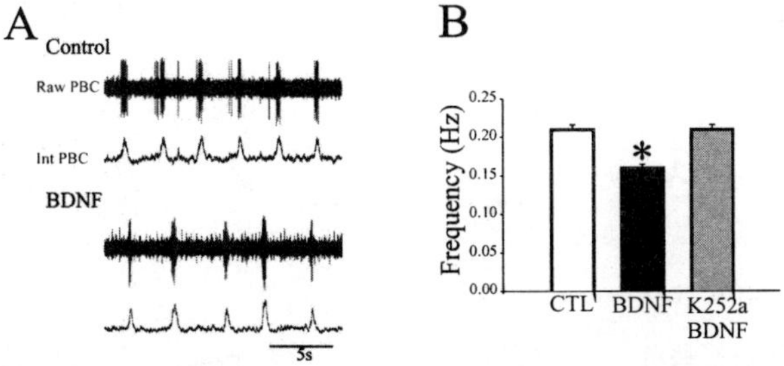

Figure 1. **BDNF decreases respiratory frequency in vitro. A**—Population activity (raw and integrated traces) recorded on the PBC in control conditions (top two traces) and after 30 min exposure to BDNF 100 ng/ml (bottom two traces). **B**—Bar histograms showing the mean frequency of rhythmic activity measured in control conditions (white bar, n = 18), in the presence of 100 ng/ml BDNF (black bar, n = 12) and in 200 nM K252a + 100 ng/ml BDNF (gray bar, n = 6). * $p < 0.05$.

activation of TrkB receptors, six slices were pre-treated with the tyrosine receptor kinase inhibitor K252a (200 nM, 1 hour). As shown in Fig. 1B, BDNF effects were abolished in the presence of the antagonist. These data indicate that BDNF modulation of population activity in the PBC is mediated through TrkB receptor activation (see ref 11).

3.2. BDNF Affects Membrane Properties Specifically on Rhythmic PBC Neurons

BDNF application did not modify membrane potentials of the nineteen PBC neurons recorded: −50.3 ± 5.2 mV in control conditions vs −49.9 ± 5.9 mV in the presence of BDNF (n = 19), suggesting that BDNF-induced frequency decrease is not likely resulting from the hyperpolarization of PBC neurons. Alternatively, BDNF might act on membrane conductances other than those controlling resting membrane potential. Within the PBC two types of neurons were recorded: neurons active in phase with population activity termed rhythmic neurons (R; Fig. $2A_1$) and neurons exhibiting tonic pattern discharge termed non-rhythmic neurons (NR; Fig. $2B_1$). In order to test whether BDNF specifically affects neurons involved in the generation of respiratory rhythm we examined membrane properties of both types of neurons in control conditions and after treatment with the neurotrophin. Average I–V curves obtained from 10 R neurons exhibited significant changes only for depolarizing pulses ranging from +10 and +30 mV (Fig. $2A_2$). In contrast, treatment with BDNF did not affect average I–V curves for 9 NR neurons (Fig. $2B_2$).

To investigate a selective BDNF action on mechanisms controlling the respiratory rhythm generation we also examined BDNF effects on the Ih current, a conductance known to be involved in the modulation of respiratory frequency[12]. The amplitude of the Ih current evoked by a series of hyperpolarzing voltage pulses significantly decreased in R neurons after BDNF exposure (Fig. $2A_3$, see ref 11). For example, for a test pulse to −120 mV the maximum current amplitude decreased from −0.24 ± 0.01 nA in control conditions to −0.14 ± 0.009 nA in BDNF. In addition, the Ih current starts to activate at voltage more hyperpolarized in the presence of BDNF (around −80 mV) compared to control (around −60 mV). In contrast, the amplitude of the Ih current recorded in NR neurons was unaffected by the treatment (Fig. $2B_3$). The Ih current maximum amplitude measured for a

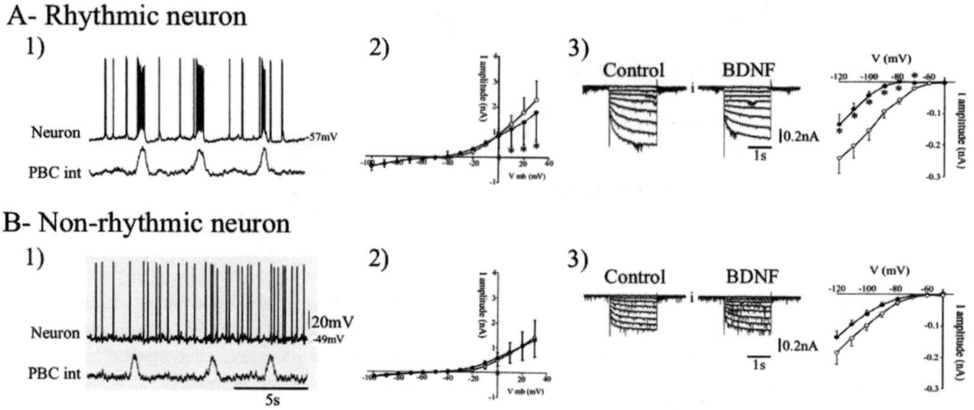

Figure 2. **BDNF modulates membrane properties specifically in rhythmic PBC neurons. A_1**—Intracellular recording (top trace) from a rhythmic neuron simultaneously with integrated population activity (bottom trace) **A_2**—Average current-voltage response curves obtained from 10 rhythmic neurons in control conditions (white circles) and in the presence of BDNF (black circles). **A_3**—Ih raw current traces recorded on a rhythmic neuron in control conditions (left) and after 30 minutes exposure to 100 ng/ml BDNF (middle). Right: graph representing the Ih current amplitude plotted against voltage. Average curves were obtained from 10 rhythmic neurons tested in control conditions (white circles) and after 30 minutes exposure to 100 ng/ml BDNF (black circles). Asterisks represent a significant difference ($p < 0.05$). **B**—Same legend than in A for 9 non-rhythmic neurons.

step at -120 mV changed insignificantly from -0.17 ± 0.001 nA in control conditions to -0.14 ± 0.006 nA in BDNF. Taken together these results demonstrate that BDNF is acting specifically on rhythmically active neurons membrane properties.

3.3. Individual PBC Neurons Express TrkB, p75 and/or BDNF mRNAs

Different responses to BDNF in R and NR neurons can result from uneven distribution of TrkB receptors inside the PBC or from the fact the BDNF/TrkB signalling controls specifically rhythm-related targets. We performed single cell multiplex RT-PCR experiments on 42 neurons (27 R neurons and 15 NR neurons) in order to examine the expression of BDNF receptors (TrkB and p75) in PBC neurons. Accordingly to what has been shown with immunolabeling directed against TrkB receptor[11] we found within the PBC 38% ($n = 16$) of the neurons that expressed either one or both of the two receptors for BDNF (Fig. 3). In each sub-group of PBC neurons we found equivallent proportion of neurons expressing BDNF receptors mRNAs: 11 out of 27 (40%) R neurons expressed either TrkB, p75 or both mRNAs. Five out of 15 (33%) NR neurons expressed TrkB mRNA. None of the 15 NR neurons tested expressed p75 mRNA.

Moreover, 29% ($n = 8$) and 26% ($n = 4$) of R and NR neurons respectively expressed BDNF mRNA. Three neurons (2 R and 1 NR) expressed both the ligand and its receptor (Fig. 3A). These results indicate that a source of BDNF exists within the PBC, with a possible autocrine and paracrine actions of BDNF on PBC neurons. These data also suggest that the expression of TrkB or p75 mRNA is not a specific marker for rhythmic neurons within the PBC.

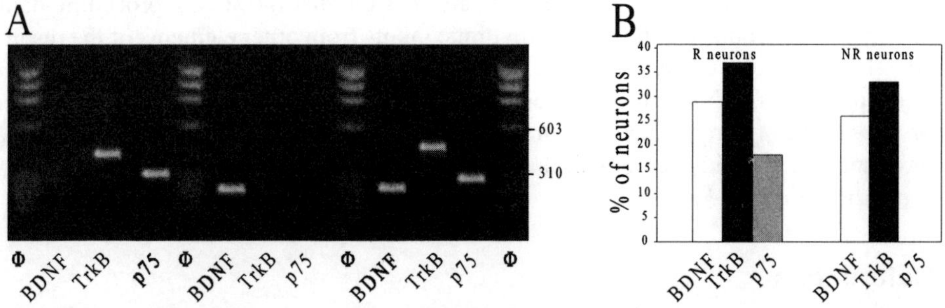

Figure 3. **PBC neurons express TrkB, p75 and BDNF mRNAs. A**—Agarose gel analysis of the multiplex RT-PCR products of three PBC neurons, one (left) expressing BDNF receptors mRNAs, one (middle) expressing BDNF mRNA and one (right) expressing both the neurotrophin and its receptors mRNAs. Φ indicates molecular weight markers. **B**—Graph representing the proportion of neurons in R (left group) and NR (right group) groups that expressed BDNF (white bar), TrkB (black bar) or p75 (gray bar) mRNAs.

4. Conclusions

The present study demonstrates that the rhythmic activity recorded at the PBC level on spontaneously active transverse brainstem slices is modulated *in vitro* by exogeneously applied BDNF[11]. The presence of the neurotrophin induces a decrease in respiratory frequency from 0.21 to 0.16 Hz. This decrease is associated with modifications of membrane properties of rhythmic neurons exclusively. Despite the fact that NR neurons express TrkB mRNA, their membrane properties were unaffected by treatment with BDNF. A small number of PBC neurons expressed both the neurotrophin and its receptor mRNAs suggesting that within the PBC some neurons (mainly rhythmic) can release the neurotrophin with a possible auto-regulation in a few subset of cells.

Among the 15 NR neurons tested in RT-PCR, 33% expressed TrkB mRNA whereas none of the 9 NR neurons intracellularly recorded exhibited modifications of membrane properties after BDNF exposure. This discrepancy can be explained by two hypothesis. First, BDNF might target in non-rhythmic neurons other membrane conductances than those tested in the present study and that would not be specifically implicated in rhythm generation. Second, the probes used to detect TrkB mRNA do not discriminate between the truncated and the complete forms of the receptor arising from differential splicing. It remains possible that NR neurons expressed truncated receptors devoid of tyrosine kinase activity. On the other hand, expression of truncated isoform of TrkB receptor might inhibit activation of Trk kinases in NR neurons by forming nonproductive heterodimers[13]. The presence of truncated receptors at the surface of non-rhythmic neurons might also limit the diffusion of BDNF to inappropriate sites of action, thereby restricting its action to the PBC region[14]. Interestingly expression of p75 receptor, known to enhance the specificity of TrkB receptor for its primary ligand BDNF[15], is found exclusively in rhythmic neurons, suggesting that R neurons might link BDNF with a higher affinity than NR neurons.

Our findings indicate that BDNF preferentially affects PBC neurons exhibiting rhythmic activity despite the fact that also non-rhythmic neurons express its receptor mRNAs. The role, if any, of the non-rhythmic neurons in the respiratory rhythm generation is unknown. Their identification is based on their location (closeby the extracellular electrode) and their discharge pattern (tonic or silent). It remains possible that the slicing procedure damaged their processes depriving them from rhythmic inputs from other elements of the respiratory neural network contained in the slice. In any case, the fact that BDNF did not change membrane properties of NR neurons argue in favor to the fact that the BDNF-induced decrease of the respiratory frequency is mainly mediated through action on membrane properties of R neurons (in addition to an enhancement of excitatory synaptic currents received by R neurons, see ref 11).

References

1. E. J. Huang and L. F. Reichardt, Neurotrophins: role in neuronal development and function, *Annu. Rev. Neurosci.* **24**, 677–736 (2001).
2. M. Barbacid, The Trk family of neurotrophin receptors, *J. Neurobiol.* **25**, 1386–1403 (1994).
3. E. J. Huang and L. F. Reichardt, Trk receptors: Roles in neuronal signal transduction, *Annu. Rev. Biochem.* **72**, 609–642 (2003).
4. R. A. Segal and M. E. Greenberg, Intracellular signaling pathways activated by neurotrophic factors, *Annu. Rev. Neurosci.* **19**, 463–489 (1996).
5. J. T. Erickson, J. C. Conover, V. Borday, J. Champagnat, M. Barbacid, G. Yancopoulos and D. M. Katz, Mice lacking brain-derived neurotrophic factor exhibit visceral sensory neuron losses distinct from mice lacking NT4 and display a severe developmental deficit in control of breathing, *J. Neurosci.* **16**, 5361–5371 (1996).
6. A. Balkowiec, and D. M. Katz, Brain-derived neurotrophic factor is required for normal development of the central respiratory rhythm in mice, *J. Physiol. (London)* **510**, 527–533 (1998).
7. J. C. Smith, H. H. Ellenberger, K. Ballanyi, D. W. Richter and J. L. Feldman, Pre-Bötzinger complex: a brainstem region that may generate respiratory rhythm in mammals, *Science* **254**, 726–729 (1991).
8. J. C. Rekling and J. L. Feldman, Pre-Bötzinger complex and pacemaker neurons: hypothesized site and kernel for respiratory rhythm generation, *Annu. Rev. Physiol.* **60**, 385–405 (1998).
9. J. M. Ramirez, U. J. A. Quellmalz and D. W. Richter, Postnatal changes in the mammalian respiratory network as revealed by the transverse brainstem slice preparation of mice, *J. Physiol. (London).* **491**, 799–812. (1996).
10. B. Cauli, E. Audinat, B. Lambolez, M. C. Angulo, N. Ropert, K. Tsuzuki S. Hestrin and J. Rossier, Molecular and physiological diversity of cortical non-pyramidal cells, *J. Neurosci.* **17**, 3894–3906 (1997).
11. M. Thoby-Brisson, B. Cauli, J. Champagnat, G. Fortin and D. M. Katz, Expression of functional tyrosine kinase B receptors by rhythmically active respiratory neurons in the pre-Bötzinger complex of neonatal mice, *J. Neurosci.* **23**, 7685–7689 (2003).
12. M. Thoby-Brisson, P. Telgkamp and J. M. Ramirez, The role of the hyperpolarization-activated current in modulating rhythmic activity in the isolated respiratory neural network of mice, *J. Neurosci.* **20**, 2994–3005 (2000).
13. F. F. Eide, E. R. Vining, B. L. Eide, K. Zang, X. Y. Wang and L. F. Reichardt, Naturally occurring truncated trkB receptors have dominant inhibiting effects on brain-derived neurotrophic factor signaling, *J. Neurosci.* **16**, 3123–3129 (1996).
14. R. M. Lindsay, S. J. Wiegand, C. A. Altar and P. S. DiStefano, Neurotrophic factors: from molecule to man, *TINS.* **17**, 182–190 (1994).
15. M. Bibel, E. Hoppe, and Y. A. Barde, Biochemical and functional Interactions between the neurotrophin receptors TrkB and p75NTR, *EMBO J.* **18**, 616–622 (1999).

Ionic Currents and Endogenous Rhythm Generation in the pre-Bötzinger Complex: Modelling and *In Vitro* Studies

Olivier Pierrefiche, Natalia A. Shevtsova, Walter M. St.-John, Julian F. R. Paton, and Ilya A. Rybak

1. Introduction

The pre-Bötzinger complex (pBC), a small area in the rostroventrolateral medulla, has been suggested to represent a "kernel" of the mammalian respiratory network[1–5]. The *in vitro* preparations from neonatal rodents containing this area can, under certain experimental conditions, generate an intrinsic rhythmic activity[4,5]. This activity does not require inhibitory neurotransmission[6] and, therefore, is likely to be generated by a population of pacemaker neurons in the pBC[1–5]. At the same time, the "decrementing" discharge pattern of rhythmic activity in the pBC recorded *in vitro* differs from the pattern of respiratory discharges observed under normal conditions *in vivo* ("eupnoea") and is similar to gasping pattern[7,8]. In order to establish possible relationships of the intrinsic rhythmic activity in the pBC to the respiratory rhythmogenesis *in vivo*, it is important to analyse the conditions in which this activity occurs *in vitro* and to compare these conditions with the rhythmogenic conditions during eupnoea and gasping *in vivo*. According to the preliminary modelling studies[9], the *in vitro* rhythmic activity in the pBC may be dependent on a relative expression of the voltage-gated potassium and persistent sodium currents in pBC neurons. Here we present the results of our combined modelling and *in vitro* studies performed to test this modelling prediction. Our studies focused on the involvement of the potassium and persistent sodium currents in the endogenous rhythmic activity in the pBC *in vitro* and on the possible relation of this activity to the genesis of the respiratory oscillations *in vivo*.

Olivier Pierrefiche • GRAP-JE-UFR de Pharmacie, 80036 Amiens, France. **Ilya A. Rybak and Natalia A. Shevtsova** • School of Biomedical Engineering, Science and Health Systems, Drexel University, Philadelphia, PA 19104. **Walter M. St. John** • Department of Physiology, Dartmouth Medical School, Lebanon, NH 03756. **Julian F. R. Paton** • Department of Physiology, School of Medical Sciences, University of Bristol, Bristol BS8 1TD, UK.

Post-Genomic Perspectives in Modeling and Control of Breathing, edited by Jean Champagnat, Monique Denavit-Saubié, Gilles Fortin, Arthur S. Foutz, Muriel Thoby-Brisson. Kluwer Academic/Plenum Publishers, 2004.

2. Modelling Intrinsic Bursting Activity in the PBC

The model of a single pBC pacemaker neuron was developed using the Hodgkin-Huxley formalism. The model was based on the previous models[2,3] and included fast sodium (I_{Naf}), persistent sodium (I_{NaP}) delayed-rectifier potassium (I_K), leakage (I_{leak}), and synaptic (I_{syn}) currents which together defined the dynamics of the neuron membrane potential. The voltage-gated and kinetic parameters for I_{Naf} and I_{NaP} were drawn from recent *in vitro* studies of isolated pBC neurons[10]. To investigate firing behaviour of a pacemaker neuron population we modelled a population of 50 neurons with all-to-all excitatory synaptic connections. Heterogeneity within the population was set by the random distribution of the maximal channel conductances ($\bar{g}_{NaP}$, $\bar{g}_K$, g_{leak} and g_{syn}). For more details see Rybak *et al.*[11]

Figure 1 shows results of our simulations. We found that the population bursting activity could be induced in the model by an elevation of the external potassium concentration $[K^+]_o$ from the physiological level (3 mM) to higher levels (Figure 1A) or by changing the maximal conductances of potassium ($\bar{g}_K$, Figure 1B) and persistent sodium ($\bar{g}_{NaP}$, Figure 1C) channels at the normal level of $[K^+]_o$. Our modelling study has demonstrated that rhythmic bursting activity in a population of pacemaker neurons may be triggered (repetition) by either (1) an increase of the extracellular potassium concentration, or (2) a suppression of the voltage-gated potassium currents, or (3) an augmentation of the persistent sodium currents (see Figure 1).

Our *in vitro* studies described below have been performed with the primarily goal to test our modelling predictions.

3. *In Vitro* Studies

Experiments were performed using transverse slices (700 μm) obtained from neonatal rats (P0-P4), deeply anaesthetized with ether. Anatomical landmarks were used for the localisation of the pre-Bötzinger region. Rhythmic activity in the slices was triggered by the elevation of $[K^+]_o$ to 5–7 mM. The population activity of the pBC and the activity of XII nerve rootlet were recorded simultaneously with suction electrodes. Raw activities were filtered, amplified and integrated. Experiments were started with aCSF-5 (aCSF containing 5 mM of potassium). With aCSF-7, all slices generated stable rhythmic activity. Then we replaced aCSF-7 with an aCSF containing 3 mM of potassium (aCSF-3) which was considered as an aCSF corresponding to the normal *in vivo* conditions. The effects of potassium current blockers (4-AP, 50–200 μM, or TEA, 2–4 mM) and of sodium cyanide (NaCN, 2 mM) were examined with aCSF-3 after rhythmic activities were completely abolished. The effects of the persistent sodium current blocker riluzole (25–50 μM) were tested using the rhythmically active slices superfused with aCSF-7.

At 7 mM $[K^+]_o$, all slices demonstrated rhythmic activity which was synchronous in pBC and XII recordings. This activity was abolished when $[K^+]_o$ was reduced to 3 mM (Figure 2A). Application of potassium channels blockers (4-AP or TEA) at 3 mM $[K^+]_o$ triggered rhythmic activity in the pBC and on XII nerve rootlet (Figure 2B). Application of riluzole (25–50 μM), a persistent sodium channel blocker, at 7 mM $[K^+]_o$ abolished rhythmic activity in the majority of slices (Figure 2C). Finally, histotoxic hypoxia induced by application of NaCN (2 mM) also triggered synchronous rhythmic activity in pBC and XII nerve rootlet at 3 mM $[K^+]_o$ (Figure 2D).

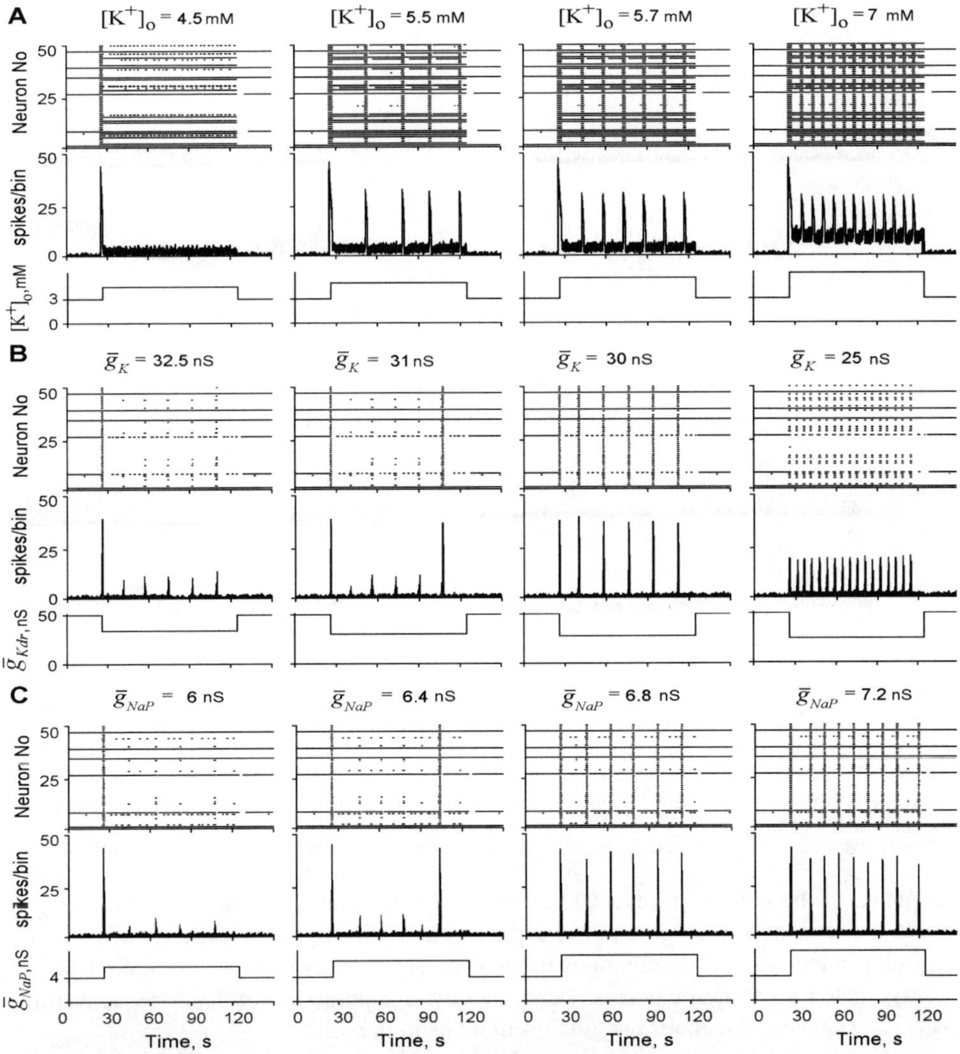

Figure 1. Triggering bursting activity in the model of pBC population of pacemaker neurons. The result of each simulation is represented by two diagrams: the top diagram is a raster plot for spike times in all 50 cells, sorted on the ordinate axis by cell index number; the bottom diagram is a corresponding integrated histogram of population activity (bin size = 10 ms). **A.** Triggering endogenous bursting activity by elevation of $[K^+]_o$ from 3 mM to higher levels from left to right. An increase of $[K^+]_o$ increases the burst frequency and the level of background asynchronous activity and decreases the burst amplitude. **B.** Triggering bursting activity by reduction of the mean value of $\bar{g}_K$ in the population. The mean value of $\bar{g}_K$ is reduced from 50 nS to lower values from left to right. **C.** Triggering bursting activity by augmentation of the mean value of $\bar{g}_{NaP}$ in the population from 4 nS to higher values from left to right.

4. Discussion

This study reveals that rhythmic activity in the pBC may be triggered by a reduction of I_K. The suppression of I_K can be obtained by either an increase in $[K^+]_o$ (via the

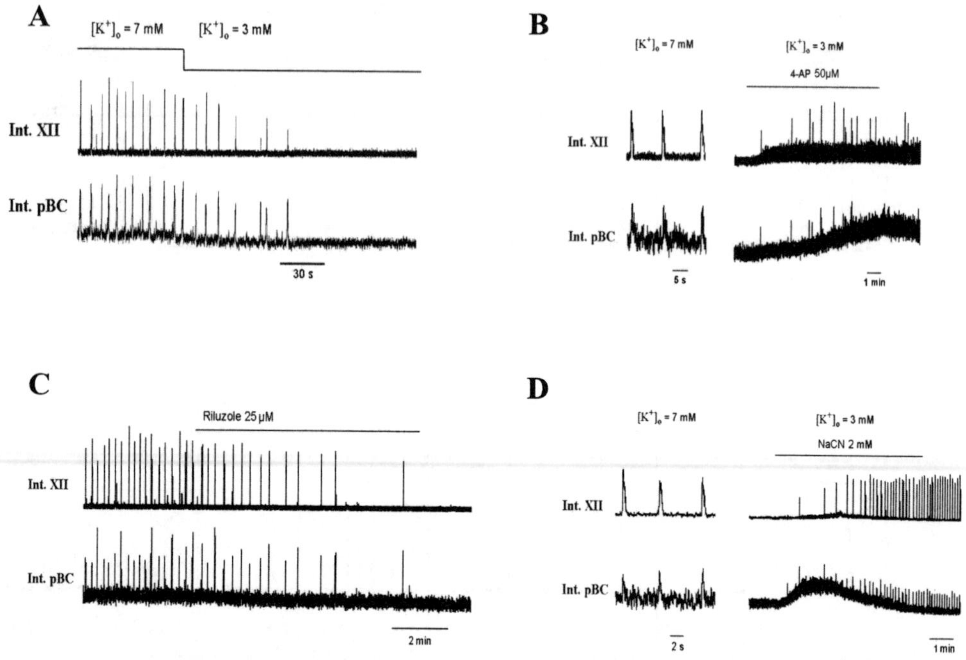

Figure 2. Results of *in vitro* studies. **A.** Reducing $[K^+]_o$ from 7 to 3 mM stopped rhythmic activity in the slice. **B.** Triggering endogenous bursting activity in the slice by application of 4-AP (3 mM) at $[K^+]_o = 3$ mM. Left column shows the intrinsic activity in the same slice at $[K^+]_o = 7$ mM before the application of the blocker. **C.** Riluzole (25 M) abolished rhythmic activity in the slice at $[K^+]_o = 7$ mM. **D.** Triggering endogenous bursting activity in the slice with NaCN (2 mM). Left column shows the activity in the same slice at $[K^+]_o = 7$ mM, before the application of NaCN.

reduction of the potassium reversal potential) or a direct suppression of the voltage-gated potassium channels. Our simulations suggest that at the physiological level of $[K^+]_o$, the I_{NaP}-dependent intrinsic oscillations in the pBC are restrained by the normally expressed voltage-gated potassium currents. Consequently, a suppression of I_K or an activation of I_{NaP} can trigger endogenous bursting oscillations in the pBC.

Our *in vitro* data confirmed our modelling predictions. We conclude that the reduction of I_K via the elevation of $[K^+]_o$ is critical for triggering and maintenance of stable rhythmic activity within the pBC. In addition, blocking the intrinsic activity in the pBC by riluzole has provided an additional support for the previous suggestions[1–3] that rhythmic activity recorded from the pBC *in vitro* is persistent-sodium dependent.

We have also demonstrated that histotoxic hypoxia induced by application of NaCN (2 mM) can also trigger rhythmic activity in the pBC. The decrementing shape of the bursts triggered by NaCN were not different from the shape of bursts triggered by the elevation of $[K^+]_o$ or application of potassium current blockers.

The role of pacemaker-driven oscillations in the pBC in the generation of the eupnoeic respiratory rhythm and its possible involvement in the generation of other patterns of breathing, such as gasping, is the subject of debate. As mentioned above, the neuronal bursts recorded *in vitro* (as seen in both the hypoglossal nerve activity and the population activity from the pBC area) have a "decrementing" pattern and differs from the "augmenting" pattern

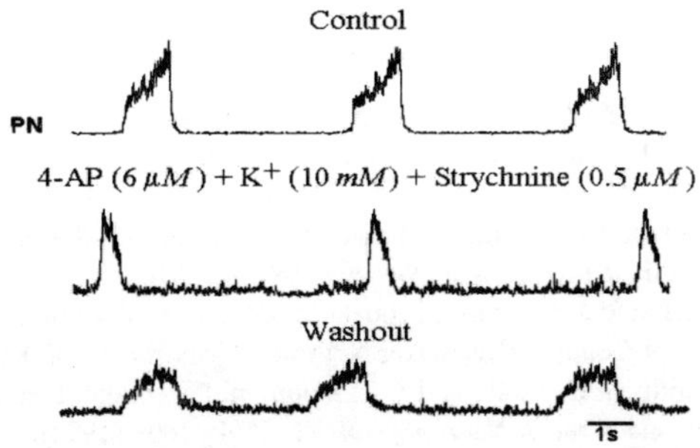

Figure 3. The results of our studies using the perfused *in situ* preparation. Application of 4-AP and strychnine (to block glycinergic inhibition) in combination with an increase of $[K^+]_o$ converted the eupnoeic pattern of phrenic nerve discharge to a decrementing discharge similar to that recorded *in vitro* and during ischemia-induced gasping *in situ*. This effect was fully reversible following washout of the drugs and returning the $[K^+]_o$ to the control levels (see "Washout").

of the phrenic nerve discharges recorded *in vivo* during eupnoea but is similar to gasping[7,8]. Similar to the *in vitro* rhythm, gasping is resistant to blockade of inhibitory transmissions, and hence it has been suggested that gasping is also generated by a pacemaker-driven mechanism[12,13]. We suggest that the pacemaker-driven activity in the pBC is not expressed during eupnoea but plays a fundamental role in the generation of gasping. According to the *switching concept* [9,13] the pBC is considered a region explicitly responsible for switching from eupnoea, generated by a network mechanism distributed in the pontomedullary region, to gasping driven by pacemaker-based oscillations in the pBC.

The decrementing high-amplitude, short-duration discharges recorded from the phrenic nerve during gasping have a similar shape to that of endogenous discharges recorded from *in vitro* preparations. Therefore, hypoxia may indeed produce a switch in the respiratory rhythm generation from a network mechanism for eupnoea to a pacemaker-driven mechanism for gasping. In support of this concept, it should be noted that hypoxia causes (1) a suppression of voltage-gated potassium currents, I_K [14–16], (2) an augmentation of the persistent sodium current, I_{NaP} [17,18], and (3) an increase of external potassium concentration $[K^+]_o$ [19], which produces an additional suppression of I_K.

In the present study, by combining *in vitro* experiments and computational modelling, we have shown that each of the above factors accompanying hypoxia may trigger endogenous bursting activity in the pBC. These results support the concept that gasping *in vivo* is driven by a pacemaker mechanism originated in the pBC.

Our recent study of the neurogenesis of gasping *in situ* provided additional experimental evidence for this concept. Specifically, application of 4-AP and strychnine (to block glycine receptor mediated inhibition) in combination with an increase of $[K^+]_o$ converted the eupnoeic pattern of phrenic nerve discharge to a decrementing discharge similar to that recorded *in vitro* and during ischemia-induced gasping *in situ*[13] (Figure 3).

5. Acknowledgements

W. M. St.-John was supported by NIH (HL26091) and a Fogarty grant. J. F. R. Paton was supported by The British Heart Foundation (BS/93003). I. A. Rybak was supported by NSF (0091942) and NIH (NS046062–02 and HL072415-01).

References

1. J. C. Smith, R. J. Butera, N. Koshiya, C. Del Negro, C. G. Wilson, and S. M. Johnson. Respiratory rhythm generation in neonatal and adult mammals: The hybrid pacemaker-network model. *Respir. Physiol.* **122**, 131–147 (2000).
2. R. J. Butera, J. R. Rinzel, and J. C. Smith, Models of respiratory rhythm generation in the pre-Bötzinger complex: I. Bursting pacemaker neurons, *J. Neurophysiol.* **82**, 382–397 (1999).
3. R. J. Butera, J. R. Rinzel, and J. C. Smith, Models of respiratory rhythm generation in the pre-Bötzinger complex: II. Populations of Coupled Pacemaker Neurons. *J. Neurophysiol.* **82**, 398–415 (1999).
4. S. M. Johnson, J. C. Smith, G. D. Funk, and J. L. Feldman, Pacemaker behavior of respiratory neurons in medullary slices from neonatal rat. *J. Neurophysiol.* **72**, 2598–2608 (1994).
5. N. Koshiya and J. C. Smith, Neuronal pacemaker for breathing visualized *in vitro*. *Nature* **400** (6742), 360–363 (1999).
6. X. M. Shao and J. L. Feldman, Respiratory rhythm generation and synaptic inhibition of expiratory neurons in pre-Bötzinger complex: differential roles of glycinergic and gabaergic neural transmission, *J. Neurophysiol.* **77**, 1853–1860 (1997).
7. W. M. St.-John, Medullary regions for neurogenesis of gasping: noeud vital or noeuds vitals? *J. Appl. Physiol.* **81**, 1865–1877 (1996).
8. W. M. St.-John, Neurogenesis of patterns of automatic ventilatory activity. *Prog. Neurobiol.* **56**, 97–117 (1998).
9. I. A. Rybak, J. F. R. Paton, R. F. Rogers, and W. M. St.-John, Generation of the respiratory rhythm: state dependency and switching. *Neurocomputing* **44–46**, 605–614 (2002).
10. I. A. Rybak, K. Ptak, N. A. Shevtsova, and D. R. McCrimmon, D R. Sodium currents in neurons from the rostroventrolateral medulla of the rat, *J Neurophysiol.* **90**, 1635–1642 (2003).
11. I. A. Rybak, N. A. Shevtsova, W. M. St.-John, J. F. R. Paton, and O. Pierrefiche, Endogenous rhythm generation in the pre-Bötzinger complex and ionic currents: Modelling and *in vitro* studies. *Eur. J. Neurosci.* **18**, 239–257 (2003).
12. W. M. St.-John, and J. F. R. Paton, Neurogenesis of gasping does not require inhibitory transmission using $GABA_A$ or glycine receptors. *Respir. Physiol. Neurobiol.*, **132**, 265–277 (2002).
13. W. M. St.-John, I. A. Rybak, and J. F. R., Paton, Potential switch from eupnea to fictive gasping after blockade of glycine transmission and potassium channels, *Am. J. Physiol.* (Integr. Comp. Physiol.) **283**, R721–R731 (2002).
14. C. Jiang and G. G. Haddad, A direct mechanism for sensing low oxygen levels by central neurons. *Proc. Natl. Acad. Sci. USA*, **91**, 7198–7201 (1994).
15. L. Conforti, L. and D. E. Millhorn, Selective inhibition of a slow–inactivating voltage-dependent K_+ channels in rat PC12 cells by hypoxia, *J. Physiol. Lond.*, **502**, 293–305 (1997).
16. Gebhardt, C. and U. Heinemann, Anoxic decrease in potassium outward currents of hippocampal cultured neurons in absence and presence of dithionite. *Brain Res.* **837**, 270–276 (1999).
17. A. K. Hammarström and P. W. Gage, Inhibition of oxidative metabolism increases persistent sodium current in rat CA1 hippocampal neurons, *J. Physiol.* **510**, 735–741 (1998).
18. E. M. Horn and T. G. Waldrop, Hypoxic augmentation of fast–inactivating and persistent sodium currents in rat caudal hypothalamic neurons. *J. Neurophysiol.*, **84**, 2572–2581 (2000).
19. J. E. Melton, S. C. Kadia, Q. P. Yu, J. A. Neubauer, and N. H. Edelman, Respiratory and sympathetic activity during recovery from hypoxic depression and gasping in cats, *J. Appl. Physiol.* **80**, 1940–1948 (1996).

Modulation of Inspiratory Inhibition of the Bötzinger Complex by Raphe Pallidus and Locus Coeruleus in Rabbits

Guimin Wang, Shuyan Yu, Fayan Zhang, Yanchun Li, Ying Cao, Qin Li, Gang Song, and Heng Zhang

1. Introduction

In mammals, the respiratory rhythm is generated by the brainstem neuronal network. The Bötzinger complex (Böt.C) is a group of inhibitory expiratory neurons at the rostral end of the ventral respiratory group (VRG) near the retrofacial nucleus. Neurons in the Böt.C have widespread inhibitory connections with other medullary respiratory neurons and phrenic motoneurons.[1,2] Long train stimulation of the Böt.C caused inhibition of inspiration and inspiratory to expiratory phase-switching.[3,4]

The nucleus raphe pallidus (RP) is a group of 5-HTergic neurons in the midline region of caudal medulla.[5] The nucleus locus coeruleus (LC) is a group of noradrenergic neurons in the dorsal pons.[6] Axons from the Böt.C, RP, and LC projected and converged to the phrenic premotor neurons in medulla and the phrenic motoneurons in spinal cord.[7–10]We hypothesize that the RP and LC could modulate the effects of the Böt.C to control the inspiratory duration and amplitude.

The present study is to observe the effect of the 5-HTergic RP and the NEergic LC on the Böt.C's inspiratory inhibition and to elucidate its role in the regulation of respiration.

2. Methods

2.1. Animal Preparation

Experiments were carried out on 52 adult rabbits (weighing 2.2–2.6 kg) of either sex. The rabbit was anesthetized with urethane (1.0 g/kg, i.v.). Tracheotomy was performed for

---

Guimin Wang, Shuyan Yu, Fayan Zhang, Yanchun Li, Ying Cao, Qin Li, Gang Song, and Heng Zhang • Institute of Physiology, School of Medicine, Shandong University, Jinan, 250012, P.R. China. phone: (86)531-8382037; email: wanggm@sdu.edu.cn.

Post-Genomic Perspectives in Modeling and Control of Breathing, edited by Jean Champagnat, Monique Denavit-Saubié, Gilles Fortin, Arthur S. Foutz, Muriel Thoby-Brisson. Kluwer Academic/Plenum Publishers, 2004.

artificial ventilation. Both cervical vagus nerves were separated for subsequent vagotomy. The animal was placed in a prone position and fixed in a stereotaxic head holder. The head was ventroflexed to facilitate recordings from the medulla. Both C_5 phrenic roots were dissected free, cut distally, and prepared for recording. The dorsal surface of the medulla was widely exposed by occipital craniotomy, and the dura and arachnoid membranes were removed. All exposed tissues were covered with warm paraffin oil (37–38°C). All rabbits were paralyzed with pancuronium bromide (Sigma; initial dose 0.5 mg/kg, supplemented by 0.1 mg/kg/hour, i.v.) and ventilated with oxygen enriched medical air (oxygen concentration at 40%). The end-tidal CO_2 was monitored (OIR-7101, Nihon Kohden) and kept at 4-5%. Blood pressure was monitored through a carotid artery cannula and maintained at 12–15 Kp by continuous infusion of 10% glucose in physiological saline through the femoral vein. Rectal temperature was kept at 37.5–39°C with a heating pad.

2.2. Stimulation and Recording

Phrenic nerve discharges were recorded with a bipolar silver electrode, amplified, filtered (50–3 k Hz), and displayed on a dual-beam memory oscilloscope (VC-10, Nihon Kohden). We defined the obex at the most rostral extent of the area postrema as a standard point of anatomic reference. For electrical stimulation of the Böt.C, the tip of an insulated monopolar tungsten microelectrode (shaft diameter 70 m, tip electrically etched to 1–3 m, impedance 3–6 MΩ) was stereotaxically inserted into the area of retrofacial nucleus, positioned 3.5–5.0 mm rostral to the obex, 2.5–3.5 mm lateral to the midline, and 4.0–4.5 mm below the dorsal medullary surface. For stimulation of the RP, another tungsten microelectrode was positioned 1.0–4.0 mm rostral to the obex, midline, and 3.0–4.5 mm below the dorsal medullary surface. For stimulation of the LC, the electrode was positioned 5.8–6.0 mm caudal to the bregma and 2.0–2.3 mm lateral to the midline, at depth of 15.5–15.8 mm from bregma surface. All signals were inputted into a computer for recording and analyzing with the Biobench software (NI Corporation, U.S.).

Rectangular constant current pulses (duration 0.1 msec, 80 Hz) from an electronic stimulator (SEN-3201, Nihon Kohden) was used for stimulation. We defined the stimulation threshold as the lowest stimulus current that produced an appreciable inhibitory (stimulating Böt.C) or excitatory (stimulating RP or LC) effects on phrenic nerve discharges during a 5-sec repetitive stimulation. Stimulation times were 10 sec for RP and LC, and 15 sec for Böt.C. The amplitude of integrated phrenic discharge (∫Phr) was calculated by averaging the peak amplitudes of 3 control respiratory cycles, or all cycles during the 15-sec stimulation. Percentage of inhibition = (control ∫Phr—∫Phr during stimulation) / control ∫Phr.

At the end of each experiment, a DC current (30 A for 30 sec) was passed through tungsten microelectrodes to produce electrolytic lesions to mark the recording and stimulation points. The animals were sacrificed with an overdose of urethane (2 g/kg) and immediately perfused with 10% formalin. The brains were removed, blocked and cut into 100 m frozen sections for the detection of stimulation and recording sites.

2.3. Statistic Analysis

Results were presented as mean ± SD. Student's t-test and F-test were used to determine the difference. Statistical significance level was set at $P < 0.05$.

3. Results

3.1. Depression of the Böt.C's Inspiratory Inhibition by Pre-stimulation of RP

As reported in previous studies, electrical stimulation of the Böt.C caused intensity-dependent depression of phrenic discharge. We found that phrenic nerve discharge began to decrease at intensity of 15 ± 5 A and was completely depressed at 30 ± 5 A (80 Hz, pulse duration 0.1 msec, 15 sec).

Electrical stimulation of the RP increased the amplitude of phrenic inspiratory discharge and the respiratory frequency. After the RP stimulation, within a short period of time, electrical stimulation of the Böt.C caused weaker inspiratory inhibition. At intensity of 20 A, stimulation of the Böt.C decreased the amplitude of integrated phrenic discharge by 59.97 ± 9.67%. Two seconds after the RP stimulation, the same stimulation at the Böt.C only decreased the integrated phrenic discharge by 26.04 ± 9.09%. The difference was highly significant ($n = 29$, $P < 0.01$). (Figure 1, Figure 3)

3.2. Depression of the Böt.C's Inspiratory Inhibition by Pre-stimulation of LC

Similar depression of Böt.C's inspiratory inhibition was observed after electrical stimulation of the LC. In another group of animals, at intensity of 20 A, stimulation of the Böt.C decreased the amplitude of integrated phrenic discharge by 61.89 ± 12.43%. Two seconds after LC stimulation, the same stimulation at the Böt.C decreased the phrenic discharge by 19.13 ± 3.12%. The difference was highly significant ($n = 17$, $P < 0.01$). Electrical stimulation at the LC increased the amplitude of phrenic discharge and the respiratory frequency. (Figure 2, Figure 3)

4. Discussion

Our results showed that long train (15 sec) electrical stimulation of the Böt.C caused intensity-dependent inhibition of phrenic nerve discharge. It also demonstrated that the 5-HTergic neurons in RP and the NEergic neurons in LC could modulate the inspiration-inhibition of the Böt.C. We suggest that the down-regulation of GABAa receptor mediated synaptic inhibition is responsible for this phenomenon.

The neurons of Böt.C projected to the spinal phrenic motor neurons and the medullary phrenic premotor neurons.[7, 10–13] Electrical and chemical stimulation of the Böt.C produced strong inhibition of phrenic discharge.[1,4,12] This inhibition could be blocked by local application of bicuculine (a specific GABAa receptor antagonist), indicating the involvement of GABAa receptors.[14, 15] Both the RP and LC projected to phrenic motor neurons and medullary respiratory neurons. In addition, axonal terminals containing 5-HT and NE were observed to form synapses with phrenic motor neurons.[8, 10, 16] Stimulation of RP and LC would cause the release of related neurotransmitters (5-HT and NE, respectively) to those neurons.

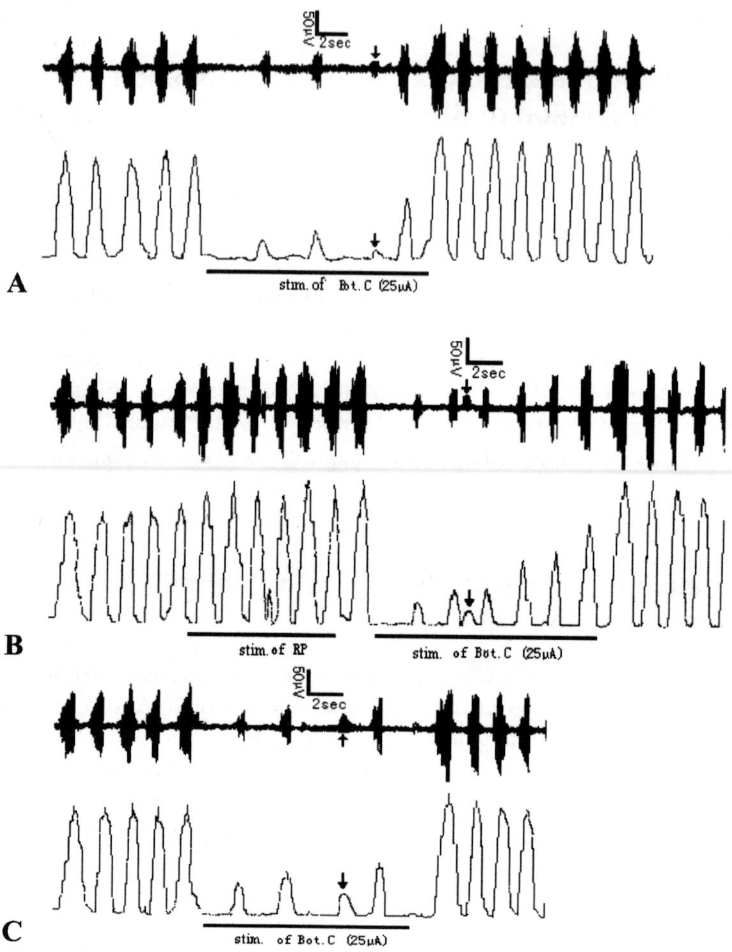

Figure 1. Pre-stimulation of the RP weakened the inspiratory inhibition caused by stimulation of the Böt.C. **A:** The phrenic discharge was almost completely abolished by electrical stimulation of the Böt.C at intensity of 25 A. **B:** Immediately after stimulating the RP, another stimulation at the Böt.C (with the same intensity as in A) only partially depressed the phrenic discharge. Stimulation at the RP (50 A) increased the amplitude of phrenic discharge and respiratory frequency. **C:** 30 seconds after the RP stimulation, stimulation of the Böt.C (25 A) produced the same degree of phrenic inhibition as in A. Arrows indicate the artifacts.

GABAa receptor is subject to phosphorylation modulation by PKA, PKC, and tyrosine kinase. The level of phosphorylation modulated the receptor current. The binding of 5-HT and NE to their respective receptors would change the activity of those protein kinases through some kinds of secondary messenger pathways and caused the down-regulation of the GABAa receptor current.[17–19] As a result, the GABAa receptor mediated inspiration-inhibition of Böt.C was compromised following pre-stimulation of the RP and LC.

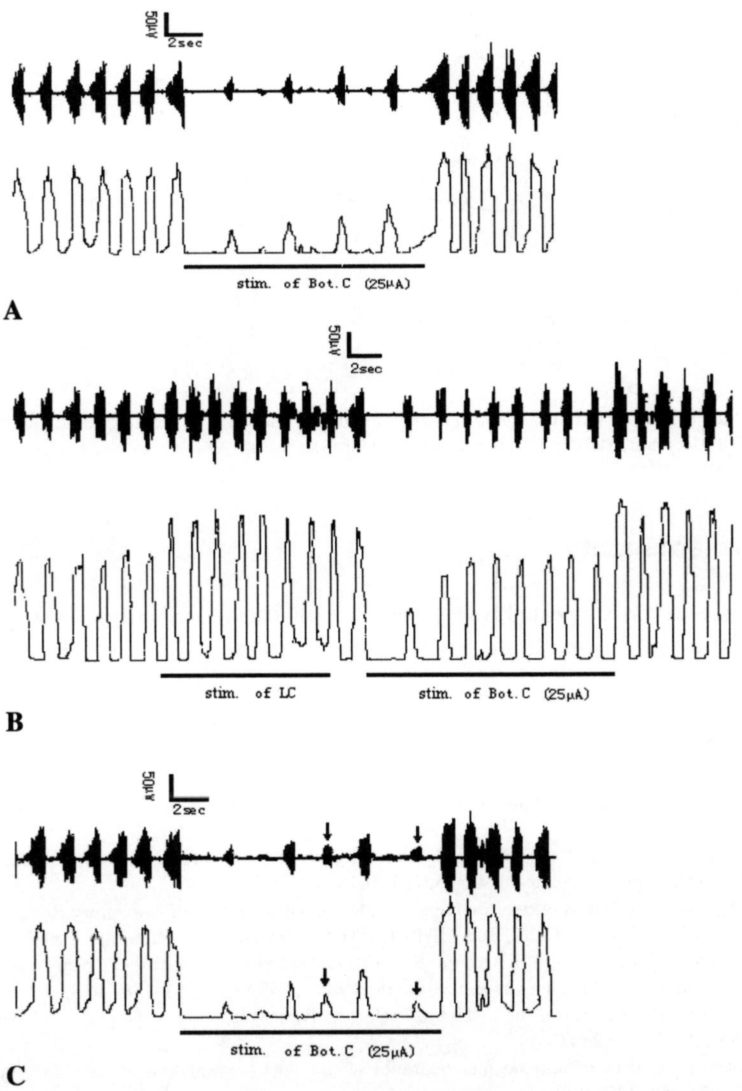

Figure 2. Pre-stimulation of the LC weakened the inspiratory inhibition caused by stimulation of the Böt.C. **A:** The phrenic discharge was almost completely abolished by electrical stimulation of the Böt.C at intensity of 25 A. **B:** Immediately after stimulating the LC, another stimulation at the Böt.C (with the same intensity as in A) caused weak depression of the phrenic discharge. Stimulation at the LC (50 A) increased the amplitude of phrenic discharge and respiratory frequency. **C:** 30 seconds after the LC stimulation, stimulation of the Böt.C (25 A) produced the same degree of phrenic inhibition as in A. Arrows indicate the artifacts.

In conclusion, the existence of such kind of "cross-talk" between different transmitter systems endows flexibility and plasticity to the respiratory neuronal network. All those properties greatly increased the ability of the respiratory system to adapt to changes in internal and external environments.

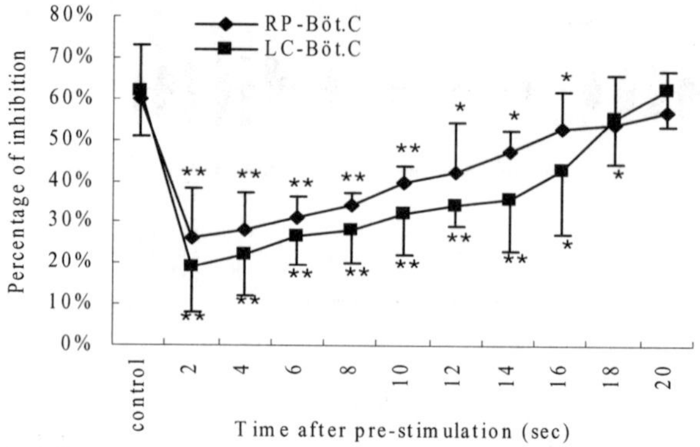

Figure 3. With the increases of time between the pre-stimulations and the succeeding stimulation of Böt.C, the weakening effects decreased rapidly. The weakening of inhibition was no longer significant after 18–20 seconds. * $P < 0.05$; ** $P < 0.01$.

5. Acknowledgments

This project was supported by a grant (No. 39970278) from China Natural Science Foundation, and a grant from China Ministry of Education.

References

1. G. Song, Bötzinger complex and the formation of respiratory rhythm, Progress Physiol. Sci. in Chinese 30, 237–240 (1999).
2. E. G. Merrill, L. Fedorko, Monosynaptic inhibition of phrenic motoneurons: a long descending projection from Bötzinger complex, J. Neurosci. 4, 2350–2353 (1984).
3. G. Song, Q. Li, M. Lü, Roles of the Bötzinger complex in the formation of respiratory rhythm, in: FRONTIERS IN MODELING AND CONTROL OF BREATHING: Integration at Molecular, Cellular and Systems Levels. Edited by C.-S. Poon and H. Kazemi (Kluwer Academic/Plenum Publishers, New York, 2001) (Advances in Experimental Medicine and Biology Series, Volume 499), pp. 153–158.
4. S. Gang, L. Liu, Reappraisal of the inspiratory effect of Bötzinger complex on phrenic nerve discharge, Respir. Physiol. 105, 17–21 (1996).
5. P. M. Lalley, Responses of phrenic motoneurones of the cat to stimulation of medullary raphe nuclei, J. Physiol. 380, 349–371 (1986).
6. Z. Y. Li, B. L. Xia, C. J. Huang, Effects of microinjection of L-glutamate into locus coeruleus complex area on respiration, J. Tongji Med. Univ. 12(4), 205–8 (1992).
7. K. Otake, H. Sassaki, H. Mannen, K. Ezure, Morphology of expiratory neurons of the Bötzinger complex: an HRP study in the cat, J. Comp. Neurol. 258, 565–579 (1987).
8. G. Song, Q. Li, F. Z. Shao, Projections of 5-HTergic fibers to the spinal phrenic nucleus and medullary phrenic premotor neurons in the cat, Acta physiologica sinica. 53 (5), 391–395 (2001).
9. S. Gang, Y. Sato, I. Kohama, M. Aoki, Afferent projections to the Bötzinger complex from upper cervical cord and other respiratory related structures in the brainstem in cats: retrograde WGA-HRP tracing, J. Auton. Nerv. Syst. 56, 1–7 (1995).
10. E. G. Dobbins, J. L. Feldman, Brainstem network controlling descending drive to phrenic motoneurons in rat, J. Comp. Neurol. 347(1), 64–86 (1994).
11. G. Song, Q. Li, F. Z. Shao, GABAergic neurons in Kolliker-Fuse nucleus and Botzinger complex with axons projecting to phrenic nucleus, Acta physiologica sinica in Chinese with English abstract. 52(2), 167–9 (2000).

12. C. Jiang, J. Lipski, Extensive monosynaptic inhibition of ventral respiratory group neurons by augmenting neurons in the Bötzinger complex in the cat, Exp. Brain Res. 81, 639–648 (1990).
13. E. G. Merrill, L. Fedorko, Monosynaptic inhibition of phrenic motoneurons: a long descending projection from Bötzinger neurons, J. Neurosci. 4, 2350–2353 (1984).
14. D. R. McCrimmon, E. J. Zuperku, F. Hayashi, Z. Dogas, C. F. Hinrichsen, E. A. Stuth, M. Tonkovic-Capin, M. Krolo, F. A. Hopp, Modulation of the synaptic drive to respiratory premotor and motor neurons, Respir. Physiol. 110(2–3), 161–76 (1997).
15. V. C. Chitravanshi, H. N. Sapru, GABA receptors in the phrenic nucleus of the rat, Am. J. Physiol. 276 (2 Pt 2), R420–8 (1999).
16. V. R. Holets, T. Hokfelt, A. Rokaeus, L. Terenius, M. Goldstein, Locus coeruleus neurons in the rat containing neuropeptide Y, tyrosine hydroxylase or galanin and their efferent projections to the spinal cord, cerebral cortex and hypothalamus, Neuroscience 24(3), 893¨C906 (1988).
17. J. P. Huidobro-Toro, C. F. Valenzuela, R. A. Harris, Modulation of GABAa receptor function by G protein-coupled 5-HT2C receptors, Neuropharmacology 35(9–10), 1355–63 (1996).
18. Z. Yan, Regulation of GABAergic inhibition by serotonin signaling in prefrontal cortex: molecular mechanisms and functional implications, Mol Neurobiol. 6(2-3), 203–16 (2002).
19. S. Matsuo, I. S. Jang, J. Nabekura, N. Akaike, alpha 2-Adrenoceptor-mediated presynaptic modulation of GABAergic transmission in mechanically dissociated rat ventrolateral preoptic neurons, J. Neurophysiol. 89(3), 1640–8 (2003).

Behavioural Control of Breathing in Mammals: Role of the Midbrain Periaqueductal Gray

Hari H. Subramanian, Ron J. Balnave and Chin Moi Chow

1. Introduction

Brainstem respiratory neurons play a critical role in the generation of basic breathing rhythm in mammals (Richter et al., 2001, Feldman et al., 2003). This basic respiratory rhythm is modified by neurons probably with behaviour roles to support development of emotional expressivity. Such modulatory neurons are thought to be hierarchically organised throughout the neuraxis from the cerebral cortex through to the brainstem and spinal cord (Holstege, 1991a,b). One area within this hierarchy, that has been shown to be involved in motor patterning of defensive behaviour (in rats and cats) and vocalization (in cats) is the midbrain periaqueductal gray (PAG) (Carrive et al., 1987, Zhang et al., 1994).

Here we focus on respiratory patterns produced by the stimulation of the PAG with excitatory amino acid (EAA): *tachypnea*, associated with behaviours such as defensive reactions in the rat and the *vocalization-breathing pattern* in the cat. We summarize recent evidence obtained from our studies to suggest a possible role for the PAG in respiratory patterning to support complex motor behaviour.

2. Studies on the Rat

The role of PAG in the behavioural control of breathing in the rat is unclear. We were interested to determine the function of PAG as a respiratory modulating centre. Nembutal-anaesthetised (70 mg/kg), spontaneously-breathing rats (Sprague-Dawley, n = 30, weighing 350–400 g of either sex) were used to examine the effect of chemical stimulation

---

Hari H. Subramanian and Ron J. Balnave • School of Biomedical Sciences, Faculty of Health Sciences, The University of Sydney, P.O. Box 170, Lidcombe, NSW 1825, Australia. **Chin Moi Chow** • School of Exercise and Sport Sciences, Faculty of Health Sciences, The University of Sydney, P.O. Box 170, Lidcombe, NSW 1825, Australia.
Corresponding author: Hari H. Subramanian: Tel: 61-2-9351 9530; Fax: 61-2-9351 9520; Email: hsub5857@usyd.edu.au

Post-Genomic Perspectives in Modeling and Control of Breathing, edited by Jean Champagnat, Monique Denavit-Saubié, Gilles Fortin, Arthur S. Foutz, Muriel Thoby-Brisson. Kluwer Academic/Plenum Publishers, 2004.

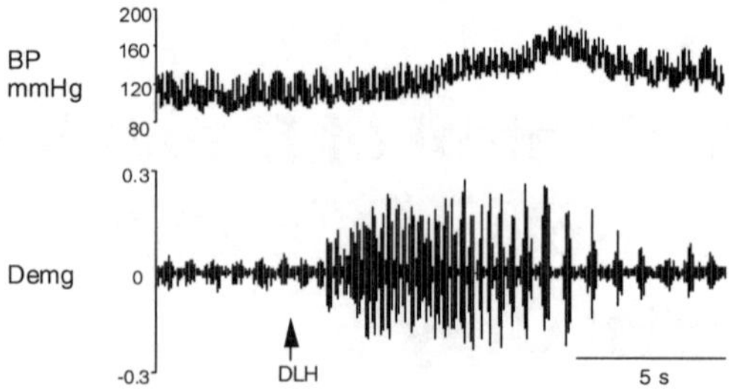

Figure 1. Effect of stimulation of dlPAG with DLH (30 nl). BP: blood pressure; Demg: diaphragm electromyogram. (from Huang et al., 2000).

(D, L-Homocysteic acid [DLH], 0.2 M 30–60 nl) of the dorsolateral PAG (dlPAG) on respiratory function. Respiratory patterns were examined through EMG recordings of various respiratory muscles.

2.1. Respiratory Patterning Following EAA Stimulation of dlPAG

Chemical stimulation of the dlPAG (7.8–8.3 mm caudal to bregma) evoked tachypnea (Fig. 1). The alteration of quiet breathing to a tachypneic pattern was site-specific and dose-dependant. Tachypnea was accompanied with significant increases in blood pressure (BP) and heart rate (HR).

The tachypnea was characterized at first by huge increases in diaphragm amplitude and increased respiratory frequency (RF). Both inspiratory (Ti) and expiratory (Te) durations were shortened. These changes were accompanied by increase in blood pressure that peaked 7.5 ms after the commencement of the respiratory effect. After 4.9 seconds RF began to decline. Ti & Te increased and Te became longer than control. In our laboratory it has been shown that abdominal muscles are recruited during tachypneic breathing along with significant increases to activity in external intercostals (Z-G Huang, personal communication). During tachypneic breathing the anaesthetized rat was observed to withdraw both fore and hind limbs and to arch its back.

2.2. dlPAG Induced Tachypnea can be Reversed by Prior Microinjection of β-Antagonist (Propranolol) into NTS

The tachypnea induced by stimulation of dlPAG could be reversed by prior microinjection of β-antagonist (propranolol 150 pmol, 30 nl, pH 7.4) into the commissural NTS; ie the DLH microinjections no longer induced tachypnea (Fig. 2). Propranolol treatment however did not alter the DLH (into dlPAG) induced cardiovascular response. Propranolol, itself, injected into NTS did not significantly alter BP, HR or RF in the absence of DLH.

The findings suggest that β-adrenoreceptors, at least in part, mediate the tachypnea, and likely through a noradrenergic link to the NTS. Li et al. (1992) reported increased respiratory activity following glutamate stimulation of the locus coeruleus. Its increase was blocked by propranolol injection into the NTS. This report together with our observation suggest that through some central respiratory circuits the effect of noradrenaline on respiratory activity is excitatory, a finding different to that of noradrenergic influences of respiratory neurons in the VRG (Champagnat et al., 1979). Propranolol pre-treatment did

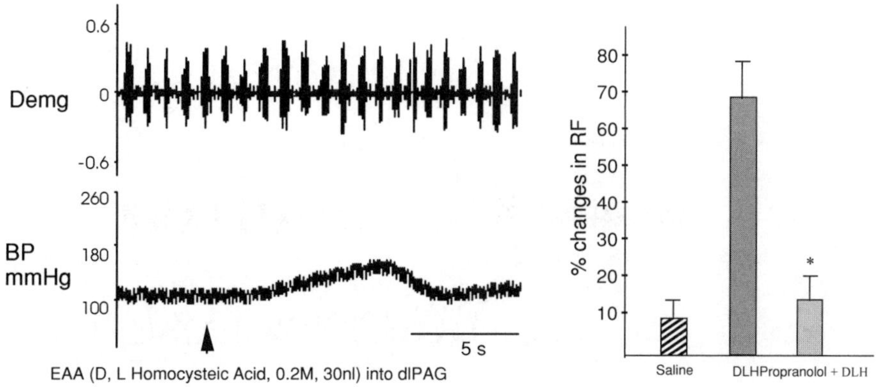

Figure 2. DLH into the dlPAG after pre-treatment of NTS with propranolol. Demg: diaphragm electromyogram; BP: blood pressure. (from Huang et al., 2000).

not alter BP. This vascular response is probably mediated by α_1/α_2 receptors (Feldman and Felder, 1989).

2.3. PAG: A Respiratory Modulator in the Rat

Production of tachypnea and abdominal breathing pattern upon stimulation of dlPAG suggests the presence of discrete respiratory neuronal populations capable of modulating quiet breathing. It is not possible for us to propose a behavioural scope for the tachypnea elicited in our anaesthetized preparation. Depaulis et al. (1992) have presented evidence of the involvement of PAG in defensive reactions in the rat. Carrive et al. (1987) have shown tachypneic breathing to be part of an integrated defense response in the cat. Perhaps the body movements observed during tachypnea in our preparation signify the elicitation of a defense reaction. As well tachypnea was often seen prior to the development of PAG induced vocalization in our cat data (below). In cats, lateral and ventrolateral PAG function as the primary center for call production (Zhang et al., 1994). In the rat, such a vocalization neuronal apparatus has not yet been demonstrated. However, abdominal patterning signifies the possibility of (fictive) phonation triggered by the PAG.

3. Studies on the Cat

3.1. Respiratory Patterning During PAG Induced Vocalization in the Cat

Analysis of data from precollicular decerebrate cats (n = 8, weighing 2.2–4.5 kg of either sex) revealed the respiratory patterns (through various respiratory and laryngeal EMG recordings) produced during vocalization elicited by DLH (0.2 M, 30–60 nl) microinjections into PAG. The patterning of breathing during vocalization was characterized by shortening of inspiration, an increase in inspiratory effort, a prolongation of expiration and marked decrease in overall RF (Fig. 3). Progressive recruitment of abdominal & cricothyroid muscles were seen. In some cases a *Double Diaphragm* pattern of breathing was also seen (Fig. 4).

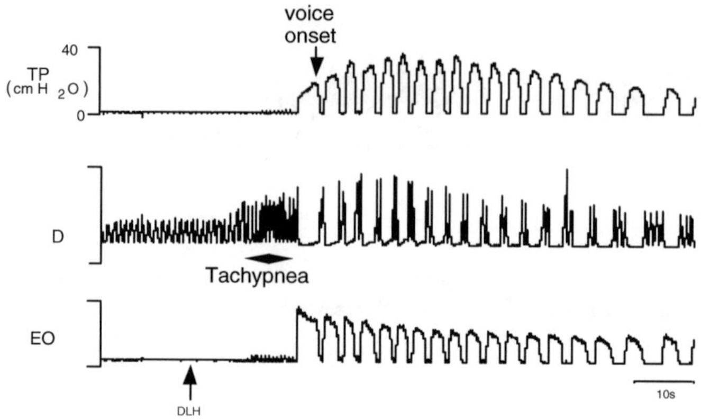

Figure 3. Vocalization breathing pattern elicited from the PAG. TP: Tracheal Pressure, D: Diaphragm EMG, EO: External Oblique EMG.

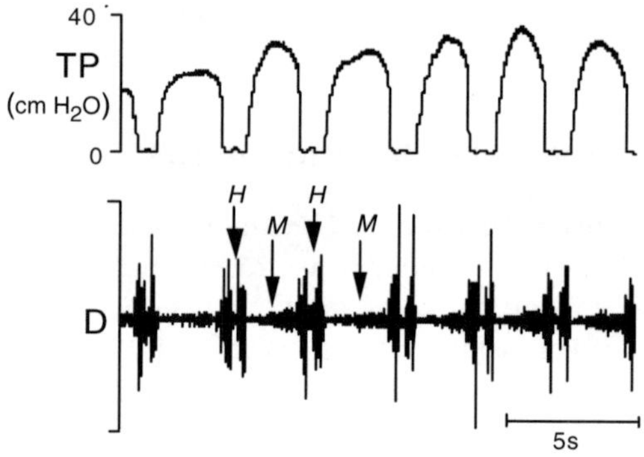

Figure 4. Characteristic *Double Diaphragm* pattern of breathing seen during mixed vocalization elicited from the PAG. TP: Tracheal Pressure, D: Diaphragm EMG. *H: Hiss, M: Mew.*

Tachypnea was observed in a number of vocalization episodes. Tachypnea occurred immediately following the DLH microinjection, prior to the elicitation of vocalization. Typically, the DLH microinjection induced a rapid rise in the respiratory rate for a few seconds, followed by a fall in RF sharply upon elicitation of vocalization. The fall ranged between 40–60% depending upon the type of vocalization (*hiss/mew/howl/growl*) elicited.

The inspiratory durations (Ti) for *voiced* vocalizations, (*mews, howls and growls*) were decreased in the range of 10–15% of its control value. The Ti preceding *unvoiced* vocalization (*hiss)* was generally larger than control. The expiratory durations (Te) were prolonged for *voiced* vocalizations. The overall Te showed an increase in the range of 170–300% as compared to that of control. The *unvoiced* vocalization had a shorter vocal expiratory phase usually lower than control. While expiratory facilitation is sublimal during quiet breathing, expiratory patterning to suit a specific vocal episode is seen during PAG-evoked vocalization. This is evident from the differential expiratory duration control for *voiced* and *unvoiced* vocalizations. The tracheal pressure (TP) increase is significant (up to 40 cmH_2O) during *voiced* vocalizations. This suggests that the prior tachypnea (increased

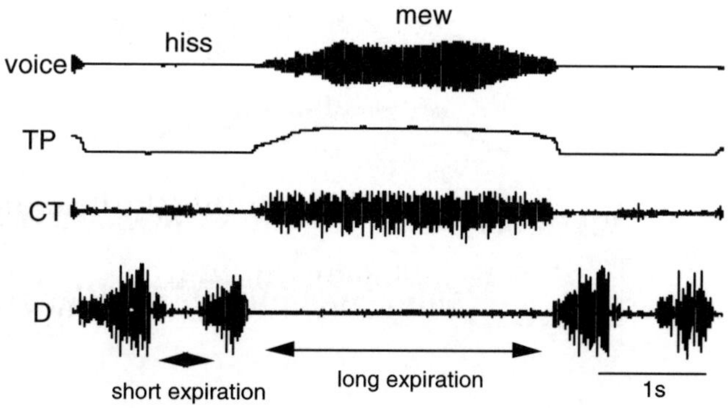

Figure 5. Short and long expirations associated with the mixed vocalization elicited from the PAG. TP: Tracheal Pressure, CT: Cricothyroid EMG, D: Diaphragm EMG.

diaphragm frequency and amplitude) may be responsible for the increased TP necessary for vocalization. Integration of expiratory and laryngeal neuronal activity may also contribute to a large modulation in TP.

3.2. Differential Inspiratory & Expiratory Patterning During *Hiss-Mew* Episodes *(Double Diaphragm Pattern)*

During a mixed vocal episode (Fig. 4) the diaphragm discharge patterns were different before a *hiss* and a *mew*. The inspiratory activity preceding the *hiss* showed an augmenting discharge pattern (Fig. 5). The inspiratory phase following the *hiss* and prior to the *mew* was similar to baseline diaphragm activity. The increase in EMG activity and the appearance of large amplitude spikes are suggestive of large motor units being recruited for mediating increased inspiratory effort and lung volume requirements. Similar control is also seen over the expiratory and laryngeal musculature. The expiratory activity is very much different for a hiss and a mew. In the case of a hiss the PAG does not produce sufficient expiratory drive to produce the vibration of the vocal folds in spite of development of sufficient air volume in the lungs. The laryngeal adductor muscles (cricothyroid) are neither active nor is there a sufficient build up of tracheal pressure (Fig. 5).

3.3. Respiratory Patterning Following Direct Chemical Stimulation of the NRA in the Caudal Brainstem

In data from six cats we looked at the respiratory patterning following DLH microinjections into the nucleus retroambigualis (NRA). The microinjections gave rise to laryngeal (CT) and abdominal (EO) muscle activation along with slight elevation of TP (up to 6 cmH_2O). A significant finding of this data was the maintenance of the respiratory frequency to such an intervention (Fig. 6). Study by Feldman and Speck (1983) has shown that stimulation of the expiratory neurons in the NRA causes changes to RF with characteristic increases in the Te. Our results are in agreement with their reporting, as lengthening of expiration is seen. However the RF does not change significantly as the expiratory lengthening is accompanied with shortening of inspiration. This opposing activity maintains a relatively constant mean RF. Whether this phenomenon is due to co-activation of NRA

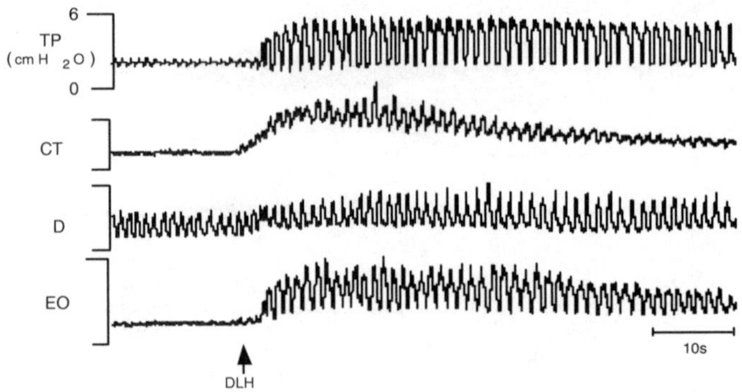

Figure 6. DLH into the NRA. TP: Tracheal Pressure, CT: Cricothyroid EMG, D: Diaphragm EMG, EO: External Oblique.

laryngeal neurons along with the premotor respiratory neurons (presumably in the ventral respiratory group [VRG]) remains to be investigated.

3.4. PAG: A Respiratory Modulator in the Cat

It is strikingly apparent that alterations to quiet breathing pattern are necessary for vocalization in the cat. Our results suggest that the PAG contains discrete neuronal populations involved in patterning of respiration during species-specific call production. Further the NRA data establish that brainstem respiratory and laryngeal neurons involved in basic respiratory function are not capable of such motor patterning, an attribute highlighting the hierarchical superiority of the PAG.

4. Conclusion

Our findings clearly show that PAG modifies the quiet breathing pattern to support complex motor behaviour in both the rat and the cat. Further investigations are required to provide understanding of the extent of sensori-motor integration happening at the PAG and the hierarchical role it plays in the behavioural control of breathing.

5. Acknowledgements

This work was supported by grants from Faculty of Health Sciences, University of Sydney, and Australian National Health and Medical Research Council.

References

1. Richter, D. W., Spyer, K. M., 2001. Studying rhythmogenesis of breathing: comparison of in vivo and in vitro models. *Trends in Neuroscience*, 24, No. 8, 464–472.
2. Feldman, J. L., Mitchell, G. S., Nattie, E. S., 2003. Breathing: Rhythmicity, Plasticity, Chemosensitivity. *Annu. Rev. Neurosci.* 26, 239–266.

3. Holstege, G., 1991a. Descending motor pathways and the spinal motor system. Limbic and non-limbic components. In *Progress in Brain Research*. Edited by G. Holstege Amsterdam: Elsevier Vol 87. p 307–421.
4. Holstege, G., 1991b. Descending pathways from the periaqueductal gray and adjacent areas. In *The Midbrain Periaqueductal Gray Matter: Functional Anatomical and Immunohistochemical Organization*. Edited by A. Depaulis and R. Bandler. New York Plenum Press, p 239–265.
5. Carrive, P., Dampney, R. A. L., Bandler, R., 1987. Excitation of neurons in a restricted region of the periaqueductal gray elicits both behavioural and cardiovascular components of defence reaction in unanaesthetised decerebrate cat. *Neurosci. Lett*, 81, 273–278.
6. Zhang, S. P., Davis, P. J., Bandler, R., Carrive, P., 1994. Brainstem Integration of Vocalization: Role of the Midbrain Periaqueductal Gray. *Journal of Neurophysiology*, 72, 1337–56.
7. Li, Z.Y., Xia, B.L., Huang, C.J., 1992. Effects of microinjection of L-glutamate into locus coeruleus complex area on respiration. *J. Tongji Medical University* 12, 205–208.
8. Champagnat, J., Denavit-Saubie, M., Henry, J., Leviel, V., 1979. Catecholaminergic depressant effects on bulbar respiratory mechanisms. *Brain Res.* 480, 57–68.
9. Feldman, P.D., Felder, R.B., 1989. Alpha 2-adrenergic modulation of synaptic excitability in the rat nucleus tractus solitarius. *Brain Res.* 480, 190–197.
10. Depaulis, A., Keay., K. A., Bandler, R. 1992. Longitudinal neuronal organization of defensive reactions in the midbrain periaqueductal gray region of the rat. *Exp. Brain. Res.* 90, 307–318.
11. Feldman, J. L., Speck, D. F., 1983. Interactions among inspiratory neurons in dorsal and ventral respiratory groups in the cat medulla. *J. Neurophysiol.* 49: 472–490.
12. Huang, Z. G., Subramanian, S. H., Balnave, R. J., Turman, A. B., Chow, C. M., 2000. Roles of periaqueductal gray and nucleus tractus solitarius in cardiorespiratory function in the rat brainstem. *Respiration Physiology*, 120, 185–195.

Breathing at Birth: Influence of Early Developmental Events

Gilles Fortin, Caroline Borday, Isabelle Germon, and Jean Champagnat

1. Introduction

Respiration is a rhythmic motor behavior that appears in the fetus and acquires a vital importance at birth. It is generated centrally, within neuronal networks of the brainstem. Recently, examination of hindbrain activities in the embryo has revealed that a central rhythm generator is active in the brainstem before fetal maturation and conforms to the segmented organization of the embryonic hindbrain at this stage of development. From physiological studies of this primordial rhythm generator in embryos, we may therefore gain an understanding of how genes govern development of neuronal networks and specify patterns of motor activities operating throughout life.

The brainstem derives from the embryonic hindbrain (rhombencephalon), one of the vesicles that appears towards the anterior end of the neural tube. When reticular neurons differentiate, the hindbrain neuroepithelium is partitioned along the antero-posterior axis into an iterated series of eight cellular compartments called rhombomeres. This segmentation process is transient and believed to specify neuronal fates by encoding positional information[1]. Electrophysiological recordings performed on an isolated preparation of chick embryo hindbrain, revealed that by the end of the segmentation period, the hindbrain neuronal network starts to exhibit a consistent and organized activity in the form of recurring episodes composed of burst discharges that occur simultaneously in the different cranial nerves. At this stage the neuronal network is already organized with distinct reticular and motor neurons[2]. When intersegmental relationships are interrupted by transverse sectioning of the hindbrain rostral and caudal to the exit of the branchiomotor nerve, the ability to generate the rhythmic pattern is preserved in each transverse slice. The network organization which is responsible for this rhythm generation conforms to the rhombomeric pattern in the hindbrain of the embryo. Specification of the future rhythmic network would therefore take

---

Gilles Fortin, Caroline Borday, Isabelle Germon, and Jean Champagnat • Neurobiologie Génétique et Intégrative, Institut de Neurobiologie Alfred Fessard, CNRS, 91198, Gif sur Yvette, France.
Supported by European grant QLG2/CT/2001-01467.

Post-Genomic Perspectives in Modeling and Control of Breathing, edited by Jean Champagnat, Monique Denavit-Saubié, Gilles Fortin, Arthur S. Foutz, Muriel Thoby-Brisson. Kluwer Academic/Plenum Publishers, 2004.

place within the segmented hindbrain of the embryo and segmentation may be determinant in conferring the ability to generate specific rhythmic patterns of activity[3].

2. Methods

For electrophysiological recording of embryos, isolated whole hindbrain preparations as well as preparations deriving from hindbrain isolated segments (see Figure 1) were performed as previously described[4] ref 4 to ref 2 and transfered into a 2 ml recording chamber superfused at 2 ml.min^{-1} with an artificial cerebrospinal fluid containing in mM: NaCl, 120; KCl, 8; NaH_2PO_4, 0.58; $MgCl_2$, 1.15; $CaCl_2$, 1.26; $NaHCO_3$, 21; glucose, 30 aerated with 5% CO_2/95% O_2 (pH7.4 at 30°C). Spontaneous electrical activity was recorded through suction electrodes positioned onto the trigeminal exit point or the seventh, ninth and tenth cranial nerve rootlets. The recording electrode was connected to a high-gain AC amplifier (Grass P511K). Amplified and filtered (bandwidth, 3 Hz-3 kHz) neurograms were rectified and integrated through an analog integrator (Grass 7P3) with a time constant of 50 ms and digitized at 100 Hz via a CED-1401 laboratory interface. Burst detection and measurements of burst duration and of interburst intervals were calculated offline using the Acquis1 software package (developed by G. Sadoc, Gif-sur-Yvette). The duration of recordings varied from 30 minutes to several hours and the reported patterns of activity were maintained throughout.

3. Results: Rhythm Generation in Embryo

3.1. Identification of a Maturation Step Leading to an HF Episodic Activity in the Chick

By recording neural rhythm generation in post-segmental chick hindbrains preparations isolated in vitro, we have identified a developmental step in which GABAergic synapses begin to exert an inhibitory action on neurons, thereby shifting rhythmic activity from low frequency (immature, figure 1, bottom left) to high (closer to mature) episodic frequency (HF, figure 1 bottom right). At the cellular level, GABAergic inhibition is responsible for the appearance of a novel neuronal phenotype (being inhibited instead of excited during burst activity) within the rhythmic network. At the membrane level, inhibition involves GABA-A receptors and an increase of neuronal input conductance; HF results from a post-inhibitory rebound regulated by the Ih voltage-dependent current[4].

3.2. Deciphering the Rhombomeric Code for HF Maturation

To understand how rhombomere identity might influence rhythm generation, we have developed in collaboration with A. Lumsden & S. Jungbluth (London), an in ovo neural tube ablation procedure that allows the later recording of rhythms (arising after the end of the segmental period) from combinations of isolated single rhombomeres (r) or doublets of adjacent rhombomeres. We found that the normal HF generator develops in r3r4 (figure 1 bottom) and r5r6 doublets, but not in isolated r2, r4 (figure 1 top) or r6, or in r1r4, r2r3 or r4r5 doublets. Therefore, turning on the HF rhythm pattern depends on the singular identity

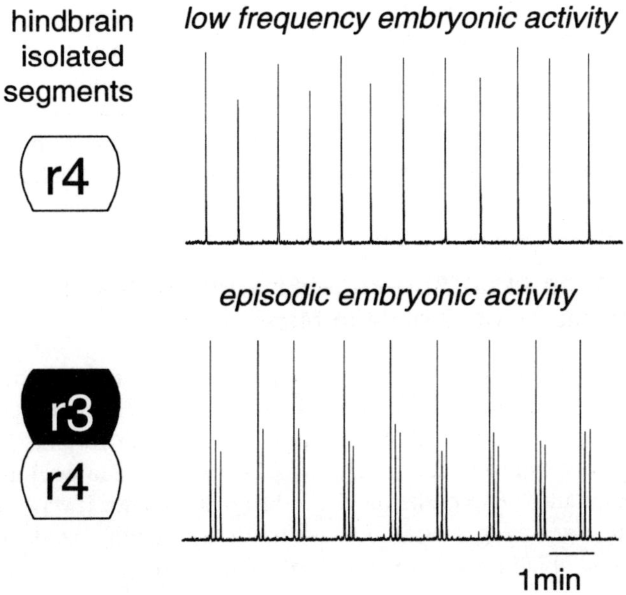

Figure 1. Experiments on chick hindbrain isolated segments demonstrate that the program of hindbrain segmentation controls formation of rhythm promoting neuronal circuits. Individual or pairs of segments are isolated at the time when they form and maintained in ovo to allow development of neuronal networks. The r3r4 segments contains the entire information required for the normal maturation of an episodic (HF) activity. The r3 segment is required, because the mature episodic network fails to develop from the isolated r4 segment.

of r3 and r5, which controls, on the basis of a two segment repeat, the later maturation of GABAergic inhibition[4]. These experiments have demonstrated a robust link between rhombomere patterning and maturation of HF reticular circuits in the avian embryo.

4. Discussion: Breathing in Transgenic Mice

A wealth of data have been accumulated on genes governing hindbrain segmentation. Before and at the onset of segmentation, genes encoding transcription factors such as *Hox*, *Krox-20*, *kreisler*, are expressed in domains corresponding to the limits of future rhombomeres[1]. Inactivation of these genes specifically disturbs the rhombomeric pattern of the hindbrain. For example, *Krox-20*, is transiently expressed within the yet unsegmented hindbrain in two stripes with sharp edges corresponding to the future rhombomeres r3 and r5. The inactivation of *Krox-20* in transgenic mice results in the deletion of r3 and r5 and lethality shortly after birth[5]. The *Krox-20* gene product acts as a direct transcription activator of other r3- and r5-related genes belonging to the Hox clusters. Concerning Hox genes themselves, expression of Hoxa1 also provides one of the earliest signs of regionalisation within the developing hindbrain. As early as 7.5 day-post-coitum in mice, the *Hoxa1* expression domain extends from the posterior end of the mouse embryo up to the presumptive r3/r4 border and is downregulated before rhombomere boundary formation. This transient expression has a profound impact on hindbrain patterning, as *Hoxa1* targeted inactivation results in severe reduction of r4 and r5 and their derived structures (e.g. the motor nucleus of the facial nerve) and in lethality shortly after birth[6]. The advent of such mutant mice in which embryonic hindbrain development is affected by the deletion of specific territories provides a potential strategy to establish a link between gene expression and breathing in intact animals. Because respiration acquires a vital importance at birth, prenatal dysfunction of neuronal networks can be responsible for functional anomalies that appear after birth.

4.1. Identification of an Anti-apneic Neuronal System Depending on the Integrity of r3 or r4 in Mice

The general strategy has been to identify phenotypic traits of the rhythmic respiratory network following loss-of-function mutation of transcription factors expressed in a rhombomere-specific manner: after the elimination of both r3 and r5, or of r5 alone, obtained respectively by inactivating *Krox-20* and *kreisler*[7] and after elimination of r4 and r5 in *Hoxa1*$^{-/-}$ mutants[8]. By comparing these different mutants, it was found that both r3- and r4-derived neurons are essential to alleviate life threatening neonatal apneas thereby maintaining normal breathing shortly after birth. This "anti-apneic" system is vital during a precise time window (the first two days after birth in mice[9]), during which survival of *Krox20*$^{-/-}$ and *Hoxa1*$^{-/-}$ mutants could be improved by blocking enkephalinergic inhibition of the respiratory rhythm in vivo. These results have established the link between neuronal loss resulting from the abnormal segmentation of the neural tube and life-threatening murine syndromes with potential importance in neonatal human pathology.

4.2. Mis-specifications in the Segmented Hindbrain Produce Novel Neuronal Controls of Breathing, Active After Birth

Abnormal expression of segmentation genes does not necessarily result in cell death. The survival and neurobiological function of mis-specified cell progeny at birth has been investigated in *Hoxa1*$-/-$ and *kreisler* $+/-$ mutants. Dominguez et al.[8] have identified and located a novel functional neuronal circuit increasing the respiratory rhythm in Hoxa1$^{-/-}$ mice but not in wild-type mice; it seems to result from the acquisition of an r2-like phenotype by progenitors located in r3–r4 (figure 2). Chatonnet et al.[7] have found an exaggeration of the anti-apneic control (as defined above[10]) indicating persistence of an abnormal control of respiration in heterozygous *kreisler* mice exhibiting mis-specifications of r3 cells[11], but neither rhombomere elimination nor massive anatomical deficits. These results demonstrate that changes in segmental gene expression pattern underlie the acquisition of neuronal circuits regulating vital adaptive behaviours and, therefore, might be implicated in the evolution of the vertebrate brainstem network.

4.3. Ontogenetic Basis for a Dual Oscillator Model of Rhythm Generation

Importantly, r4 is the neural tube territory from which the facial branchio-motor nucleus and adjacent nuclei are derived. Experiments illustrated in figure 1 show that

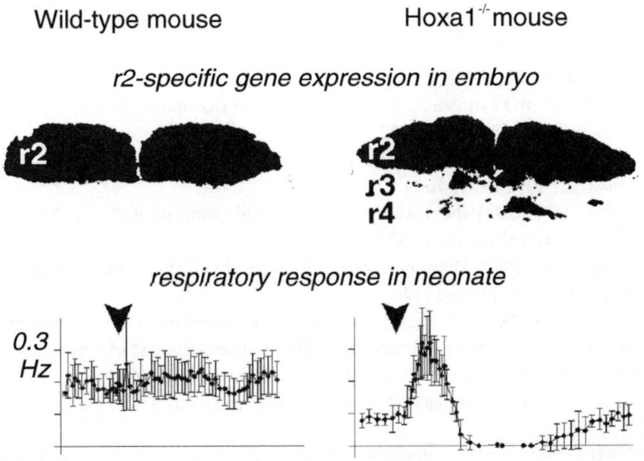

Figure 2. Experiments on *Hoxa1*$^{-/-}$ mutant mice demonstrate that genes orchestrating hindbrain segmentation control formation of rhythm promoting neuronal circuits. In situ hybridization of the segmented hindbrain (top) identifies mis-specification of progenitors in rhombomeres r3 and r5, a region of the neural tube giving rise to the caudal Pons. Slice preparations have been developed in the mutant neonates to investigate function of pontine structures and generation of the respiratory rhythm *in vitro*. Stimulation of the caudal Pons (arrowhead) in neonatal Hoxa1−/− mutants reveals an abnormal connection that influences respiratory frequency.

rhythm generation in r4-derived structure is higly dependent on proper interactions with the more rostral, *Krox-20*-expressing r3[4]. In *Hoxa1* homozygous mutant mice[8], supplementary rhythm-promoting cell are induced from r4 (figure 2) and located at the facial antero-posterior level of the brainstem, indicating the important role of r4 in the generation of the anti-apneic system. In contrast, the elimination of r5 in *kreisler* homozygous mutants[7] does not affect respiratory rhythm generation. These results show that there is a gap of at least a one rhombomere length between a rostral, r4-derived oscillator and the more caudal generators of the Pre-Bötzinger complex[12,13]. The analysis of early developmental events therefore provides support for a dual oscillator model of respiratory rhythm generation, by which pre-inspiratory neurons from the parafacial neuronal group[14,15] interact with Pre-Bötzinger neurons to generate a normal respiratory drive after birth[16].

5. Conclusions

Overall, studies on embryos and neonates demonstrate the role of hindbrain segmentation in the formation of rhythm-promoting circuits. Developmental mechanisms orchestrating the early organogenesis of the brainstem appear to be crucial in establishing the postnatal breathing pattern. Experiments performed after birth in transgenic mice indicate that, although expression of these genes and hindbrain segmentation are transient events of the early embryonic development, they are important for the process of respiratory rhythm generation. Therefore, early developmental processes have to be taken into account to understand normal and pathological diversity of the vertebrate breathing behaviours during postnatal life.

References

1. Lumsden, A. and Krumlauf, R., 1996, Patterning the vertebrate neuraxis. *Science* 274: 1109–1115.
2. Fortin, G., Champagnat, J. and Lumsden, A., 1994, Onset and maturation of branchiomotor activities in the chick hindbrain. *NeuroReport*, 5, 1149–1152.
3. Champagnat, J. and Fortin, G., 1997, Primordial respiratory-like rhythm generation in the vertebrate embryo. *Trends in Neurosciences*, 20: 119–124.
4. Fortin, G. et al., 1999, Segmental specification of GABAergic inhibition during development of hindbrain neural networks. *Nature Neuroscience* 2: 873–877.
5. Schneider-Maunoury, et al., 1993, Disruption of Krox-20 results in alteration of rhombomere 3 and 5 in the developing hindbrain. *Cell* 75: 1199–1214.
6. Mark, et al., 1993, Two rhombomeres are altered in *Hoxa-1* mutant mice. Development 119:319–338.
7. Chatonnet, F. et al., 2002, Different respiratory control systems are affected in homozygous and heterozygous kreisler mutant mice, 2002, *European Journal of Neuroscience* 15: 684–692.
8. Dominguez, et al., 2001, Generation of a novel functional neuronal circuit in Hoxa1 mutant mice. *J Neurosci.* 2001 Aug 1; 21(15): 5637–42.
9. Fortin, G. et al., 2000, Genetic and Developmental Models for the Neural Control of Breathing in Vertebrates. *Respiration Physiology*, 122: 247–257.
10. Jacquin, T. D., et al., 1996, Reorganization of pontine rhythmogenic neuronal networks in *Krox-20* knockout mice. *Neuron* 17: 747–758.
11. Manzanares, et al., 1999, The role of *kreisler* in segmentation during hindbrain development. *Developmental Biology* 211 (2), 220–237.
12. Smith, J.C., Ellenberger, H. H., Ballanyi, K., Richter, D. W., and Feldman, J. L., 1991, Pre-Bötzinger complex: a brainstem region that may generate respiratory rhythm in mammals, *Science* 254, 726–729.
13. Thoby-Brisson, M., Cauli, B., Champagnat, J., Fortin, G., and Katz, D. M., 2003, Expression of functional tyrosine kinase B receptors by rhythmically active respiratory neurons in the pre-Bötzinger complex of neonatal mice. *J. Neurosci.* 23, 7685–7689.
14. Onimaru, H. Arata, A., and Homma, I., 1990, Inhibitory synaptic inputs to the respiratory rhythm generator in the medulla isolated from newborn rats. *Pflügers Arch.* 417, 425–432.
15. Onimaru H. and Homma I., 2003, A novel functional neuron group for respiratory rhythm generation in the ventral medulla. *J Neurosci.* 23: 1478–86.
16. Chatonnet, F., Dominguez del Toro E., Thoby-Brisson M., Champagnat, J., Fortin, G., Rijli F. M., Thaeron-Antono C., 2003, From hindbrain segmentation to breathing after birth: developmental patterning in rhombomeres 3 and 4. *Mol Neurobiol.* 28: 277–94.

4

From Molecular to Integrated Neural Control of Breathing

Synaptic Transmission

A Dual-Role Played by Extracellular ATP in Frequency-Filtering of the Nucleus Tractus Solitarii Network

Fusao Kato, Eiji Shigetomi, Koji Yamazaki, Noriko Tsuji, and Kazuo Takano

1. Introduction

ATP is now identified to be an important signaling molecule in the CNS[1–3]. However, there are only few brain regions in which the function of ATP-mediated signaling is demonstrated from the molecular to whole animal levels. The caudal part of the nucleus of the solitary tract (cNTS) is such a rare structure. In the cNTS, neuronal ATP release[4,5], hypoxia-induced increase in purine concentration[6], abundant expression of P2X and P2Y receptors[7–12], extracellular hydrolysis of ATP to adenosine[4,13], and rich expression of adenosine transporters[14] have been demonstrated. In addition, a microinjection of agonists for P2X and adenosine receptors into cNTS in anesthetized rats exerts profound cardiorespiratory effects[15–17]. Taken together, ATP, in tandem with its extracellular metabolite adenosine, is thought to be involved in the neuronal signaling in the cNTS, where various visceral signals including those from pulmonary stretch receptors and peripheral chemoreceptor converge.

The cNTS network presents two distinct types of activities: tonic and phasic. The tonic activity is generated by regularly firing properties of the cNTS neurons[18], re-excitatory connections between cNTS excitatory neurons[19,20] and spontaneous action potential-independent release of transmitters from terminals[13]. The phasic activity arises from short-term responses to the primary afferent inputs[19,20]. These tonic and phasic activities may represent DC (low-frequency) and AC (high-frequency) responses, respectively, of the input-output characteristics of cNTS network. Here we describe two examples, *in vitro* and *in vivo*, showing that the distinct frequency components of the cNTS activity are regulated by distinct purinoceptors.

Fusao Kato, Eiji Shigetomi, Koji Yamazaki, and Noriko Tsuji • Laboratory of Neurophysiology, Department of Neuroscience. **Kazuo Takano** • Department of Pharmacology, Jikei University School of Medicine, Tokyo 105-8461, Japan.

Post-Genomic Perspectives in Modeling and Control of Breathing, edited by Jean Champagnat, Monique Denavit-Saubié, Gilles Fortin, Arthur S. Foutz, Muriel Thoby-Brisson. Kluwer Academic/Plenum Publishers, 2004.

2. Methods

2.1. *In vitro*

Coronal brainstem slices (400-m thickness) of Wistar rats (2–5 weeks old) were continuously perfused with ACSF (in mM: NaCl 125, KCl 3, $CaCl_2$ 2, $MgCl_2$ 1.3 or 3, NaH_2PO_4 1.25, D-glucose 12.5, L-ascorbic acid 0.4, $NaHCO_3$ 25, picrotoxin 0.1; 95% O_2 + 5% CO_2)[13]. The transmembrane current was recorded from small (soma diameter <15 m) second-order cNTS neurons visually identified with IR-DIC videomicroscopy with a patch-pipette containing (in mM): 120 gluconic acid potassium, 6 NaCl, 5 $CaCl_2$, 2 $MgCl_2$, 2 ATP Mg, 0.3 GTP Na, 10 EGTA, 10 HEPES. The membrane potential was held at −70 mV. The solitary tract (TS) was stimulated every 5 s at a suprathreshold intensity (0.01–3 mA; 100 s) with a concentric bipolar electrode placed on the ipsilateral TS. Recordings were made at room temperature.

2.2. *In vivo*

Adult Japanese white rabbits were anesthetized with urethane (0.5 g/kg, ip and 0.5 g/kg, iv). After tracheal cannulation, the animals were immobilized with gallamine triethiodide (6 mg/kg, iv; then, 3 mg/kg/h) and artificially ventilated with room air under monitoring of the end-tidal CO_2 concentration (4.0%–4.5%). The bilateral phrenic nerves were sectioned in the neck, and the central cut end of a nerve was hooked on a bipolar platinum electrode for recording. The A and A afferent fiber groups in the vagus nerve arising from the SA-PSR were selectively stimulated (see METHODS in Ref. 22) with a bipolar platinum electrode (0.5 V; 100 s).

3. Results

The second-order cNTS neurons showed both spontaneously occurring and stimulation-evoked EPSCs (sEPSCs and eEPSCs, respectively). For a period of <500 ms following the TS stimulation, sEPSC frequency increased to ∼150% of the basal level (Fig. 1A). This "short-term excitation" of the cNTS network following TS stimulation is likely to result from the "re-excitatory processing" between cNTS neurons[19]. ATP (100 M) exerted two distinct effects (Fig. 1, middle). First, ATP increased the frequency of sEPSC (filled bars in Fig. 1A and inward events in Fig. 1B). Second, ATP decreased the amplitude of eEPSC (Fig. 1B and C). The short-term excitation was no more observed following TS stimulation during the ATP effect (Fig. 1A, filled bars). The following analyses suggest that distinct classes of purinoceptors mediated these distinct effects. Adenosine (100 M) markedly decreased the eEPSC amplitude (Fig. 1B and C), whereas it did not increase the sEPSC frequency (Fig. 1A and B). In contrast, ATP applied in the presence of 8-cyclopentyltheophylline (CPT), an adenosine A_1 receptor antagonist, did not affect eEPSC amplitude despite a marked increase in sEPSC frequency (Fig. 1A–C, right). The increase in spontaneous activity by ATP was completely abolished by a selective P2X receptor blocker, pyridoxal-phosphate-6-azophenyl-2′,4′-disulphonic acid (PPADS, 40 M; data not shown).

These results indicate that ATP (1) promotes spontaneous tonic activity by activating P2X receptors, and (2) depresses afferent-evoked phasic activities by activating A_1 receptors

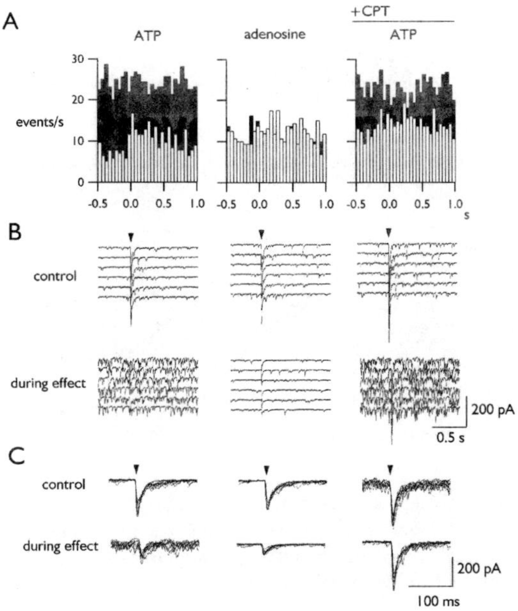

Figure 1. Effect of ATP and adenosine on the spontaneous and evoked EPSCs. A, peristimulus histogram of EPSC event frequency. Open bars, control; filled bars, during drug effect. B, peristimulus traces of the membrane current. The solitary tract was stimulated at the arrowheads. control, traces before application of agonists; during effect, traces at the period during which the effect of agonists (shown on the top) was almost maximal. C, time-expanded superimposition of the traces (10 consecutive stimuli).

after extracellular breakdown to adenosine. Analyses of the effect of P2X agonists on the miniature EPSC frequency recorded in the presence of tetrodotoxin (1 M) and the effects of ATP and adenosine on the paired-pulse depression of the eEPSC (data not shown) revealed that the receptors underlying these responses are located presynaptically. The amplitude of eEPSC represents the gain of the phasic, reflexogenic and dynamic activity of the cNTS network while the frequency of sEPSC reflects the basal level of the tonic, spontaneous and static activity of the network. It is therefore expected that, in cNTS network, the lower-frequency components are more affected by activation of P2 receptors whereas the higher-frequency components are more sensitive to activation of P1 receptors.

To examine whether this frequency-dependent distinction exists in the *in vivo* preparation, we analyzed the effect of P2 receptor blockade on the "vagal inspiratory promotion reflex", a respiratory reflex highly dependent on the afferent input frequency[21,22].

The vagus nerve stimulation at a low-frequency (10–40 Hz) in vagotomized rabbits or maintaining the lung volume near the functional residual capacity (FRC) in rabbits with intact vagi accelerated the respiratory rhythm, whereas that at a high-frequency (80–160 Hz) or lung inflation markedly slowed it (Fig. 2A, left). The latter response is the well-described "vagal inspiratory off-switch"[23], whereas the former is the "vagal inspiratory promotion" originally described in detail by Refs. 21, 22. After unilateral injection of PPADS (5 nmole/rabbit) into the cNTS in bilaterally vagotomized rabbits, the low-frequency stimulation of the vagus afferent ipsilateral to the injection no more accelerated respiratory

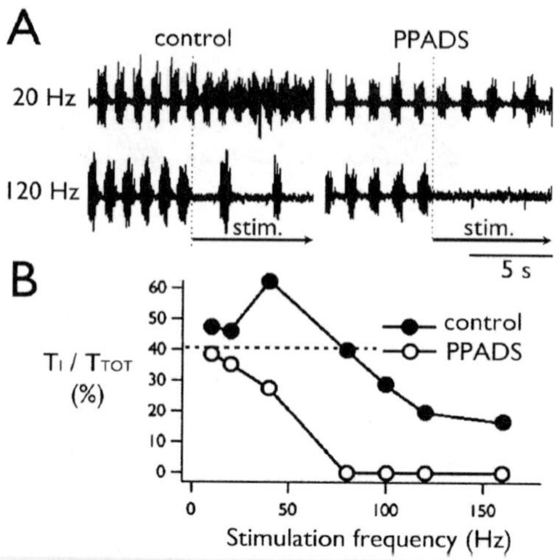

Figure 2. Effect of PPADS on the vagal inspiratory promotion reflex. A, the phrenic nerve discharges in response to 20 Hz and 120 Hz stimulation of the vagus nerve before and after PPADS microinjection into cNTS. B, The inspiratory time over total respiratory cycle duration during continuous stimulation of the vagus nerve of the side ipsilateral to the PPADS injection. Broken line indicates the T_I/T_{TOT} without stimulation.

rhythm (Fig. 2A right and open circles in 2B). In contrast, the vagal stimulation at a high-frequency still slowed the respiratory rhythm and suppressed the inspiratory activity (Fig. 2A and 2B), indicating the inhibitory effect of PPADS was limited to low-frequency responses. The PPADS injection did not affect the effects of vagal stimulation contralateral to the injection. These results indicate that a blockade of P2X receptors in the cNTS suppresses exclusively the responses provoked by low-frequency afferent signals and leaves those by high-frequency signals intact.

4. Discussion

ATP is a unique neuromessenger because it plays a dual-role[13]: first, after being released, it activates ionotropic P2X and/or metabotropic P2Y receptors, then ATP is broken-down to adenosine in the extracellular milieu by ecto-nucleotidases, and activates adenosine receptors. An important finding of the present study is that ATP and its metabolite adenosine modulate the input-output relationship of the cNTS network in distinct manners depending on the frequency-range of the activity, through activation of distinct classes of receptors.

Lines of evidence indicate that the synaptic transmission in the cNTS is highly dependent on frequency[24,25]. In addition, the synaptic transmission from the TS to the cNTS second-order neurons shows strong short-term plasticity, which is profoundly modulated by adenosine[26]. The present results suggest that ATP promotes tonic DC/low-frequency activity and adenosine inhibits phasic AC/high-frequency activities, by activating P2X and A_1 receptors, respectively. Fig. 3 summarizes a hypothetical model to explain these frequency-dependent effects of ATP and adenosine. It has been shown that an electrical stimulation of the hypothalamic defense area triggers ATP release, which is followed by a rapid adenosine

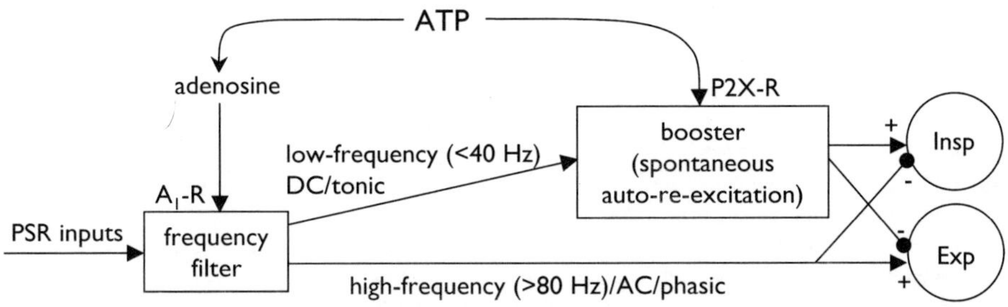

Figure 3. A hypothetical model to explain frequency-dependent effects of ATP and its metabolite adenosine on the frequency filtering in the cNTS network. A1-R, adenosine A_1 receptors; P2X-R, P2X receptors; Insp, inspiratory pattern generator; Exp, expiratory pattern generator; PSR, pulmonary stretch receptor.

production, in the cNTS[4]. ATP released in the cNTS might promote the tonic endogenous activity, while, at the same time, attenuate phasic reflexogenic activity after being broken-down to adenosine in the cNTS network, which may underlie complex cardiorespiratory responses including the region-dependent changes in the vascular tone and lowered baroreflex gain during the defense reactions.

5. Acknowledgments

This work was supported by Grants-in-Aid from the Ministry of Education, Culture, Sports, Science and Technology, Japan (Nos. 13680902 and 15650071) and Grants for the Research on Health Sciences focusing on Drug Innovation from The Japan Health Sciences Foundation (KH21014) to F. K. The expert assistance in the experiments and during preparation of the manuscript by T. Matsuo is acknowledged.

References

1. S. J. Robertson, S. J. Ennion, R. J. Evans and F. A. Edwards, Synaptic P2X receptors, *Curr. Opin. Neurobiol.*, **11**, 378–386 (2001).
2. R. A. North, Molecular Physiology of P2X Receptors, *Physiol. Rev.*, **82**, 1013–1067 (2002).
3. B. S. Khakh, Molecular physiology of P2X receptors and ATP signalling at synapses, *Nat. Rev. Neurosci.*, **2**, 165–174 (2001).
4. N. Dale, A. V. Gourine, E. Llaudet, D. Bulmer, T. Thomas and K. M. Spyer, Rapid adenosine release in the nucleus tractus solitarii during defence response in rats: real-time measurement in vivo, *J. Physiol. (Lond)*, **544**, 149–160 (2002).
5. J. H. St Lambert, T. Thomas, G. Burnstock and K. M. Spyer, A source of adenosine involved in cardiovascular responses to defense area stimulation, *Amer. J. Physiol.*, **272**, R195–R200 (1997).
6. R. A. Barraco, G. A. Walter, P. M. Polasek and J. W. Phillis, Purine concentrations in the cerebrospinal fluid of unanesthetized rats during and after hypoxia, *Neurochem. Int.*, **18**, 243–248 (1991).
7. I. von Kögelgen and A. Wetter, Molecular pharmacology of P2Y-receptors, *Naunyn-Schmiedeberg's Arch. Pharmacol.*, **362**, 310–323 (2000).
8. G. Collo, R. A. North, E. Kawashima, E. Merlo-Pich, S. Neidhart, A. Surprenant and G. Buell, Cloning of P2X5 and P2X6 receptors and the distribution and properties of an extended family of ATP-gated ion channels, *J. Neurosci.*, **16**, 2495–2507 (1996).

9. L. Vulchanova, M. S. Riedl, S. J. Shuster, G. Buell, A. Surprenant, R. A. North and R. Elde, Immunohistochemical study of the P2x2 and P2x3 receptor subunits in rat and monkey sensory neurons and their central terminals, *Neuropharmacol.*, **36**, 1229–1242 (1997).
10. R. Kanjhan, G. D. Housley, L. D. Burton, D. L. Christie, A. Kippenberger, P. R. Thorne, L. Luo and A. F. Ryan, Distribution of the P2X2 receptor subunit of the ATP-gated ion channels in the rat central nervous system, *J. Comp. Neurol.*, **407**, 11–32 (1999).
11. L. Atkinson, T. F. C. Batten and J. Deuchars, P2X2 peceptor immunoreactivity in the dorsal vagal complex and area postrema of the rat, *Neuroscience*, **99**, 683–696 (2000).
12. S. T. Yao, J. A. Barden and A. J. Lawrence, On the immunohistochemical distribution of ionotropic P2X receptors in the nucleus tractus solitarius of the rat, *Neuroscience*, **108**, 673–685 (2001).
13. F. Kato and E. Shigetomi, Distinct modulation of evoked and spontaneous EPSCs by purinoceptors in the nucleus tractus solitarii of the rat, *J. Physiol. (Lond)*, **530**, 469–486 (2001).
14. J. C. Bisserbe, J. Patel and P. J. Marangos, Autoradiographic localization of adenosine uptake sites in rat brain using [^{3}H]nitrobenzylthioinosine, *J. Neurosci.*, **5**, 544–550 (1985).
15. J. W. Phillis, T. J. Scislo and D. S. O'Leary, Purines and the nucleus tractus solitarius: Effects on cardiovascular and respiratory function, *Clin. Exp. Pharmacol. Physiol.*, **24**, 738–742 (1997).
16. K. M. Spyer, J. H. St Lambert and T. Thomas, Central nervous system control of cardiovascular function: neural mechanisms and novel modulators, *Clin. Exp. Pharmacol. Physiol.*, **24**, 743–747 (1997).
17. T. J. Scislo, A. M. Kitchen, R. A. Augustyniak and D. S. O'Leary, Differential patterns of sympathetic responses to selective stimulation of nucleus tractus solitarius purinergic receptor subtypes, *Clin. Exp. Pharmacol. Physiol.*, **28**, 120–124 (2001).
18. J. F. R. Paton, W. T. Rogers and J. S. Schwaber, Tonically rhythmic neurons within a cardiorespiratory region of the nucleus tractus solitarii of the rat, *J. Neurophysiol.*, **66**, 824–838 (1991).
19. G. Fortin and J. Champagnat, Spontaneous synaptic activities in rat nucleus tractus solitarius neurons in vitro: evidence for re-excitatory processing, *Brain Res.*, **630**, 125–135 (1993).
20. Y. Kawai and E. Senba, Organization of excitatory and inhibitory local networks in the caudal nucleus of tractus solitarius of rats revealed in In Vitro slice preparation, *J. Comp. Neurol.*, **373**, 309–321 (1996).
21. K. Takano and F. Kato, Inspiration-promoting vagal reflex under NMDA receptor blockade in anaesthetized rabbits, *J. Physiol. (Lond)*, **516**, 571–582 (1999).
22. K. Takano and F. Kato, Inspiration-promoting vagal reflex in anaesthetized rabbits after rostral dorsolateral pons lesions, *J. Physiol. (Lond)*, **550**, 973–983 (2003).
23. M. Denavit-Saubié and A. S. Foutz, Neuropharmacology of respiration, in "*Neural Control of the Respiratory Muscles*" (A. D. Miller, A. L. Bianchi and B. P. Bishop, Eds.), CRC Press, Boca Raton, New York (1996).
24. R. Miles, Frequency dependence of synaptic transmission in nucleus of the solitary tract in vitro, *J. Neurophysiol.*, **55**, 1076–1090 (1986).
25. Z. Liu, C.-Y. Chen and A. C. Bonham, Frequency limits on aortic baroreceptor input to nucleus tractus solitarii, *Am. J. Physiol. Heart Circ. Physiol.*, **278**, H577–H585 (2000).
26. N. Tsuji, E. Shigetomi, K. Yamazaki and F. Kato, Activation of presynaptic adenosine A_1 receptors in the nucleus of the solitary tract slows transmitter release adaptation during repeated stimulation of the primary afferents in the rat, *2003 Abstract Viewer/Itinerary Planner*, Society for Neuroscience, Washington, DC (2003). 686.9.

Role of GABA in Central Respiratory Control Studied in Mice Lacking GABA-Synthesizing Enzyme 67-kDa Isoform of Glutamic Acid Decarboxylase

Shun-ichi Kuwana, Yasumasa Okada, Yoshiko Sugawara, and Kunihiko Obata

1. Introduction

In *in vivo* adult mammals, the inhibitory neurotransmitter gamma-aminobutyric acid (GABA) has been shown to play an essential role in the termination of the respiratory phase in the central respiratory rhythm generator.[1–3] On the other hand, works with *in vitro* brainstem-spinal cord preparations have revealed that the respiratory rhythm of neonatal rats is unaffected by blockade of GABAergic and glycinergic receptors.[4–6] These results suggest that either GABAergic or glycinergic synaptic inhibition is not essential for the generation of respiratory rhythm in neonatal mammals. We recently analyzed the role of GABA in the generation of respiratory rhythm and pattern and reported that GABA plays an important role in the maintenance of regular respiratory rhythm and normal inspiratory pattern in neonatal mice[7]. However, the precise role of GABA in the generation of respiratory rhythm and pattern in neonatal mammals is not well understood at the level of respiratory neurons.

A new function of GABA has been revealed recently in the developing brain. In various regions of the developing brain GABA serves as an excitatory neurotransmitter,[8–10] and as a neuronal trophic factor.[11–13] Therefore, we hypothesize that GABA plays a role in the development of the respiratory control system during the perinatal period in mammals. However, the role of GABA in the development of central respiratory control cannot be analyzed by simply applying GABA receptor antagonists in fetal or neonatal mice, because

Shun-ichi Kuwana and Yoshiko Sugawara • Department of Physiology, Teikyo University School of Medicine, Tokyo 173-8605 Japan. **Yasumasa Okada** • Department of Medicine, Keio University Tsukigase Rehabilitation Center, Shizuoka 410-3215 Japan. **Kunihiko Obata** • Neural Circuit Mechanisms Research Group, RIKEN Brain Science Institute, Wako 351-0198 Japan.

Post-Genomic Perspectives in Modeling and Control of Breathing, edited by Jean Champagnat, Monique Denavit-Saubié, Gilles Fortin, Arthur S. Foutz, Muriel Thoby-Brisson. Kluwer Academic/Plenum Publishers, 2004.

short-term GABA-blocking experiments cannot elucidate the contribution of GABA over the two-week period of brainstem development.

Recently, gene-targeting techniques have been effectively employed in studies of several ontogenetic processes, including those of the respiratory system.[14–17] We also succeeded to apply neonatal mice lacking GABA synthesizing enzyme 67-kDa isoform of glutamic acid decarboxylase (GAD67−/−) in the functional analysis of GABA in central respiratory control.[7] In the present study, we looked at the role of GABA in the ontogenesis of the central respiratory neuronal network using GAD67−/− mice.

2. Methods

2.1. Genotyping

Establishment of a GAD67 mutant mouse line has been reported previously.[18] Homozygous GAD67−/− newborn mice were produced by crossing GAD67 heterozygous (+/−) mice on a C57BL/6 strain background. The homozygous GAD67−/− genotype is easily identified by the presence of a cleft palate. All animals were anesthetized with ether after the experiment, and a piece of tail was sampled from each and frozen. Tail DNA was later extracted and genotyped by PCR analysis.[18]

2.2. Isolated Brainstem-Spinal Cord Preparation

The isolated brainstem-spinal cord preparation has been described in detail.[19,20] Twenty-four neonatal mice (8 GAD67+/+ and 16 GAD67−/−) were used. Mice were deeply anesthetized with ether, and the brainstem with cervical spinal cord was isolated. The rostral pons and cerebellum were ablated. The preparation was transferred to a recording chamber and fixed with the ventral side up. The preparation was perfused at 26°C with control mock cerebrospinal fluid (mock CSF), which was equilibrated with a gas mixture of 5% CO_2 in O_2 (pH = 7.4). The composition of the control mock CSF was as follows (in mM): 126 NaCl, 5 KCl, 1.5 $CaCl_2$, 1.3 $MgSO_4$, 1.25 NaH_2PO_4, 26 $NaHCO_3$ and 30 glucose. Central respiratory activity was recorded from the C4 ventral roots. After the preparation was perfused for at least 30 minutes with control mock CSF, the perfusate was switched to the test mock CSF containing 10 μM GABA or 1–10 μM bicuculline. The signals were recorded on a thermal array recorder and stored on a digital tape for subsequent analysis.

2.3. Neuronal Recording

The activity of neurons in the superficial (<200 μm) rostral ventrolateral medulla was recorded intracellularly using a perforated patch technique. The method of recording was based on our previous experiment.[20] Briefly, a glass pipette was pulled with a horizontal puller (PA-91, Narishige, Tokyo) to a tip size of approximately 2 μm. Electrode resistance ranged from 12 to 18 MΩ when filled with a solution containing (in mM) 140 K-gluconate, 10 EGTA, 10 HEPES, 1 $CaCl_2$, 1 $MgCl_2$ and nystatin (100 μg/ml). The micropipette was inserted into the rostral ventrolateral medulla with a manual hydraulic micromanipulator. Membrane potentials were recorded using a single electrode voltage clamp amplifier (CEZ 3100; Nihon Koden, Tokyo).

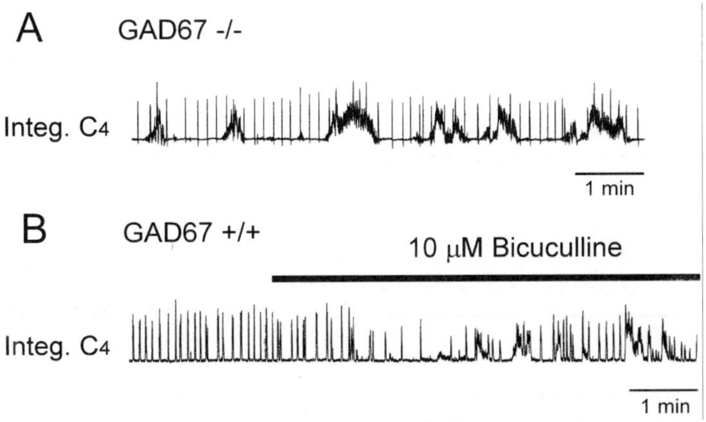

Figure 1. Typical rhythmic activity recorded from C4 ventral roots in isolated brainstem-spinal cord preparations from GAD67−/− (**A**) and GAD67+/+ (**B**) mice. Rhythm and amplitude of integrated C4 (Integ. C4) bursts of GAD67+/+ subjects were regular and constant compared with those of GAD67−/− mice. Two-thirds of the preparations from GAD67−/− mice showed a large amount of non-respiratory activity of long duration. In GAD67+/+ preparations, respiratory rhythm became irregular and substantial non-respiratory activity appeared during perfusion with 10 μM bicuculline.

3. Results

3.1. Respiratory Rhythm and Pattern

Examples of C4 recordings obtained from GAD67+/+ and GAD67−/− preparations are shown in Figure 1. In these recordings, rhythmic burst discharges in C4 and amplitudes of integrated C4 activities were relatively constant in GAD67+/+ preparations. On the other hand, in GAD67−/− preparations, inspiratory discharges appeared in C4, but the rhythms and amplitudes were more variable than in GAD67+/+ preparations. Most notably, large non-respiratory discharges accompanied inspiratory discharges in 10 out of 16 GAD67−/− mice. These discharges were characterized by bursts of long duration (10–30 sec), appearing every 2–10 min. Although these discharges were observed also in other mouse genotypes, the amplitudes of these non-respiratory discharges were lower than those of inspiratory discharges.

In three GAD67+/+ preparations, application of 1–10 μM bicuculline produced disturbance of inspiratory rhythm and induced large non-respiratory discharges (Fig. 1B) similar to the discharge pattern of GAD67−/− preparations (Fig. 1A).

3.2. Effect of Exogenous GABA on Respiratory Neurons

We performed whole cell recordings successively in 6 inspiratory neurons in GAD67+/+ preparations and in 6 inspiratory neurons in GAD67−/− preparations. Typical inspiratory neuron activity obtained from a GAD67+/+ mouse was characterized by a large inspiratory drive potential and prolonged bursting during the inspiratory phase (Fig. 2A). In contrast, inspiratory neurons of GAD67−/− mice exhibited weak and short inspiratory drive potentials, accompanied by only a few spikes during the inspiratory phase (Fig. 2B).

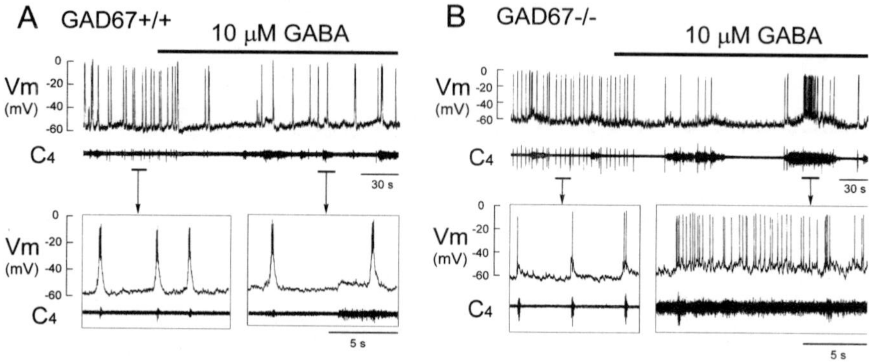

Figure 2. Effect of GABA application on inspiratory neurons. **A:** In GAD67+/+ preparations, respiratory rhythm was slowed during perfusion with mock CSF containing 10 μM GABA. Lower panels in boxes indicate the inspiratory neuron activity on an enlarged time scale before and after perfusion with GABA. The trajectory of membrane potentials was not affected by GABA application. **B:** In GAD67−/− preparations, perfusion with GABA induced a slow respiratory rhythm, prolongation of the inspiratory phase, and an increase in non-respiratory activity in C4. Simultaneously, exogenous GABA produced increments in inspiratory drive potentials and non-respiratory EPSPs, followed by a notable increase in spike frequency.

We examined the effect of GABA on central respiratory activity in GAD67+/+ and GAD67−/− preparations. In GAD67+/+ preparations, the respiratory frequency decreased with 10 μM GABA (Fig. 2A). Membrane potential trajectories of inspiratory neurons were not affected by perfusion with GABA. In contrast, application of the same concentration of GABA to GAD67−/− preparations induced a marked decrease in respiratory frequency. Fig. 2B shows a representative recording of an inspiratory neuron, in which the resting membrane potential fluctuated and firing was synchronized with inspiratory C4 burst activities. Non-respiratory long burst activity was also augmented by exogenous GABA. The neuron was activated and tonic firing was synchronized with C4 activity by perfusion with GABA. In 4 out of 6 inspiratory neurons from GAD67−/− preparations, this kind of exogenous GABA induced-activation was observed, and the resting membrane potentials of these GABA-activated neurons fluctuated. The rest of the inspiratory neurons had stable resting membrane potentials and were insensitive to 10 μM GABA.

In two expiratory neurons from GAD67−/− preparations, responses of membrane potentials to GABA and bicuculline were successively recorded. Fig. 3 shows an expiratory neuron that received inhibitory synaptic input during the inspiratory phase in a GAD67−/− preparation. Respiratory rhythm and amplitudes of IPSPs during the inspiratory phase were not affected during perfusion of 1 μM bicuculline in the GAD67−/− preparations. In contrast, 10 μM GABA induced an increase in firing frequency during the expiratory phase in expiratory neurons.

4. Discussion

We demonstrated that reduction of GABA by genetic disruption of GAD67 resulted in irregular respiratory rhythm with notable non-respiratory activity in the brainstem-spinal cord of neonatal mutant mice. This abnormal respiratory pattern was confirmed at the level of respiratory neurons. The abnormal respiratory activity in GAD67−/− preparations

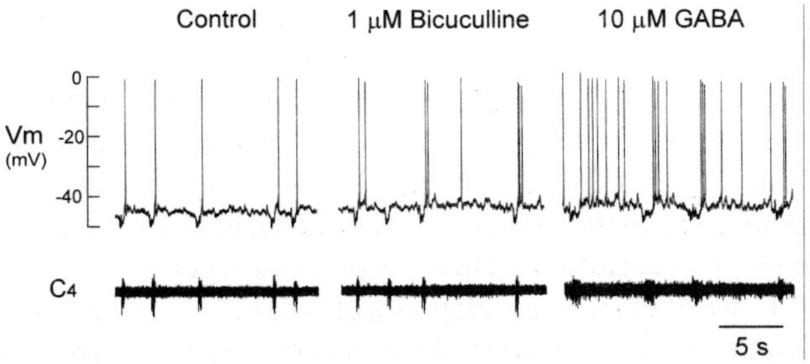

Figure 3. Effects of bicuculline or GABA on an expiratory neuron. Inspiration-related IPSPs were observed in expiratory neurons in GAD67−/− mice. The IPSPs and firing pattern of expiratory neurons were not affected by perfusion with 1 μM bicuculline. Perfusion with 10 μM GABA induced an increase in firing frequency during the expiratory phase.

was similar to the respiratory activity of GAD67+/+ preparations during application of bicuculline. This suggests that GABAergic synaptic transmission plays an important role in the maintenance of stable respiratory rhythm and pattern in neonatal mice. We also showed that the response of the respiratory neural network to exogenous GABA differed between GAD67+/+ and GAD67−/− mice. These results suggest that deficiency of GABA during the perinatal period induces a disturbance in the functioning and/or development of the central respiratory neuronal network that generates respiratory rhythm and pattern.

4.1. Role of GABA in Inspiratory Pattern Generation

In the previous study we showed that the respiratory pattern in GABA-deficient mice is characterized by irregular rhythm with a short inspiratory duration.[7] In the present study we analyzed the membrane potential trajectories of inspiratory neurons in GAD−/− mice and found that inspiratory drive potentials are weak and short in these neurons. This suggests that the GABAergic system plays an important role in inspiratory pattern generation by modulating the activity of inspiratory neurons. Therefore, adequate GABA levels in the medulla oblongata must be necessary for the stabilization of respiratory rhythm and pattern during the perinatal period.

Our results are consistent with the suggestion by Feldman and Smith[5] and Onimaru *et al.*[6] that an inhibitory synaptic network is not required for basic respiratory rhythm generation in neonatal mammals. We have shown the existence of IPSPs during the inspiratory phase in expiratory neurons of GAD67−/− mice, even during the blockade of $GABA_A$ receptors. This result suggests that there exists an inhibitory transmitter other than GABA and that glycine produces these IPSPs. Indeed, it has been reported that inspiration-related IPSPs in the medullary expiratory neurons of neonatal rats are abolished by blockade of glycine receptors with strychnine.[21,22]

4.2. Role of GABA in Development of the Respiratory Neural Network

The low firing rate of inspiratory neurons in GAD67−/− mice during the inspiratory phase (Fig. 2) suggests that the interconnection between respiratory neurons is weak.

This weak inspiratory activity in the medulla oblongata would directly affect the phrenic motoneuron pool and produce a short inspiratory duration.

It is well-known that GABAergic responses change from excitatory to inhibitory during the perinatal period.[8–10] As shown in Fig. 2, exogenous GABA induced prolongation of inspiratory C4 burst duration and excitation in both the inspiratory and expiratory neurons of GAD67−/− neonatal mice. In contrast, GABA did not induce excitation in the inspiratory neurons of GAD67+/+ preparations. This may be due to a delay in the transformation from excitation to inhibition in the GABAergic system of GAD67−/− mice. Ganguly *et al.*[23] have shown recently that the switch of neural GABAergic responses from excitation to inhibition is delayed by chronic blockade of GABA receptors and accelerated by GABA receptor activation. They concluded that GABA acts as a self-limiting trophic factor during neural development. Our results suggest the possibility that the development of the respiratory neuronal network is delayed by lowered levels of GABA in the brain of GAD67−/− mice. Further studies will be necessary to elucidate the role of GABA in the development of the central respiratory neuronal network.

4.3. Non-Respiratory Activity in GAD67−/− Mice

In our *in vitro* preparations, large non-respiratory rhythmic discharges were often observed in GAD67−/− mice (Figs. 1 and 2). These non-respiratory rhythmic discharges were similar to those seen with the application of bicuculline to GAD67+/+ preparations. This similarity suggests that the non-respiratory rhythmic discharge of GAD67−/− preparations is induced by blockade of $GABA_A$ receptors. Onimaru *et al.*[6] showed that application of GABA or glycine antagonists induced seizure-like discharges in C4 motoneurons in the brainstem-spinal cord of neonatal rats. They suggested that such discharges affected respiratory rhythm but that medullary respiratory neurons were not the origin of seizure-like activity. In GAD67−/− preparations, however, non-respiratory activities in medullary inspiratory neurons were synchronized with seizure-like discharges in C4 and intensified by application of GABA. It thus appears that the seizure-like discharges in C4 originated in the respiratory neural network in the medulla of GAD67−/− mice.

In summary, we have reduced GABA levels in mice during the perinatal period by disrupting the GAD67 gene and demonstrated malfunctioning of the central respiratory control system in these neonatal mice. We suggest that GABAergic transmission is not essential for basic respiratory rhythm generation, but that it plays an important role in the maintenance of regular respiratory rhythm and a normal inspiratory pattern in neonatal mice by modulating the activity of inspiratory neurons. We also suggest that GABA plays a role in the development of the respiratory neuronal network in the perinatal period.

5. Acknowledgements

This work was supported by a Grant-in-Aid for Scientific Research from the Ministry of Education, Science, Sports and Culture of Japan (SK), a Research Grant for Specific Diseases from the Japanese Ministry of Health and Welfare (YO) and a Grant-in-Aid for Exploratory Research from the Japan Society for the Promotion of Science (YO).

References

1. D.W. Richter, K. Ballanyi, and S. Schwarzacher, Mechanisms of respiratory rhythm generation, *Curr. Opin. Neurobiol.* **2**, 788–793 (1992).
2. D.W. Richter, S.L. Mironov, D. Busselberg, P.M. Lalley, A.M. Bischoff, and M.H. Wilkinson, Respiratory rhythm generation: Plasticity of a neuronal network, *Neuroscientist* **6**, 181–198 (2000).
3. A. Haji, R. Takeda, and M. Okazaki, Neuropharmacology of control of respiratory rhythm and pattern in mature mammals, *Pharmacol. Ther.* **86**, 277–304 (2000).
4. T. Murakoshi, and M. Otsuka, Respiratory reflexes in an isolated brainstem-lung preparation of the newborn rat: possible involvement of gamma-aminobutyric acid and glycine, *Neurosci. Lett.* **62**, 63–68 (1985).
5. J.L. Feldman, and J.C. Smith, Cellular mechanisms underlying modulation of breathing pattern in mammals, *Ann. N.Y. Acad. Sci.* **563**, 114–130 (1989).
6. H. Onimaru, A. Arata, and I. Homma, Inhibitory synaptic inputs to the respiratory rhythm generator in the medulla isolated from newborn rats, *Pflügers Arch.* **417**, 425–432 (1990).
7. S. Kuwana, Y. Okada, Y. Sugawara, N. Tsunekawa, and K. Obata, Disturbance of neural respiratory control in neonatal mice lacking GABA synthesizing enzyme 67-kDa isoform of glutamic acid decarboxylase, *Neuroscience* **120**, 861–870 (2003).
8. K. Obata, M. Oide, and H. Tanaka, Excitatory and inhibitory actions of GABA and glycine on embryonic chick spinal neurons in culture, *Brain Res.* **144**, 179–184 (1978).
9. E. Cherubini, J.L. Gaiarsa, and Y. Ben-Ari, GABA: an excitatory transmitter in early postnatal life. *Trends Neurosci.* **14**, 515–519 (1991).
10. Y. Ben-Ari, R. Khazipov, X. Leinekugel, O. Caillard, and J.L. Gaiarsa, $GABA_A$, NMDA and AMPA receptors: a developmentally regulated 'menage a trois', *Trends Neurosci.* **20**, 523–529 (1997).
11. K. Obata, Excitatory and trophic action of GABA and related substances in newborn mice and organotypic cerebellar culture, *Dev. Neurosci.* **19**, 117–119 (1997).
12. X.B. Gao, and A.N. van den Pol, GABA release from mouse axonal growth cones. *J. Physiol. Lond.* **523 Pt 3**, 629–637 (2000).
13. X.B. Gao, and A.N. van den Pol, GABA, not glutamate, a primary transmitter driving action potentials in developing hypothalamic neurons. *J. Neurophysiol.* **85**, 425–434 (2001).
14. T.D. Jacquin, V. Borday, S. Schneider-Maunoury, P. Topilko, G. Ghilini, F. Kato, P. Charnay, and J. Champagnat, Reorganization of pontine rhythmogenic neuronal networks in Krox-20 knockout mice, *Neuron* **17**, 747–758 (1996).
15. G.D. Funk, S.M. Johnson, J.C. Smith, X.W. Dong, J. Lai, and J.L. Feldman, Functional respiratory rhythm generating networks in neonatal mice lacking NMDAR1 gene, *J. Neurophysiol.* **78**, 1414–1420 (1997).
16. S. Shirasawa, A. Arata, H. Onimaru, K.A. Roth, G.A. Brown, S. Horning, S. Arata, K. Okumura, T. Sasazuki, and S.J. Korsmeyer, Rnx deficiency results in congenital central hypoventilation, *Nat. Genet.* **24**, 287–290 (2000).
17. J. Zhao, H. Chen, J.J. Peschon, W. Shi, Y. Zhang, S.J. Frank, and D. Warburton, Pulmonary hypoplasia in mice lacking tumor necrosis factor-α converting enzyme indicates an indispensable role for cell surface protein shedding during embryonic lung branching morphogenesis, *Dev. Biol.* **232**, 204–218 (2001).
18. H. Asada, Y. Kawamura, K. Maruyama, H. Kume, R.G. Ding, N. Kanbara, H. Kuzume, M. Sanbo, T. Yagi, and K. Obata, K. Cleft palate and decreased brain gamma-aminobutyric acid in mice lacking the 67-kDa isoform of glutamic acid decarboxylase, *Proc. Natl. Acad. Sci. U.S.A.* **94**, 6496–6499 (1997).
19. Y. Okada, A. Kawai, K. Mückenhoff, and P. Scheid, Role of the pons in hypoxic respiratory depression in the neonatal rat, *Respir. Physiol.* **111**, 55–63 (1998).
20. S. Kuwana, Y. Okada, and T. Natsui, Effects of extracellular calcium and magnesium on central respiratory control in the brainstem-spinal cord of neonatal rat, *Brain Res.* **786**, 194–204 (1998).
21. X.M. Shao, and J.L. Feldman, Respiratory rhythm generation and synaptic inhibition of expiratory neurons in pre-Bötzinger complex: differential roles of glycinergic and GABAergic neural transmission, *J. Neurophysiol.* **77**, 1853–1860 (1997).
22. J. Brockhaus, and K. Ballanyi, Synaptic inhibition in the isolated respiratory network of neonatal rats, *Eur. J. Neurosci.* **10**, 3823–3839 (1998).
23. K. Ganguly, A.F. Schinder, S.T. Wong, and M. Poo, GABA itself promotes the developmental switch of neuronal GABAergic responses from excitation to inhibition, *Cell* **105**, 521–532 (2001).

Breathing Without Acetylcholinesterase

Fabrice Chatonnet, Eliane Boudinot, Arnaud Chatonnet, Jean Champagnat, and Arthur S. Foutz

1. Introduction

Acetylcholine (ACh) mediates neurotransmission at the neuromuscular junction and is involved in respiratory control[1], notably chemosensitivity[2] of central and peripheral origin. The level of ACh at the synaptic cleft and neuromuscular junction is regulated by the enzyme acetylcholinesterase (AChE). Blockade of AChE by organophosphorus compounds produces death by respiratory failure[3], but despite the absence of AChE activity in all tissues, AChE (−/−) mice knockout for the gene coding for AChE develop to term[4] and survive to adulthood if provided special care[5]. However, they show many aspects of a cholinergic syndrome, such as pinpoint pupils and muscle tremors[5].

We compared respiration at rest and ventilatory responses to hypoxic and hypercapnic challenges in adult AChE (−/−) mice and littermate wild-type (+/+) mice, and we found that (−/−) mice have a distinct respiratory phenotype and altered chemosensitivity. We then investigated the mechanisms through which these mice have adapted to excess ACh. To distinguish whether the adaptation processes occur in brainstem rhythm-generating structures or peripherally at the neuromuscular junction, we used a combined *in vivo/in vitro* approach. We found that the survival of AChE knockout mice is dependent on an enzyme usually considered irrelevant, butyrylcholinesterase (BChE) acting peripherally, and on a down-regulated response of respiratory neurons and motoneurons to cholinergic stimulation.

2. Methods

Original mice founders (provided by Prof. O. Lockridge, Eppley Institute, Omaha, NE) were maintained in a 129SVJ strain background. Feeding the weaned pups with liquid enriched food[5,6] allowed some AChE (−/−) mice to survive several months. Mice of either sex, genotyped by PCR[6], were used between post-natal days 0 to 3 for both *in vivo* and *in vitro* experiments and between 1–5 months of age for *in vivo* studies. Ventilatory

---

Fabrice Chatonnet, Eliane Boudinot, Jean Champagnat and Arthur S. Foutz • NGI—Institut de Neurobiologie A. Fessard—CNRS, Gif sur Yvette, France. Supported by DGA: DSP/STTC 00/34/077.
Arnaud Chatonnet • Dept of animal physiology, INRA, Montpellier, France.

Post-Genomic Perspectives in Modeling and Control of Breathing, edited by Jean Champagnat, Monique Denavit-Saubié, Gilles Fortin, Arthur S. Foutz, Muriel Thoby-Brisson. Kluwer Academic/Plenum Publishers, 2004.

activity in neonate and adult mice was measured using a barometric method[6]. Each animal was placed in the plethysmograph chamber (15 to 700 ml and maintained at 28–30°C or 27–28°C, depending on animal size), flushed with fresh humidified air and hermetically sealed during 2–3 min recordings. Chamber and animal temperatures (oral in neonates, rectal in adults) were measured. In the chamber, adult animals were partially restrained by the tail with a flexible probe permanently inserted in the rectum[6]. The spirogram was stored on a PC, and a program (Acquis1) was used to measure the durations of inspiration (T_I) and expiration (T_E), breathing frequency (f_R) and tidal volume (V_T). Baseline ventilation at rest was measured, then chemosensitivity was tested with brief (5 min) exposures to gas mixtures: 3% or 5% CO_2, 21% O_2 balance N_2 (hypercapnia protocols) and 10% O_2 balance N_2 (hypoxia protocol).

To dissociate effects occurring at the diaphragmatic neuromuscular junction from those of central origin, we recorded respiratory activity generated in 0–3 day-old isolated neonatal mouse brainstem preparations *in vitro*, using standard procedures. The rostral section was made at the level of the inferior cerebellar arteries to eliminate pontine and cerebellar structures, and the caudal section was made at a low cervical level. The preparations were bathed in a Ringer solution[6] equilibrated with 5% CO_2, 95% O_2 at 26°C, pH 7.4. Inspiratory motor activity was recorded from C1–C2 spinal rootlets by means of a suction electrode, and was amplified, filtered (3 Hz- 3 kHz), full wave rectified, integrated (40 ms time constant), sampled at 1 kHz and stored on a PC. The results (means ± SEM) were analyzed using analysis of variance and post-hoc tests.

3. Results

3.1. Ventilation at Rest and Chemosensory Control

AChE (−/−) adult mice were smaller than their wild-type (+/+) littermates[4,5] (16.3 g vs 24.9 g), but no difference was detected in their body temperature.

AChE (−/−) animals presented a 70% increase in minute-ventilation (V_E), compared with their (+/+) littermates (Fig. 1A). This increase was due to by an elevated V_T, which was nearly twice as high as in (+/+) mice (Fig. 1B), whereas breathing frequency was not significantly different (Fig. 1C). Expiratory time (T_E) was longer (+49%) and inspiratory time (T_I) shorter (−43%) than in (+/+) mice, which resulted in a T_I/Ttot ratio decreased by one half. The ratio V_T/T_I, which expresses the average inflow during inspiration, was extremely high in AChE−/− mice (3.5 times higher than in (+/+) mice), a consequence of the concomitant increase in V_T and decrease in T_I.

There was a robust ventilatory response to hypoxia in both (+/+) and (−/−) animals, but the response was slightly lower in the latter (82% vs 95% increase of V_E over baseline, $P < 0.02$). No significant interactions were observed for f_R, which increased in the same proportion in the two genotypes, nor for V_T.

The ventilatory response to hypercapnia was stronger in (−/−) mice than in (+/+) mice for the two concentrations of CO_2 tested (Fig. 1D, $P < 0.001$ for genotype by treatment interaction). This ventilatory response was entirely due to an enhanced frequency response which was twice as large as in (+/+) mice (71% vs 31% increases with 3% CO_2 and 101% vs 50% increases with 5% CO_2, $P < 0.01$).

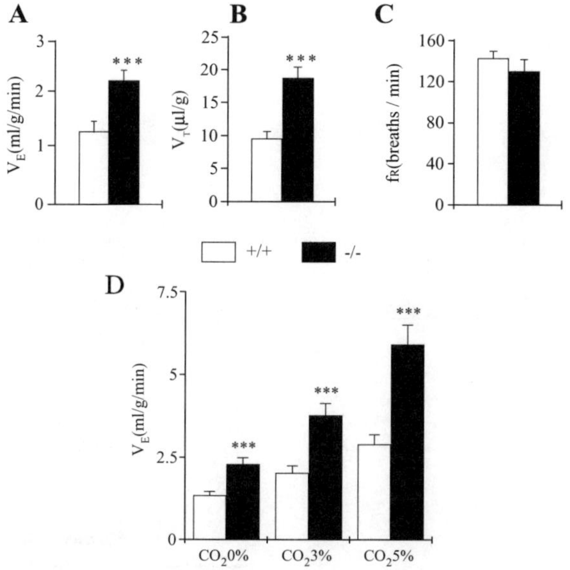

Figure 1. Ventilation at rest (A), tidal volume (B) and breathing frequency (C) in wild-type (white bars) and AChE (−/−) mice (black bars). Ventilatory response to hypercapnia (D) is enhanced in AChE(−/−) mice. ***: $P < 0.001$ versus wild-type mice. Nine (+/+) and 9 (−/−) mice were recorded.

3.2. Critical Role of Peripheral but not Central Butyrylcholinesterase

In adult mice a low dose of the BChE inhibitor bambuterol (50 µg/kg) decreased V_T and produced apnea (Fig. 2A). Thus bambuterol was used to screen AChE−/− mice at birth. A dose of 0.5 mg/kg killed all AChE−/− mice within a few min but had no effect on littermates later genotyped as (+/+) or (+/−). Plethysmographic recordings revealed a phase of disorganized, labored breathing followed by a gasp-like activity and progressive slowing of breathing accompanied with a bluish coloring of the skin, revealing asphyxia (Fig. 2B, *in vivo* traces). When breathing movements ceased the animals were cold-anesthetized and dissected for *in vitro* recording of the isolated brainstem.

After bambuterol-treated AChE −/− animals presented apnea *in vivo* and their brainstem was isolated *in vitro*, a regular inspiratory bursting activity was recorded from spinal nerves rootlets, with no difference in frequency compared to (+/+) brainstems (Fig. 2B, right traces). *In vitro*, however, neither bambuterol (1–100 µM), nor the irreversible BChE inhibitor iso-OMPA (10–100 µM) affected inspiratory burst frequency in isolated brainstems from (−/−) mice or from the other genotypes. Hence the lethal effect of BChE inhibition *in vivo* does not originate in central rhythm-generating structures.

3.3. Reduced Effects of Cholinergic Agonists in the Isolated Brainstem of AChE−/− Mice

In both (+/+) and (+/−) brainstems, muscarine application (50–100 µM) produced distinct dose-dependent effects (Fig. 3A): firstly, a tonic activity, which reached rapidly

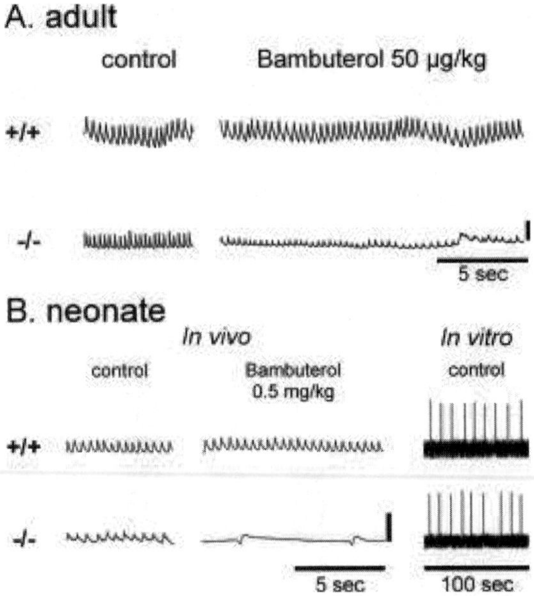

Figure 2. Effects of BChE inactivation in adult and neonate mice *in vivo* and *in vitro*. A. Plethysmographic recordings of adult mice before (control) and after injection of bambuterol. Note the ineffectiveness of bambuterol in the wild-type (+/+) mouse breathing air (upper trace) and the profound decrease of tidal volume in the AChE (−/−) mouse. Inspiration: upward deflections. B, left: *in vivo* recordings of both wild-type (+/+) and AChE (−/−) newborn littermates in control and after injection of bambuterol. Note the apnea in the AChE −/− mouse. B, Right: integrated inspiratory bursting activity recorded *in vitro* from the C2 motor rootlets in the brainstems of the same mice, isolated immediately after apnea in the (−/−) mouse and 12 min after bambuterol injection in the (+/+) mouse. Note the recovery of a rhythmic activity in the AChE (−/−) brainstem similar to the activity recorded in the wild-type. Calibration bars: 250 µL in (A), 100 µL in (B). Inspiration up.

a maximal value then declined despite the continued muscarine application; secondly, an increase in respiratory frequency and burst duration, and a decrease in burst amplitude. These latter effects outlasted the muscarine application (Fig. 3A). Muscarine effects were all greatly reduced in (−/−) mice. Tonic activity was nearly absent, increase of respiratory frequency was smaller than in the other genotypes (Fig. 3C), decrease of burst amplitude was also smaller and burst duration was unchanged.

Nicotine (0.5 to 10 µM) produced the same dose-dependent effects as muscarine (Fig. 3B). In (+/+) and (+/−) mice a tonic activity appeared then declined, inspiratory burst frequency and duration increased, and burst amplitude decreased. Recovery of burst frequency and amplitude was not complete after 30 min of wash (Fig. 3B). As with muscarine, (−/−) mice showed a very weak tonic activity, increase of respiratory frequency was smaller than in the other genotypes (Fig. 3B and D), and there were no significant effects on burst duration or amplitude.

4. Conclusions

Mice knockout for the AChE gene present an increased ventilation and an increased tidal volume compared with wild-type littermates. Chemosensory ventilatory responses are

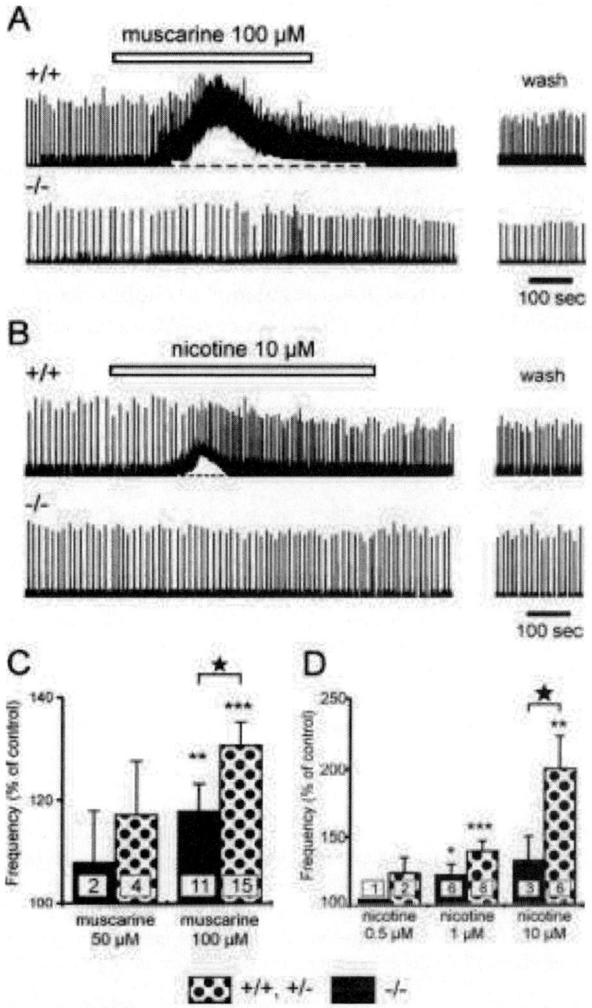

Figure 3. Effects of cholinergic agonists on the isolated brainstem *in vitro*. Effects of muscarine (A) and nicotine (B) applications on integrated respiratory bursting activity in wild-type (+/+) and AChE (−/−) mouse brainstems from littermate animals recorded simultaneously in the same bath. Note the appearance of a rapidly desensitizing tonic activity (above the dashed line) and an increase in respiratory frequency in (+/+) brainstems, whereas in (−/−) brainstems tonic activity is barely visible and frequency increase is small. Wash: 30 min after muscarine application. Maximal changes of respiratory frequency over pre-drug control induced by increasing concentrations of muscarine (C) and nicotine (D). The number of animals in each experiment is indicated in the bar graphs. (+/+) and (+/−) mice were pooled. Asterisks: significant vs pre-drug control, white stars: comparison between genotypes. * $P < 0.05$; ** $P < 0.01$; *** $P < 0.001$.

altered in opposite directions, with an augmented ventilatory response to hypercapnia and a slightly blunted response to hypoxia.

AChE−/− mice have a phenotype presenting similarities with symptoms of poisoning with anticholinesterase agents[5]. The striking respiratory phenotype (increased tidal volume and ventilation) is also suggestive of excess ACh in neural structures, because it can be reproduced to a large extent by pharmacological means[7,8,9]. The shortened T_I *in vivo* is opposite to the increased burst duration produced by cholinergic agonists in wild-type brainstems or slices[10,11] *in vitro*, and might reflect the inability of diaphragmatic neuromuscular junctions to sustain evoked endplate potentials during repetitive stimulation in (−/−) animals[12]. The increased ventilatory response to hypercapnia is in keeping with the involvement of muscarinic mechanisms in central chemosensitivity[13,14].

As BChE controls transmitter release[12], we speculate that BChE inhibition in (−/−) mice has a lethal effect by decreasing ACh release on a post-synaptic membrane that has been adapted to high levels of ACh through down-regulating nicotinic receptors, resulting in a decreased V_T, whereas AChE inhibition in wild-type mice increases V_T.

The present study suggests that down-regulation of cholinergic receptors[15] is essential to rhythm generation and transmission in the absence of AChE, and is likely the mechanism of recovery after organophosphate intoxication.

References

1. A. Haji, R. Takeda, and M. Okazaki, Neuropharmacology of control of respiratory rhythm and pattern in mature mammals, *Pharmacol. Ther.* **86**, 277–304 (2000).
2. M. D. Burton, and H. Kazemi, Neurotransmitters in central respiratory control, *Respir. Physiol.* **122**, 111–121 (2000).
3. R. W. Brimblecombe, Drugs acting on cholinergic mechanisms and affecting respiration, in: *Respiratory Pharmacology*, edited by J. Widdicombe (Pergamon Press, Oxford, 1981), pp. 175–184.
4. W. Xie, J. A. Stribley, A. Chatonnet, P. J. Wilder, A. Rizzino, R. D. McComb, P. Taylor, S. H. Hinrichs, and O. Lockridge, Postnatal developmental delay and supersensitivity to organophosphate in gene-targeted mice lacking acetylcholinesterase, *J. Pharmacol. Exp. Ther.* **293**, 896–902 (2000).
5. E. G. Duysen, J. A. Stribley, D. L. Fry, S. H. Hinrichs and O. Lockridge, Rescue of the acetylcholinesterase knockout mouse by feeding a liquid diet; phenotype of the adult acetylcholinesterase deficient mouse, *Dev. Brain Res.* **137**, 43–54 (2002).
6. F. Chatonnet, E. Boudinot, A. Chatonnet, L. Taysse, S. Daulon, J. Champagnat, and A. S. Foutz, respiratory survival mechanisms in acetylcholinesterase knockout mouse, *Eur. J. Neurosci.* **18**, 1419–1427 (2003).
7. A. S. Foutz, E. Boudinot, and M. Denavit-Saubié, Central respiratory depression induced by acetylcholinesterase inhibition: involvement of anaesthesia, *Eur. J. Pharmacol.* **142**, 207–213 (1987).
8. A. S. Foutz, I.S. Delamanche, and M. Denavit-Saubié, Persistence of central respiratory rhythmogenesis after maximal acetylcholinesterase inhibition in unanaesthetized cats, *Can. J. Physiol. Pharmacol.* **67**, 162–166 (1989).
9. M. D. Burton, D. C. Johnson, and H. Kazemi. The central respiratory effects of acetylcholine vary with CSF pH, *J. Auton. Nerv. Syst.* **62**, 27–32. (1997).
10. X. M. Shao, and J. L. Feldman, Acetylcholine modulates respiratory pattern: effects mediated by M3-like receptors in pre-Bötzinger complex inspiratory neurons, *J. Neurophysiol.* **83**, 1243–1252 (2000).
11. X. M. Shao, and J. L. Feldman, Mechanisms underlying regulation of respiratory pattern by nicotine in preBötzinger complex, *J. Neurophysiol.* **85**, 2461–2467 (2001).
12. J. Minic, A. Chatonnet, E. Krejci, and J. Molgó, Butyrylcholinesterase and acetylcholinesterase activity and quantal transmitter release at normal and acetylcholinesterase knockout mouse neuromuscular junctions, *Br. J. Pharmacol.* **138**, 177–187 (2003).
13. E. E. Nattie, and A. Li, Ventral medulla sites of muscarinic receptor subtypes involved in cardiorespiratory contro, *J. Appl. Physiol.* **69**, 33–41 (1990).
14. A. Haji, S. Furuichi, and R. Takeda, Effects of iontophoretically applied acetylcholine on membrane potential and synaptic activity of bulbar respiratory neurones in decerebrate cats, *Neuropharmacology* **35**, 195–203 (1996).
15. V. Bernard, C. Brana, I. Liste, O. Lockridge, and B. Bloch, Dramatic depletion of cell surface m2 muscarinic receptor due to limited delivery from intracytoplasmic stores in neurons of acetylcholinesterase-deficient mice, *Mol. Cell. Neurosci.* **23**, 121–133 (2003).

In-Silico Model of NMDA and Non-NMDA Receptor Activities Using Analog Very-Large-Scale Integrated Circuits

Guy Rachmuth and Chi-Sang Poon

1. Introduction

Computational modeling is a useful analytical tool that (with valid data) may help to explain how complex brain systems work. In respiratory control, various neural network models have been proposed to explore the possible mechanisms of respiratory rhythmogenesis based on intrinsic neuronal properties as well as interactions at the network level[1,2]. However, the large parameter space that must be tested for such neuromorphic models makes it rather cumbersome and time consuming to simulate even a modest size network with discrete respiratory neurons. For large-scale neural network models involving multiple populations of respiratory neurons, such simulations become computationally prohibitive on general-purpose digital computers.

In addition, the respiratory rhythm is known to be modulated by a complex cascade of excitatory and inhibitory processes in the afferent and efferent pathways. The diversity of neurotransmitters and neuromodulators acting at multiple respiratory-related brainstem sites with varying forms of plasticity[3] further complicate the modeling and simulation of the closed-loop respiratory control system using computer software.

An alternative analog hardware approach to neuromorphic modeling first suggested by Mead[4,5] involves the implementation of neuronal models by means of metal-oxide-silicon (MOS) transistor circuits. Such electronic models perform computations using intrinsic transistor physics in analog mode, making it possible to simulate neuronal activities at a speed orders-of-magnitude faster than digital simulation. This analog modeling approach allows detailed investigation of brain processes whose basic building blocks are ionic channels and synapses on neuronal somas and dendrites.

Guy Rachmuth and Chi-Sang Poon • Harvard-MIT Division of Health Sciences and Technology, MIT, Cambridge, MA 02139. **Guy Rachmuth** • Division of Engineering and Applied Sciences, Harvard University, Cambridge, MA 02138.

Post-Genomic Perspectives in Modeling and Control of Breathing, edited by Jean Champagnat, Monique Denavit-Saubié, Gilles Fortin, Arthur S. Foutz, Muriel Thoby-Brisson. Kluwer Academic/Plenum Publishers, 2004.

Furthermore, by using very large scale integration (VLSI) technology, it is possible to implement a large-scale neural network with thousands of model neurons on a single MOS substrate. Analog VLSI circuits offer several advantages over traditional software simulations in terms of speed and power. Using this approach, a large network of artificial synapses and neurons may be implemented and such detailed neuromorphic models may be simulated in virtually real time.

As a first step toward the application of this novel methodology toward the modeling of the respiratory system, we have obtained MOS transistor-based circuit realizations that emulate NMDA and non-NMDA receptor gated ion channel dynamics, intracellular calcium dynamics and synaptic learning rules.

2. Methods

2.1. Basic Building-Block Circuits

The analog neuromorphic modeling approach relies on generating MOS-based circuit models of ionic channels, such as voltage-dependent Na^+ or K^+ channels or ligand-dependent AMPAR and NMDAR channels, and a circuit that mimics the current integration on a patch of membrane. The post-synaptic dendritic membrane current is given by the following equations:

$$I(t) = g_{syn}(t, V_{MEM}) * (V_{MEM} - E_{syn}) \quad (1)$$

$$g(t, V_{MEM}) = g_{\max}(V_{MEM}) * te^{-t/\tau_{peak}} \quad (2)$$

where g_{syn} is the time-dependent and/or voltage-dependant conductance of the channel, V_{MEM} is the membrane potential, and E_{syn} is the reversal potential of the patch of membrane. Three parameters governing these equations are g_{max}, the maximum conductance of the channel, which can be voltage dependent; τ_{peak}, the opening rate time constant of the channels, which determines the shape of the alpha function; and E_{SYN}, the reversal potential.

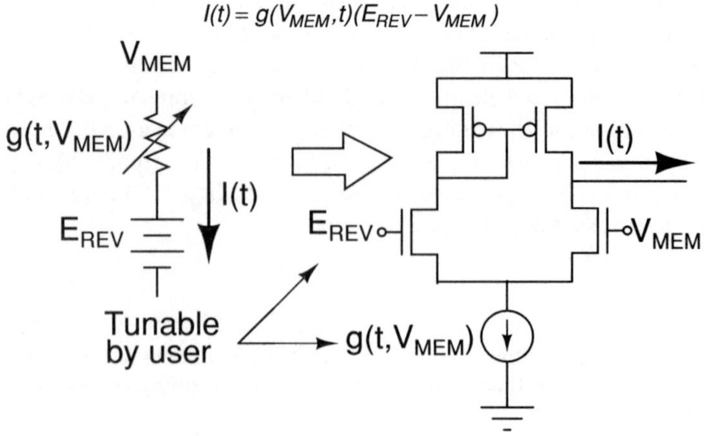

Figure 1. Equivalent circuit models of equation 1, shown as a passive resistive circuit (*left*) and electronic implementation of ion channels (*right*).

All three parameters can be set externally by the user, and the effect of changing them can be observed in real time. The circuits can be configured to model synaptic ligand-gated channels such as AMPAR channels, as well as the more complex NMDAR channels, whose gating is both voltage and ligand dependent. This ability allows for tremendous flexibility when simulating a large network.

An important factor in determining the instantaneous value of the output current of a given ionic channel population is the value of V_{MEM}. The membrane potential is in turn strongly dependent on the currents which charge the membrane capacitor, as shown mathematically by equation 3:

$$C_{MEM}\frac{dV_{MEM}}{dt} = g_{leak}(V_{MEM} - E_{LEAK}) + \sum I_{EXC}(t) - \sum I_{INH}(t) \qquad (3)$$

where C_{MEM} is the membrane capacitance, g_{leak} is the leak conductance of the membrane patch, and I_{EXC} and I_{INH} are excitatory and inhibitory current sources, respectively, which are impinging onto the membrane node and are described by equation 1.

This important feedback is crucial to the correct descriptions of voltage gated channels and other cellular processes. However, its implementation entails a significant computational cost associated with computing equation 3 at each time step in digital simulation platforms. When simulating large scale networks, it is sometimes omitted and the driving force of the current ($E_{REV}-V_{MEM}$) is modeled as a constant for some time and is updated only occasionally. However, by using circuits to model these equations, the feedback is designed into the system directly with virtually no computational cost.

Having the ability to tune channel parameters, and the implementation of the important feedback in the current—voltage relationship of the cell membrane, allows the inclusion of any population of excitatory or inhibitory channels with any desired dynamics in the simulations. The circuits that implement these basic building blocks are relatively small, and consist of only a few dozen transistors. Consequently, the use of VLSI technology allows the implementation of a large number of channels, synapses and neurons, each with tunable dynamics, reversal potentials, and conductance values.

2.2. Synaptic Plasticity

Several learning rules, such as Hebbian[6] and correlation based rules[7] have been proposed to underlie plasticity in different parts of the brain. Using analog VLSI circuits, a mathematically or graphically described learning rule can be implemented and integrated on the same chip as part of a network of artificial neurons and synapses. This added flexibility allows different learning rule to be assessed in a large-scale simulation in virtually real time.

3. Results

A hardware model of a hippocampal synaptic plasticity system has been designed as a proof-of-principle system[8]. The model consists of AMPAR channels and NMDAR channels, as well as calcium dynamics circuits which control other circuits that implement a calcium-dependent Hebbian learning rule.

In order to take advantage of the superior speed of silicon circuits compared to neurons, the design uses a temporal scaling factor of 100 for time constants in the equations:

$$\tau_{\text{BIOLOGICAL}} = 100 \times \tau_{\text{ELECTRONIC}} \tag{4}$$

Additionally, the voltage range is multiplied by a factor of 10 as well as level-shifted (with 0mV biological set to a bias voltage of 3V), i.e.:

$$V_{\text{ELECTRONIC}} = 10 \times V_{\text{BIOLOGICAL}} + 3V \tag{5}$$

The output currents are adjusted to create signals which are virtually identical to biological signals (with respect to the scaling). The system is able to emulate long-term synaptic plasticity of hippocampal neurons, such as long-term potentiation (LTP) or depression (LTD) induced by high-frequency or low-frequency afferent stimulation. By implementing a Hebbian learning rule, such as those that are thought to govern synaptic plasticity in the hippocampus, the system can now provide a neuromorphic model of learning synapses.

4. Conclusions

A neuromorphic model of glutamatergic transmission and intracellular calcium dynamics has been implemented using analog VLSI circuits. This neuromorphic synapse model has been shown to emulate classic forms of synaptic plasticity such as LTP and LTD. Similar circuit design principles can be readily extended to in-silico modeling of inhibitory transmission in the future. These VLSI circuit models of excitatory and inhibitory transmission and neuronal action potential will form the basic building blocks for the implementation of large-scale neural network models of the respiratory system consisting of populations of respiratory neurons with distributed neuronal properties, rather than discrete neurons with stereotypic properties.

In addition to respiratory modeling, such hardware-based models of cellular plasticity may find applications in the design and implementation of brain-machine interfaces[9,10], such as a real-time adaptive control mechanism capable of interacting with neural tissue and induce reorganization following injury[11], or machine learning devices that emulate human intelligence[12].

5. Acknowledgment

The VLSI chips were fabricated with the support of the MOSIS Education Program. G.R. was a recipient of the National Defense Science and Engineering Graduate Fellowship. This work was supported in part by NIH grant HL60064.

References

1. Rybak, I.A., Shevtsova, N.A., St-John, W.M., Paton, J.F.R., Pierrefiche, O. Endogenous rhythm generation in the pre-Botzinger complex and ionic currents: modeling and in vitro studies, *European Journal of Neuroscience* **18** (**2**): 239–257 (2003).

2. Butera, R.J., Johnson, S.M., DelNegro, C.A., Rinzel, J., Smith, J.C. Dynamics of excitatory networks of bursting pacemaking neurons: Modeling and experimental studies of the respiratory central pattern generator. *Neurocomputing*: **32**: 323–330 (2000).
3. Poon, C.-S. and M. Siniaia. Plasticity of cardiorespiratory neural processing: classification and computational functions. *Respirat. Physiol.* Special Issue on Modeling and Control of Breathing **122**: 83–109 (2000).
4. Mead., C. *Analog VLSI and Neural Systems.* 1989. Addison-Wesley. Reading Massachusetts.
5. Mahowald, M. and R. Douglas, A Silicon neuron. *Nature* **354**: pp. 515–518, (1991).
6. Hebb, D. *The Organization of Behavior.* 1949, New York Wiely.
7. Young, D., and C.S. Poon. A Hebbian feedback covariance learning paradigm for self-tuning optimal control. *IEEE Tran Systems and Cybernetics* **31**: 173–186 (2001).
8. Rachmuth, G. and C.S. Poon. Design of a neuromorphic hebbian synapse using analog VLSI. *First International IEEE EMBS Conference on Neural Engineering, Capri, Italy, 20–22 March 2003 Conference Proceedings.*, pp. 221–224 (2003).
9. Mussa-Ivalidi, F.A., and Miller, L.E. Brain-Machine interfaces: computational demands and clinical needs meet basic neuroscience. *Trends in Neuroscience.* **26**: 329–334 (2003).
10. Jung, R., Brauer, E.J., Abbas, J.J. Real-time interaction between a neuromorphic electronic circuit and the spinal cord. *IEEE Transactions on Neural Systems and Rehabilitation Engineering.* **9**: 319–326 (2001).
11. Krebs, H. Hogan, N., Aisen, M.L., and Volpe, T. Robot-aided neurorehabilitation. *IEEE Trans Rehabil Engineering*, **6**: 75–87 (1998).
12. Sutton, R.S., and Barto, A.G. *Reinforcement Learning.* (1998). Cambridge, Massachusetts, MIT Press.

Respiratory Role of Ionotropic Glutamate Receptors in the Rostral Ventral Respiratory Group of the Rabbit

Donatella Mutolo, Fulvia Bongianni, and Tito Pantaleo

1. Introduction

Neuronal mechanisms responsible for generating the eupneic pattern of breathing are localized within the pons and the medulla oblongata.[1–3] Excitatory amino acids (EAAs) are known to be involved in the generation of rhythmic respiratory drive in mammals; both N-methyl-D-aspartic acid (NMDA) and non-NMDA receptors are present within the medullary respiratory neuronal aggregates, and in particular in the ventral respiratory group (VRG).[2,4] Recently, we have provided evidence[5] that EAA-mediated neurotransmission controls the intensity of eupneic inspiratory activity within the intermediate, almost purely inspiratory VRG (iVRG) mainly through NMDA receptors, whilst it does not appear to have a significant role in shaping the pattern of breathing at the level of caudal VRG.

The present study was undertaken to investigate the respiratory role played by ionotropic glutamate receptors located within the Bötzinger complex (Böt.c.) and the pre-Bötzinger complex (pre-Böt.c.), *i.e.* the rostral VRG subregions probably subserving respiratory rhythm generation,[1,6–8] by using microinjections of a broad-spectrum EAA receptor antagonist as well as selective antagonists acting on either NMDA or non-NMDA receptors.

2. Methods

Experiments were carried out on 38 male New Zealand white rabbits (2.7–3.2 kg) anesthetized with a mixture of α-chloralose (40 mg/kg i.v.; Sigma, St. Louis, MO, USA) and urethane (800 mg/kg i.v.; Sigma), supplemented when necessary (4 mg/kg and 80 mg/kg, respectively). The adequacy of anesthesia was assessed by the absence of reflex withdrawal of the hindlimb in response to noxious pinching of the hindpaw. All animal care and experimental procedures were conducted in accordance with the Directives of the European

---

Donatella Mutolo, Fulvia Bongianni, and Tito Pantaleo • Dipartimento di Scienze Fisiologiche, Università di Firenze, Viale G.B. Morgagni 63, I-50134 Firenze, Italy.

Post-Genomic Perspectives in Modeling and Control of Breathing, edited by Jean Champagnat, Monique Denavit-Saubié, Gilles Fortin, Arthur S. Foutz, Muriel Thoby-Brisson. Kluwer Academic/Plenum Publishers, 2004.

Community as well as with the Italian legislation. Animal preparation and experimental procedures were similar to those described in previous reports.[5,8,9] In particular, the animals were vagotomized, paralysed (gallamine triethiodide 4 mg/kg i.v., supplemented with 2 mg/kg every 30 min, Sigma) and artificially ventilated. Intratracheal pressure and arterial blood pressure as well as end-tidal CO_2 partial pressure (P_{CO2}) were monitored. In paralysed animals, the depth of anesthesia was assessed by monitoring a stable and regular pattern of phrenic activity as well as the absence of fluctuations in arterial blood pressure whether spontaneous or in response to somatic nociceptive stimulation. End-tidal P_{CO2} was maintained at desired levels (28–32 mmHg) by adjusting the frequency and stroke volume of the respiratory pump. Efferent phrenic nerve activity was recorded with bipolar platinum electrodes from desheathed C_5 phrenic roots, amplified, full-wave rectified and integrated (RC filter, time constant 100 ms).

To localize the different VRG subregions, extracellular recordings were made with tungsten microelectrodes (5–10 MΩ impedance at 1 kHz) as already described;[8] we encountered an almost purely expiratory population of neurons in the Böt.c., and a mix of inspiratory and expiratory neurons in the pre-Böt.c.. The localization of the investigated areas was confirmed by evaluating the respiratory effects induced by microinjections of DL-homocysteic acid (DLH).[6,9,10] Integrated phrenic nerve activity as well as the signals of all variables studied were recorded on an eight-channel rectilinearly writing chart recorder.

Microinjection procedures have been fully described in previous reports.[5,8,9] The following drugs (Tocris Cookson, Bristol, UK) were used: kynurenic acid (KYN, 50 mM), D(-)-2-amino-5-phosphonopentanoic acid (D-AP5; 1, 10 and 20 mM) and 6-cyano-7-nitroquinoxaline-2, 3-dione (CNQX; 1, 10 and 20 mM). DLH (20 mM) was obtained from Sigma. Only one ionotropic receptor antagonist was tested in each preparation. Microinjections (30–50 nl) were performed via a glass micropipette (tip diameter 10–25 m) by applying pressure using an air-filled syringe connected to the micropipette by polyethylene tubing. Control injections of equal volumes of the vehicle solution were also performed.

To evaluate chemical sensitivity during respiratory responses induced by NMDA and/or non-NMDA receptor antagonism, hypercapnic and hypoxic stimuli were employed. Hypercapnia was produced by allowing the animal to inspire appropriate mixtures of CO_2 and O_2 from a large bag (150 l); P_{CO2} was adjusted at approximately 30 mmHg higher than the level of spontaneous breathing (range 60–65 mmHg). After a stable level of P_{CO2} had been achieved, hypercapnic stimulation was maintained for at least 3 min. Hypoxia was induced by allowing the animal to inspire a gas mixture containing 6% O_2 and 94% N_2 for about 2 min.

The histological control of injection sites was performed on serial frozen coronal sections (20 m thick) stained with Cresyl violet. The atlas of Shek et al.[11] was used for comparison.

3. Results

Bilateral microinjections of 50 mM KYN ($n = 3$) into the Böt.c. induced within 3–5 min a pattern of breathing characterized by low-amplitude, high-frequency irregular oscillations superimposed on tonic phrenic nerve activity. An intense tonic activity developed within 10 min associated with the disappearance of respiratory rhythmicity ("tonic apnea").

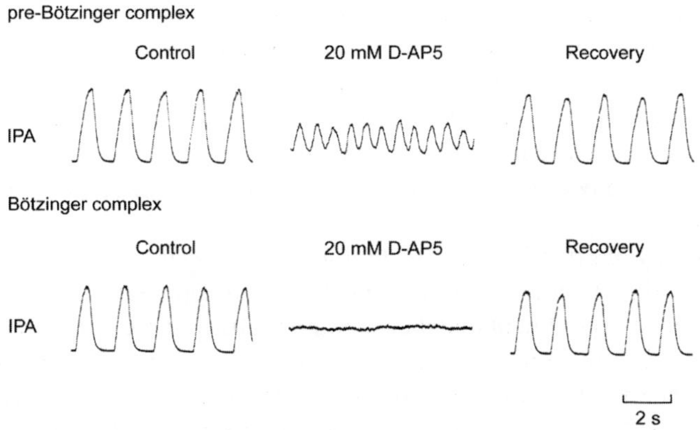

Figure 1. Respiratory responses induced by bilateral microinjections (30 nl) of 20 mM D-AP5 into the rostral subregions of the VRG. Integrated phrenic nerve activity (IPA) under control conditions and 10 min after the completion of the bilateral microinjections of D-AP5 performed into the pre-Bötzinger complex and the Bötzinger complex. Recovery 60 min following the completion of the microinjections in the right panels.

Similar microinjections ($n = 3$) into the pre-Böt.c. caused a progressive increase in the respiratory frequency and rate of rise of inspiratory activity, associated with reductions in peak phrenic amplitude and the appearance of a low level of tonic discharges (maximum effect within 10 min).

Bilateral microinjections of 1 mM ($n = 9$) and 10 mM ($n = 9$) D-AP5 within the Böt.c. induced dose-dependent increases in respiratory frequency and rate of rise of inspiratory activity, accompanied by reductions in peak phrenic amplitude. Respiratory responses quite similar to those induced by KYN were observed following microinjections of 20 mM D-AP5 ($n = 9$). Neither hypoxia nor hypercapnia were able to restore respiratory rhythmicity during tonic apnea. Similar changes in the respiratory pattern, although less pronounced, were seen in response to microinjections of 1 mM ($n = 9$), 10 mM ($n = 9$) and 20 mM ($n = 9$) D-AP5 into the pre-Böt.c.; in no case loss of respiratory rhythmicity was observed. Examples of maximum respiratory responses to NMDA receptor blockades within the Böt.c. and the pre-Böt.c. have been reported in Fig. 1.

Blockades of non-NMDA receptors with CNQX (1, 10 and 20 mM) within the Böt.c. ($n = 9$ for each concentration) and the pre-Böt.c. ($n = 9$ for each concentration) caused much less pronounced effects mainly consisting of increases in respiratory frequency. These respiratory responses were associated with the development of low levels of tonic activity and reductions in peak phrenic amplitude following microinjections of 10–20 mM CNQX within the pre-Böt.c..

Control injections of the vehicle solution performed at responsive sites caused no appreciable effects. All the observed respiratory responses were not accompanied by significant changes in arterial blood pressure.

4. Discussion

The results of this study show that ionotropic glutamate receptors within the Böt.c. and the pre-Böt.c. of the rabbit *in vivo* are involved in the control of both the intensity and frequency of inspiratory activity. The respiratory responses to microinjections of EAA receptor antagonists into these regions argue for an endogenous activation of these receptors during eupnea; NMDA receptors appear to play a major role. Apneic responses induced by KYN and D-AP5 microinjections into the Böt.c., along with the observation that respiratory rhythmicity cannot be restored by hypoxic or hypercapnic stimuli, suggest that EAA receptor antagonism disrupts important connections in the neuronal network subserving respiratory rhythm generation.

The reliability of microinjection techniques as well as the localization of the injection sites and the spread of the injectate have been discussed in previous reports;[5,8,9,12] reported considerations make us confident that the respiratory effects were confined to circumscribed areas localized within the Böt.c. and the pre-Böt.c.. The accuracy of micropipette placements was ensured by extracellular recordings, DLH-induced respiratory responses and the histological control.[5,6,8–10] Antagonist concentrations required to achieve effective blockade of ionotropic EAA receptors were selected in preliminary trials. They were in the same range as those reported previously in *in vivo* studies on medullary respiration-related regions (see e.g. Ref. 5, 12 also for further Refs.).

In agreement with previous findings and the results of a recent study (see e.g. Ref. 5 also for further Refs.) showing that NMDA receptors play a prominent role within the iVRG, present findings do not support the hypothesis that a depolarization mediated by non-NMDA receptors is required to overcome the voltage dependent Mg^{2+}-block of ion channels linked to NMDA receptors.[13] As already discussed,[5] there is indeed good evidence that NMDA receptors can mediate synaptic transmission without a preceding non-NMDA-mediated depolarization.

It could be reasoned that blockade of EAA-mediated fast excitatory synaptic transmission and, therefore, the removal of a prominent excitatory input to these VRG subregions may correspond, to some extent, to their ablation. Previous results obtained with blockades or ablations of neuronal activities within the investigated subregions have been extensively discussed in a previous report.[8] Noticeably, respiratory responses induced by blockade of ionotropic glutamate receptors are generally consistent with our previous findings obtained in the rabbit following bilateral kainic acid (KA) lesions.[8] KA microinjections into either the Böt.c. or the pre-Böt.c. transiently eliminated respiratory rhythmicity and caused tonic apnea; rhythmic activity resumed as low-amplitude, high-frequency oscillations superimposed on tonic activity and displayed progressive, but incomplete recovery. The finding that not only the amplitude, but also the frequency of respiratory bursts changes in response to KA lesions or EAA receptor blockades suggests that these subregions are important elements of the respiratory central pattern generator.[8,14]

The importance of inhibitory interactions in the respiratory rhythm generation has been addressed in several models of respiratory rhythmogenesis (see e.g. Refs. 2, 7, 15–17). These models are based on reciprocal inhibitory interactions between different types of propriobulbar neurons and as more recently suggested[7] on a combination of a network of inhibitory interneurons and excitatory pacemaker-like neurons with intrinsic oscillatory bursting properties (hybrid pacemaker-network model). It has been proposed[7] that a respiratory rhythm can be generated by interactions within the neuronal network in which kernel neurons are embedded, even when these neurons have been silenced or destroyed. On the basis of these models, we can speculate that removal of inhibitory VRG interneurons[2,7,14] may produce breathing patterns characterized by decreases in peak phrenic amplitude and increases in respiratory frequency up to the arrest of respiratory rhythm. These patterns of breathing are possibly due to the loss of part of inhibitory interneurons and reduced strength of inhibitory synapses[16] which cause desynchronization of the respiratory network. The apneic responses are probably due to very important deficits in inhibitory mechanisms and might represent a level of network desynchronization beyond which the respiratory rhythm cannot be generated. EAA receptor blockades cause increases in respiratory frequency and decreases in peak phrenic amplitude with parallel increases in tonic activity; tonic apnea as extreme possible outcome of such blockades only occurred when they are performed at the level of the Böt.c.. However, the maximal respiratory responses evoked from either rostral VRG subregion are characterized by ineffective, life threatening, patterns of breathing (see Fig. 1); they may be subserved by similar neural mechanisms. It is not possible to revert tonic apnea by increasing chemical drive inputs. We cannot rule out that apnea ensued because the rostral VRG is involved in chemical reception or in the transmission and integration of chemical drive inputs.[8] As already fully discussed elsewhere,[8] it seems more plausible that essential connections of the neural network underlying respiratory rhythm generation have been disrupted by our experimental manoeuvres. In particular, an important role in the genesis of tonic inspiratory activity and tonic apnea could be ascribed to a failure in the inhibitory input from rostral expiratory neurons that are mainly concentrated in the Böt.c., but are also present within the pre-Böt.c..[8,9,18–21] In this connection, we can speculate that blockades of ionotropic glutamate receptors within the pre-Böt.c., although complete, do not cause a loss of inhibitory mechanisms sufficient to suppress the respiratory rhythm.

5. Acknowledgements

This studies was supported by grants from the Ministero dell'Istruzione, Università e Ricerca of Italy.

References

1. J. L. Feldman, G. S. Mitchell, and E. E. Nattie, Breathing: Rhythmicity, Plasticity, Chemosensitivity, *Annu. Rev. Neurosci.* **26**, 239–266 (2003).
2. A. L. Bianchi, M. Denavit-Saubie, and J. Champagnat, Central control of breathing in mammals: neuronal circuitry, membrane properties, and neurotransmitters, *Physiol. Rev.* **75**, 1–45 (1995).
3. W. M. St John, Neurogenesis of patterns of automatic ventilatory activity, *Prog. Neurobiol.* **56**, 97–117 (1998).
4. A. Haji, R. Takeda, and M. Okazaki, Neuropharmacology of control of respiratory rhythm and pattern in mature mammals, *Pharmacol. Ther.* **86**, 277–304 (2000).
5. F. Bongianni, D. Mutolo, M. Carfi, and T. Pantaleo, Respiratory responses to ionotropic glutamate receptor antagonists in the ventral respiratory group of the rabbit, *Pflugers Arch.* **444**, 602–609 (2002).
6. A. Monnier, G. F. Alheid, and D. R. McCrimmon, Defining ventral medullary respiratory compartments with a glutamate receptor agonist in the rat, *J. Physiol.* **548**, 859–874 (2003).
7. J. C. Smith, R. J. Butera, N. Koshiya, C. Del Negro, C. G. Wilson, and S. M. Johnson, Respiratory rhythm generation in neonatal and adult mammals: the hybrid pacemaker-network model, *Respir. Physiol.* **122**, 131–147 (2000).
8. D. Mutolo, F. Bongianni, M. Carfi, and T. Pantaleo, Respiratory changes induced by kainic acid lesions in rostral ventral respiratory group of rabbits, *Am. J. Physiol.* (*Regul. Integr. Comp Physiol.*) **283**, R227–R242 (2002).

9. F. Bongianni, D. Mutolo, and T. Pantaleo, Depressant effects on inspiratory and expiratory activity produced by chemical activation of Bötzinger complex neurons in the rabbit, *Brain Res.* **749**, 1–9 (1997).
10. I. C. Solomon, N. H. Edelman, and J. A. Neubauer, Patterns of phrenic motor output evoked by chemical stimulation of neurons located in the pre-Bötzinger complex in vivo, *J. Neurophysiol.* **81**, 1150–1161 (1999).
11. J. W. Shek, G. Y. Wen, and H. M. Wisniewski. Atlas of the Rabbit Brain and Spinal Cord. Karger, Basel, 1986.
12. F. Bongianni, D. Mutolo, M. Carfi, and T. Pantaleo, Area postrema glutamate receptors mediate respiratory and gastric responses in the rabbit, *Neuroreport.* **9**, 2057–2062 (1998).
13. L. Nowak, P. Bregestovski, P. Ascher, A. Herbet, and A. Prochiantz, Magnesium gates glutamate-activated channels in mouse central neurones, *Nature.* **307**, 462–465 (1984).
14. O. Pierrefiche, S. W. Schwarzacher, A. M. Bischoff, and D. W. Richter, Blockade of synaptic inhibition within the pre-Bötzinger complex in the cat suppresses respiratory rhythm generation in vivo, *J. Physiol.* **509**, 245–254 (1998).
15. M. D. Ogilvie, A. Gottschalk, K. Anders, D. W. Richter, and A. I. Pack, A network model of respiratory rhythmogenesis, *Am. J. Physiol.* **263**, R962–R975 (1992).
16. I. A. Rybak, J. F. Paton, and J. S. Schwaber, Modeling neural mechanisms for genesis of respiratory rhythm and pattern. II. Network models of the central respiratory pattern generator, *J Neurophysiol.* **77**, 2007–2026 (1997).
17. D. W. Richter, D. Ballantyne, and J. E. Remmers, How is the respiratory rhythm generated? A model, *News Physiol. Sci.* **1**, 109–112 (1986).
18. C. A. Connelly, E. G. Dobbins, and J. L. Feldman, Pre-Bötzinger complex in cats: respiratory neuronal discharge patterns, *Brain Res.* **590**, 337–340 (1992).
19. S. W. Schwarzacher, J. C. Smith, and D. W. Richter, Pre-Bötzinger complex in the cat, *J. Neurophysiol.* **73**, 1452–1461 (1995).
20. R. St Jacques and W. M. St John, Transient, reversible apnoea following ablation of the pre-Bötzinger complex in rats, *J. Physiol.* **520**, 303–314 (1999).
21. Q. J. Sun, A. K. Goodchild, J. P. Chalmers, and P. M. Pilowsky, The pre-Bötzinger complex and phase-spanning neurons in the adult rat, *Brain Res.* **809**, 204–213 (1998).

Serotonergic Receptors and Effects in Hypoglossal and Laryngeal Motoneurons

Semi-Quantitative Studies in Neonatal and Adult Rats

Denys V. Volgin, Victor B. Fenik, Richard Fay, Shinichi Okabe, Richard O. Davies, and Leszek Kubin

1. Introduction

Respiratory motoneurons of the brainstem express multiple receptors for serotonin (5-HT).[1–7] Some of these receptors mediate excitatory effects (e.g., type 2), whereas others (e.g., type 1) are inhibitory, yet the net effect of 5-HT applied onto motoneurons of mature animals is excitation, and antagonism of endogenous serotonergic effects suppresses the activity in brainstem respiratory motoneurons.[8–10] Since all brainstem serotonergic neurons consistently exhibit decreased activity during slow-wave sleep and are silenced during rapid eye movement (REM) sleep, it has been proposed that a withdrawal of serotonergic excitation from brainstem respiratory motoneurons may play an important role in sleep-related decrements of their activity.[8] In individuals with anatomical abnormalities of their upper airway, such decrements may cause nocturnal upper airway obstructions, leading to the obstructive sleep apnea syndrome (reviewed in ref. 11).

While multiple evidence supports such a state-dependent function of serotonergic modulation of upper airway motoneuronal activity, studies in immature animals suggest that the role of 5-HT in motoneurons may vary with the stage of development and motor pool (e.g., ref. 12), and that the complement of 5-HT receptors present in upper airway motoneurons undergoes changes in the early postnatal and juvenile period (e.g., ref. 5). In mature animals, the magnitude of sleep-related decrements in activity also varies greatly among different pools of upper airway motoneurons, with those innervating pharyngeal muscles more depressed than those innervating laryngeal muscles.[11,13] These

Leszek Kubin • Department of Animal Biology 209E/VET, Univ. of Pennsylvania, Philadelphia, PA 19104-6046, USA. Tel.: (1) 215-898-1893; Fax: (1) 215-573-5186; E-mail: lkubin@vet.upenn.edu

Post-Genomic Perspectives in Modeling and Control of Breathing, edited by Jean Champagnat, Monique Denavit-Saubié, Gilles Fortin, Arthur S. Foutz, Muriel Thoby-Brisson. Kluwer Academic/Plenum Publishers, 2004.

differences may be related to the magnitude of serotonergic effects in distinct motoneuronal pools.

The goals of these studies were to conduct a quantitative assessment of postnatal development for two major 5-HT receptor subtypes, type 1B and 2A in upper airway (hypoglossal—XII) motoneurons, and to compare the magnitude of serotonergic excitatory effects in two motoneuronal pools, XII and laryngeal.

2. Methods

Experiments were performed on developing Sprague-Dawley rats (receptor mRNA and immunohistochemical studies), and on adult cats (iontophoretic studies). All experimental procedures were approved by the Institutional Animal Care and Use Committee of the University of Pennsylvania.

2.1. Detection of 5-HT Receptor mRNA by Single Cell RT-PCR

Hypoglossal motoneurons were retrogradely labeled with rhodamine dextran, and acutely dissociated, as described previously.[14,15] Brainstem slices containing the XII nucleus were obtained following decapitation under deep isoflurane anesthesia. Motoneurons were individually collected, ultrasonicated, their DNA digested, and RNA subjected to reverse transcription (RT). Uniform aliquots of cDNA from each cell were subjected to two stages of semi-nested polymerase chain reaction (PCR).[14,15] For each cell, we assessed the presence of mRNA for 5-HT_{1B} and 5-HT_{2A} receptors (cells negative for both were tested for neuron-specific enolase to verify that their cDNA was properly generated; see ref. 15 for primers used). An average of 29 ± 8 (SD) motoneurons per animal for rats sacrificed on nearly every other postnatal day (P) from P3 through P33 were studied. Such a large number of cells allowed us to characterize the level of expression for a given receptor by the proportion of positive motoneurons detected on a given postnatal day, and to calculate the confidence interval (SE) of this estimate (Analyse-It Software, UK). To validate this approach, in selected motoneurons, we also studied the expression of 5-HT_{1A} receptor mRNA because receptor binding and *in situ* hybridization show a decline in the XII nucleus during the postnatal period.[5]

2.2. Immunohistochemistry for 5-HT_{2A} Receptors

Rats P5 and older were deeply anesthetized with pentobarbital and perfused with saline followed by 4% paraformaldehyde. Coronal sections of the medulla were cut (35 m) and processed free-floating. 5-HT_{2A} receptor-like immunoreactivity was visualized with avidin-biotin-horseradish peroxidase with heavy metal intensification, as described previously.[6,15] Importantly, sections from rats of different ages were processed together in different age combinations and then mounted and digitally photographed under identical conditions. Staining in the XII nucleus was measured densitometrically (ImageQuant Software; Molecular Dynamics); background staining was also determined for each section and subtracted from that in the XII nucleus.

2.3. Quantitative 5-HT Iontophoresis onto Hypoglossal and Laryngeal Motoneurons

Adult, decerebrate, paralyzed and artificially ventilated cats were used to assess the sensitivity of XII and laryngeal motoneurons to iontophoretic application of 5-HT.[16] Cells recorded in the XII and ambiguus nuclei were identified as motoneurons by spike-triggered averaging of the activity simultaneously recorded from the ipsilateral XII and vagus nerve, respectively. In preliminary experiments, we determined that XII motoneurons are exquisitely sensitive to 5-HT, so that a current of 20–50 nA ejected from a barrel filled with 20 mM 5-HT in H_2O for 2–3 min produced a long-lasting and abnormally erratic firing. Therefore, in order to compare the sensitivity of different motoneurons to the excitatory effect of 5-HT in a quantitative manner, we studied spontaneously active motoneurons only, and we minimized and standardized the 5-HT application. 5-HT was ejected from a barrel filled with 10 mM 5-HT creatinine sulphate and 150 mM NaCl using 15 nA current that was applied for 3 min. The changes in firing rate observed during the third minute of 5-HT application were measured and subjected to statistical analysis.[16]

3. Results

As reported previously,[5] the percentage of cells expressing 5-HT_{1A} receptor mRNA was determined in 35 motoneurons from three rats. The proportion of positive motoneurons decreased from 50% ± 15(SE) on P9 to 8.3% ± 8(SE) on P23 ($p < 0.05$) (Fig. 1A).

The expression of 5-HT_{1B} and 5-HT_{2A} receptor mRNA was assessed in 414 XII motoneurons. The 5-HT_{1B} receptor mRNA was detected in 50% of the motoneurons on P3, increased to 74% on P7, and then transiently decreased to 27% on P14. The last was significantly lower than in younger (P5, 7 or 10) or older (P17 or 21) rats ($p < 0.05$). The percentage of XII motoneurons expressing the 5-HT_{1B} receptor mRNA was maximal on P21 (85%) and then declined to 50–63%. The 5-HT_{2A} receptor mRNA was expressed significantly less frequently in P3 and P5 than in older animals ($p < 0.05$). An increase from 65 to 100% of cells expressing the 5-HT_{2A} receptor mRNA occurred between P6 and P9 (Fig. 1B). The slight dip in expression on day 14 was not significant.

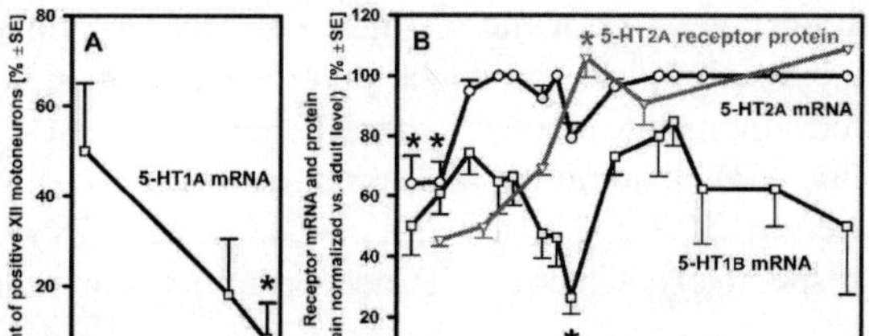

Figure 1. A: Significant developmental decline in the proportion of XII motoneurons expressing 5-HT_{1A} receptor mRNA validates the single-cell RT-PCR approach. B: Developmental changes in the 5-HT_{1B} and 5-HT_{2A} mRNA and 5-HT_{2A} receptor protein expression in XII motoneurons point to P7-15 as a period of major reorganization.

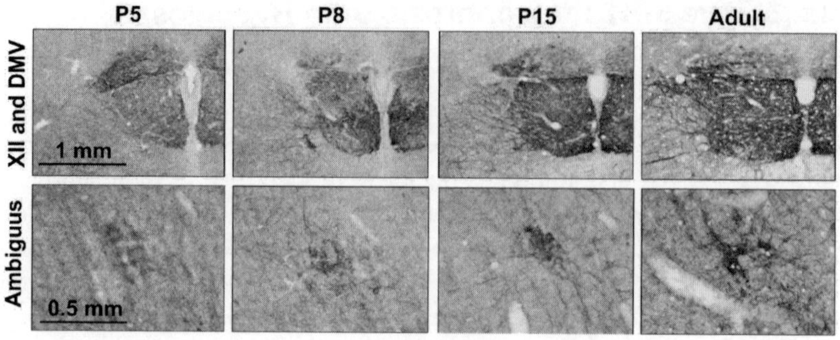

Figure 2. 5-HT_{2A} receptor protein expression gradually increases in the XII and ambiguus nuclei between P5 and P15. The top and bottom rows show coronal medullary sections of the opposite sides of the medulla at a caudal and rostral levels of the area postrema, respectively. (DMV—dorsal motor nucleus of the vagus).

The 5-HT_{2A} receptor-like immunoreactivity in the medulla of adult rats is most intense in the motor nuclei.[6] However, its level on P5 and P8 is much lower than in adults. In the XII nucleus, it then increases gradually until about P15 (gray line in Fig. 1B), even though all XII motoneurons express 5-HT_{2A} mRNA as early as P9. 5-HT_{2A} receptor protein becomes prominent in the dendrites of XII motoneurons around P12–P15 (Fig. 2, top row). Laryngeal motoneurons of the nucleus ambiguus show the same trend (Fig. 2, bottom row), although their less dense packing precluded densitometric measurements like those obtained from the XII nucleus.

In adult cats, 5-HT applied iontophoretically excited 97% of XII and 84% laryngeal motoneurons. This effect was blocked by methysergide ejected from another barrel of the electrode. The standardized iontophoretic application of 5-HT to 13 spontaneously active inspiratory XII motoneurons, and 11 inspiratory and 11 expiratory laryngeal motoneurons revealed different sensitivity of these motoneurons to 5-HT. The average relative increases in activity were to $220\% \pm 24$ ($p < 0.001$), $147\% \pm 23$ ($p < 0.001$), and $148\% \pm 9$ ($p < 0.001$) of control, respectively. XII motoneuronal activity was enhanced more than the activity of either inspiratory or expiratory laryngeal motoneurons ($p < 0.05$) (Fig. 3A).

In a separate study also performed in decerebrate cats, we assessed the magnitude of the REM sleep-like depression of inspiratory and expiratory activity in recurrent laryngeal (RL) nerve and inspiratory activity of the XII nerve.[13] This was done by microinjecting a cholinergic agonist, carbachol, into the dorsal pontine tegmentum.[17] The magnitudes of the depression differed among the nerves, with RL nerve activities depressed less than XII nerve activity (Fig. 3B). Inspiratory and expiratory activity of the recurrent laryngeal nerve were reduced by $21\% \pm 3.1$ ($n = 13, p < 0.001$) and $28\% \pm 4.7$ of the control ($n = 10, p < 0.01$), respectively, whereas XII nerve activity was reduced by $85\% \pm 3.2$ ($n = 15, p < 0.001$).

4. Discussion

In mature animals, 5-HT_{2A} receptors mediate a major portion of the serotonergic excitatory drive in brainstem respiratory motoneurons. In contrast, little is known about

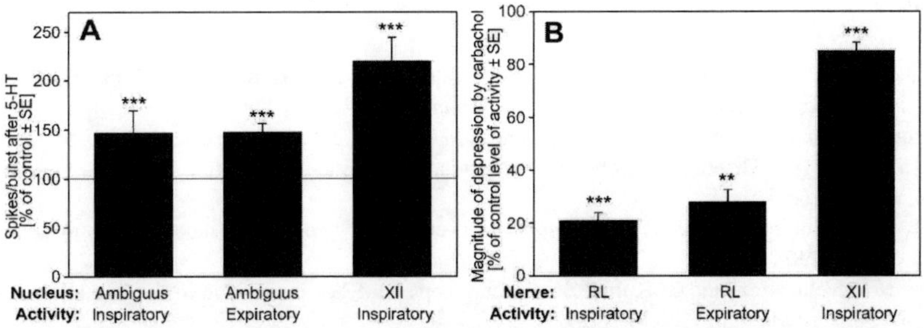

Figure 3. A: The mean 5-HT-induced increase in the number of action potentials per respiratory cycle relative to control was larger in XII than laryngeal motoneurons. B: Similarly, the magnitude of the REM sleep-like depression of activity produced by pontine carbachol is larger in XII than laryngeal motoneurons (data from ref. 13). This correlation is consistent with the hypothesis that the relative magnitude of the serotonergic excitatory drive that motoneurons of different groups may receive during wakefulness is an important determinant of the magnitude of the depression that they exhibit during REM sleep. ***-$p < 0.001$ and **-$p < 0.01$ vs. control conditions (paired t-test).

the role of 5-HT_{1B} receptors. To date, their pre-, rather than postsynaptic, modulation of excitatory and inhibitory inputs to XII motoneurons was investigated in neonatal and juvenile rats. However, our data show that XII motoneurons themselves express the mRNA for these receptors, and recent immunohistochemical study shows that XII motoneurons also express 5-HT_{1B} receptor-like protein.[19] In XII motoneurons of juvenile rats, stimulation of these receptors causes an increase in input resistance,[20] and in adult rats microinjections of a 5-HT_{1B} receptor agonist into the XII nucleus cause a depression of XII nerve activity.[3] No developmental changes were detected in the magnitude of presynaptic effects of 5-HT_{1B} receptors on XII motoneurons;[19] whereas the development of postsynaptic effects has not been studied. The timing of developmental changes in 5-HT receptor expression suggests a relationship to postnatal changes in feeding behavior. It is possible that common cellular processes underlie the transient decline in 5-HT_{1B} mRNA and the increase in 5-HT_{2A} receptor protein.

In addition to these developmental changes, our data show that the relative magnitude of the excitatory drive mediated by 5-HT may be larger in XII than in laryngeal motoneurons. This difference was detected in iontophoretic *in vivo* studies even though it was not clearly reflected in immunohistochemical staining of the two motor pools for 5-HT_{2A} receptors. This functional difference correlates with the magnitude of sleep-related depression observed in XII and laryngeal motoneurons,[13] which is consistent with the concept that 5-HT mediates an important portion of the wakefulness-related excitatory drive in brainstem respiratory motoneurons.[8]

5. Acknowledgments

The studies were supported by grants from the Heart, Lung and Blood Institute of the National Institutes of Health (HL-42236, HL-6287, HL-47600), and the University of Pennsylvania Research Foundation. RF was supported by Parker B. Francis Fellowship.

References

1. S. Manaker and P. C. Zucchi, Autoradiographic localization of neurotransmitter binding sites in the hypoglossal and motor trigeminal nuclei of the rat, *Synapse* **28**, 44–59 (1998).
2. G. Mengod, M. Pompeiano, M. I. Martínez-Mir, and J. M. Palacios, Localization of the mRNA for the 5-HT_2 receptor by in situ hybridization histochemistry. Correlation with the distribution of receptor sites, *Brain Res.* **524**, 139–143 (1990).
3. S. Okabe and L. Kubin, Role of 5HT_1 receptors in the control of hypoglossal motoneurons in vivo, *Sleep* **19**, S150–S153 (1996).
4. S. Okabe, M. Mackiewicz, and L. Kubin, Serotonin receptor mRNA expression in the hypoglossal motor nucleus, *Respir. Physiol.* **110**, 151–160 (1997).
5. D. A. Bayliss, F. Viana, E. M. Talley, and A. J. Berger, Neuromodulation of hypoglossal motoneurons: cellular and developmental mechanisms, *Respir. Physiol.* **110**, 139–150 (1997).
6. R. Fay and L. Kubin, Pontomedullary distribution of 5-HT_{2A} receptor-like protein in the rat, J. Comp. Neurol. **418**, 323–345 (2000).
7. G. Zhan, F. Shaheen, M. Mackiewicz, P. Fenik, and S. C. Veasey, Single cell laser dissection with molecular beacon polymerase chain reaction identifies 2A as the predominant serotonin receptor subtype in hypoglossal motoneurons, *Neuroscience* **113**, 145–154 (2002).
8. L. Kubin, H. Tojima, R. O. Davies, and A. I. Pack, Serotonergic excitatory drive to hypoglossal motoneurons in the decerebrate cat, *Neurosci. Lett.* **139**, 243–248 (1992).
9. H. Arita, K. Ichikawa, and M. Sakamoto, Serotonergic cells in nucleus raphepallidus provide tonic drive to posterior cricoarytenoid motoneurons via 5-hydroxytryptamine2 receptors in cats, *Neurosci. Lett.* **197**, 113–116 (1995).
10. P. Fenik and S. C. Veasey, Pharmacological characterization of serotonergic receptor activity in the hypoglossal nucleus, *Am. J. Respir. Crit. Care Med.* **167**, 563–569 (2003).
11. L. Kubin and R. O. Davies, Mechanisms of airway hypotonia, in: *Sleep Apnea. Pathogenesis, Diagnosis, and Treatment*, edited by A. I. Pack (Dekker, New York, 2002), pp. 99–154.
12. D. Morin, R. Monteau, and G. Hilaire, Compared effects of serotonin on cervical and hypoglossal inspiratory activities: an in vitro study in the newborn rat, *J. Physiol. (Lond.)* **451**, 605–629 (1992).
13. V. Fenik, R. O. Davies, A. I. Pack, and L. Kubin, Differential suppression of upper airway motor activity during carbachol-induced, REM sleep-like atonia, *Am. J. Physiol.* **275**, R1013–R1024 (1998).
14. D. V. Volgin, M. Mackiewicz, and L. Kubin, α_{1B} receptors are the main postsynaptic mediators of adrenergic excitation in brainstem motoneurons, a single-cell RT-PCR study, *J. Chem. Neuroanat.* **22**, 157–166 (2001).
15. D. V. Volgin, R. Fay, and L. Kubin, Postnatal developm ent of serotonin 1B, 2A and 2C receptors in brainstem motoneurons, *Eur. J. Neurosci.* **17**, 1179–1188 (2003).
16. V. Fenik, L. Kubin, S. Okabe, A. I. Pack, and R. O. Davies, Differential sensitivity of laryngeal and pharyngeal motoneurons to iontophoretic application of serotonin, *Neuroscience* **81**, 873–885 (1997).
17. H. Kimura, L. Kubin, R. O. Davies, and A. I. Pack, Cholinergic stimulation of the pons depresses respiration in decerebrate cats, *J. Appl. Physiol.* **69**, 2280–2289 (1990).
18. J. H. Singer, M. C. Bellingham, and A. J. Berger, Presynaptic inhibition of glutamatergic synaptic transmission to rat motoneurons by serotonin, *J. Neurophysiol.* **76**, 799–807 (1996).
19. A. J. Berger and P. Huynh, Activation of 5HT_{1B} receptors inhibits glycinergic synaptic inputs to mammalian motoneurons during postnatal development, *Brain Res.* **956**, 380–384 (2002).
20. V. A. Bouryi and D. I. Lewis, The modulation by 5-HT of glutamatergic inputs from the raphe pallidus to rat hypoglossal motoneurons in vitro, *J. Physiol. (Lond.)*, **533**, 1019–1031 (2003).

Key words: development; motoneurons; serotonin receptors; sleep; single-cell RT-PCR.

Modelling Respiratory Rhythmogenesis: Focus on Phase Switching Mechanisms

Ilya A. Rybak, Natalia A. Shevtsova, Julian F. R. Paton, Olivier Pierrefiche, Walter M. St.-John, and Akira Haji

1. Introduction

It has been established that the normal respiratory pattern ("eupnoea") in mammals is generated in the lower brainstem[1,2] and may involve several medullary and pontine regions. Although some researchers suggest that a smaller region within the medulla (e.g., the pre-Bötzinger Complex (pre-BötC)) may be sufficient for the respiratory rhythm generation[3–5], the eupnoeic respiratory rhythm (as well as apneustic breathing) has never been reproduced in reduced medullary preparations without the pons. At the same time, the specific ponto-medullary interactions related to genesis, shaping and control of the respiratory pattern have not been well characterized so far. Here we present a preliminary computational model of the ponto medullary respiratory network that is considered a basis for the future interactive modeling-experimental studies. The model has been developed using a series of assumptions. Specifically, we have suggested that, under normal conditions *in vivo*, the eupnoeic respiratory rhythm is generated by a ponto-medullary network. Hence, although the pre-BötC is a necessary part of this network, the intrinsic oscillations in this region are suppressed during eupnoea by ponto-medullary interactions. These endogenous oscillations, however, may be released under some specific conditions, e.g., *in vitro*, because of the lack of the pons, or during hypoxia *in vivo*[6]. We have also assumed that the medullary part of the respiratory network contains special neural circuits performing the respiratory phase switching. Moreover, these circuits are also targets for pulmonary feedback and

---

Ilya A. Rybak and **Natalia A. Shevtsova** • School of Biomedical Engineering, Science and Health Systems, Drexel University, Philadelphia, PA 19104. **Julian F. R. Paton** • Department of Physiology, School of Medical Sciences, University of Bristol, Bristol BS8 1TD, UK. **Olivier Pierrefiche** • GRAP-JE-UFR de Pharmacie, Amiens 80036, France. **Walter M. St.-John** • Department of Physiology, Dartmouth Medical School, Lebanon, NH 03756. **Akira Haji** • Department of Pharmacology, Faculty of Medicine, Toyama Medical and Pharmaceutical University, 2630 Sugitani, Toyama 930-0194, Japan.

Post-Genomic Perspectives in Modeling and Control of Breathing, edited by Jean Champagnat, Monique Denavit-Saubié, Gilles Fortin, Arthur S. Foutz, Muriel Thoby-Brisson. Kluwer Academic/Plenum Publishers, 2004.

inputs from the pons and major afferent nerves, which use the same medullary switching circuits to regulate the timing of phase transitions and modulate the respiratory motor pattern[7].

2. Model of the Ponto-medullary Respiratory Network

The model (Figure 1) contains several interacting populations of respiratory neurons characterized in the rostroventolateral medulla and pons *in vivo* including the rostral ventral respiratory group (rVRG), pre-BötC, and Bötzinger Complex (BötC). The following neural populations were included in the model: ramp-I, and late-I (both in rVRG); early-I and pre-I (in pre-BötC); post-I and aug-E (in BötC); I-modulated, E-modulated and two distinct IE-modulated, IE_1 and IE_2, (in the rostral pons, rPons, that was supposed to include the dorsolateral (NPMB/KF) and ventrolateral pontine regions); tonic (in the caudal pons, cPons). Network interactions within the rVRG (i.e., between the ramp-I, early-I, and late-I populations) and between rVRG and BötC populations define the basic circuitry for the inspiratory off-switching (IOS) mechanism with the late-I population playing the key role in the inspiratory off-switching[2,7–9]. Interactions among the rVRG populations, post-I population and pre-I population of pre-BötC define the basic circuitry for the expiratory off-switching (EOS) mechanism with the pre-I population performing the inspiratory on-switching function[8]. Importantly, IOS and EOS mechanisms operate under control of both pontine input and pulmonary feedback which both are excitatory to the late-I, ramp-I and post-I populations[7,9].

The model generates a stable "eupnoeic" respiratory rhythm with "augmenting" phrenic discharges and demonstrates realistic firing patterns and membrane potential trajectories of individual respiratory neurons (Figure 2A). In addition, the model reproduces a number of known physiological phenomena. For example, the disconnection of vagal feedback produces an increase in the amplitude and duration of phrenic discharges (Figure 2B), whereas a moderate vagal stimulation shortens inspiration and prolongs expiration. Both these effects are consistent with the Hering-Breuer (HB) reflex[2,10]. A relatively strong stimulation of the vagus produces "postinspiratory apnea"[11,12] (Figure 2C1). Short stimulation of the vagus delivered during inspiration can terminate the inspiratory phase, and the threshold for inspiratory termination decreases during inspiration[2,10,13] (Figure 2C2). Stimuli delivered during post-inspiration prolong expiration[10,12], whereas the late part of expiration is insensitive to vagal stimulation[10,14] (Figure 2C3). Short stimulation of pontine IE_1 population terminates inspiration[1,2,15] (Figure 3B). Continuous pontine stimulation shortens inspiration and prolongs expiration[16] (Figure 3C), which suggests the existence of ponto-medullary (PM) reflexes. The removal of rPons converts the eupnoeic pattern to apneusis[1,2,6,15,17] (Figure 3D2). Short vagal stimulation may terminate the apneustic burst (Figure 3E); continuous stimulation shortens the apneustic inspiratory discharges[15] (Figure 3F). The complete removal of the pons releases a pacemaker-driven rhythm in the pre-BötC and converts the eupnoeic pattern to a gasping-like (*in vitro*-like) "decrementing" pattern[12,6,18] (Figure 3D3). The latter is characterized by significant reduction of postinspiratory activity and shortening the delay between hypoglossal and phrenic discharges[6,19] (compare Figure 3B3 with Figures 1 and 3B1).

3. Conclusions and Model Predictions

Our modeling studies support the following suggestions: (1) The eupnoeic respiratory rhythm is generated by a ponto-medullary network mechanism; (2) The pontine inputs

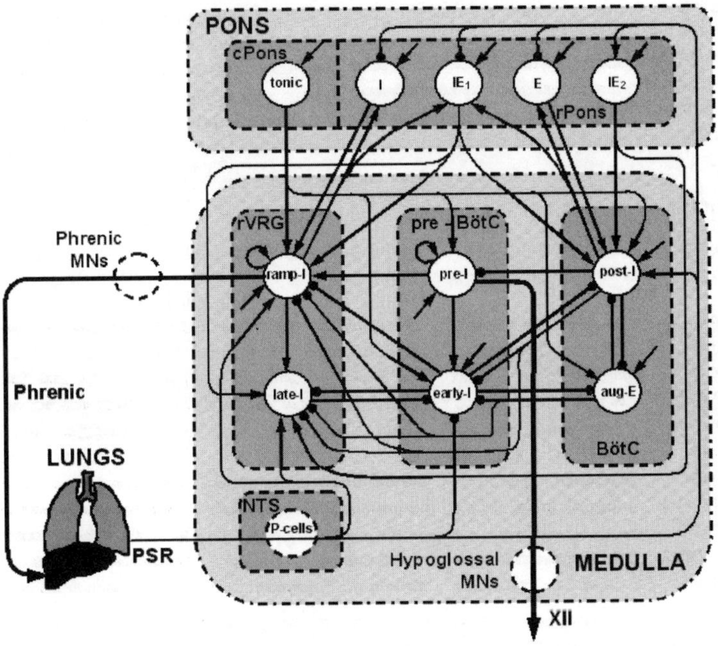

Figure 1. Schematic of the model of the ponto-medullary respiratory network. The white circles represent populations of different respiratory neurons (the populations of phrenic and hypoglossal motoneurons and NTS pump (P) cells are not present in the current model). Each population consisted of 50 neurons. All neurons were modeled in the Hodgkin-Huxley style and incorporated biophysical properties and channel kinetics characterized in respiratory neurons *in vitro*. Specifically, the fast sodium currents in all neurons and the persistent sodium currents in the pre-I population were described using recent experimental data on rat RVLM neurons[20]; the high-threshold and low-threshold calcium currents were described using data of Elsen and Ramirez[21]; intracellular calcium dynamics was based on data of Frermann et al.[22]; other cellular parameters were accepted from the previous models[23–25]. Neuronal parameters were randomly distributed within each neural population. All populations received tonic excitatory drive. The excitatory (arrows) and inhibitory (small black circles) connectivities among the neural populations within the medulla were assigned using the excising direct or indirect physiological data. Some additional connectivities were assigned based on the suggested IOS and EOS mechanisms. Reciprocal excitatory connections were assigned between the medullary ramp-I and the pontine I and IE_1 populations, and between the medullary post-I and the pontine IE_1 and E populations. These connections provided I, IE or E modulation of activity of the corresponding pontine populations. We also assumed that reticular neurons from caudal pons (tonic population) provided an additional excitatory tonic drive to the medullary respiratory populations. Simplified models of the lungs and slowly adapting pulmonary stretch receptors (PSR) were included in the model to provide pulmonary feedback to the respiratory network. The pulmonary feedback controlled activity of the key neural populations involved in IOS and EOS circuits (see text for details) and hence provided regulation of the duration of the respiratory phases and HB reflex. In addition, this feedback inhibited the activity of the pontine neural populations that received excitation from the medullary populations (I, IE_1, E) and hence suppressed the pontine control of the respiratory pattern. At the same time, one pontine population (IE_2) received excitation from the feedback and contributed to HB reflex. Integrated activities of ramp-I and pre-I populations were considered as, respectively, phrenic and hypoglossal outputs.

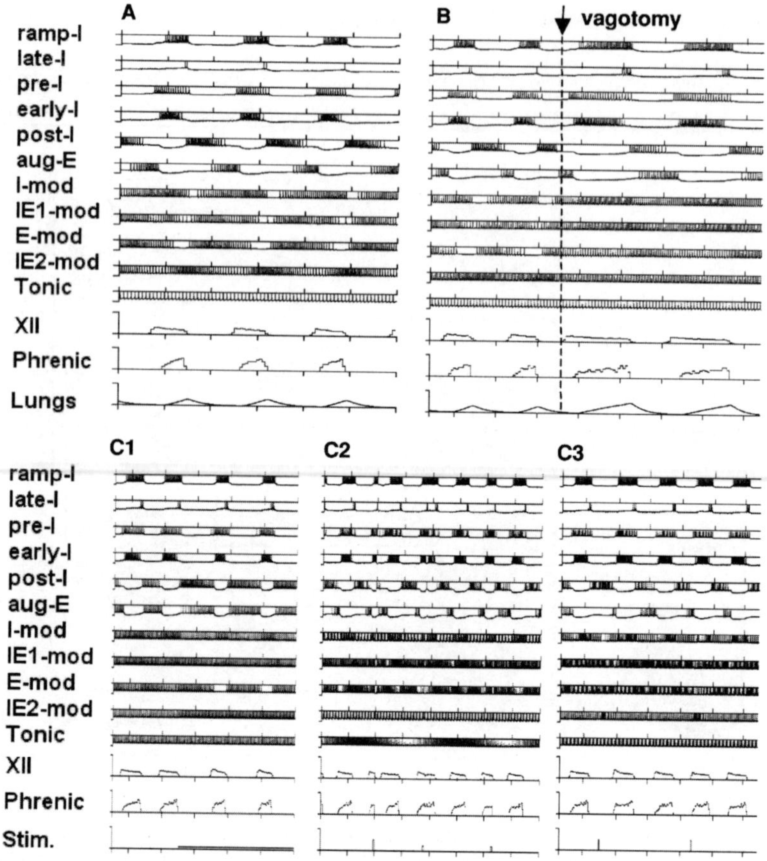

Figure 2. **Model performance.** Activity of each neural population is represented by the trace of the membrane potential of one representative neuron from the population. Tick marks on the time axes correspond to seconds. **A.** The eupnoeic respiratory pattern generated by the model under the normal conditions. **B.** Disconnection of vagal feedback ("vagotomy") produces an increase in amplitude and duration of phrenic discharges. **C1–C3:** The effects of vagus nerve stimulations. **C1.** Continuous vagal stimulation shortens inspiration and prolongs expiration reflecting the HB reflex. **C2.** Simulation of the effects of short vagal stimulation applied during inspiration. The results of simulation demonstrate a decrease of the threshold for inspiratory termination during the inspiratory phase. The first high-amplitude stimulus being applied at the beginning of inspiration terminates current inspiration. The second stimulus has smaller amplitude. Although applied at the same phase, it cannot terminate inspiration. The third stimulus has the same amplitude as the second one, but being delivered at a later time during inspiration terminates this inspiration. **C3.** Short vagal stimulation delivered in the middle of expiration (first stimulus) prolongs expiration. The same stimulus delivered at the end of expiration has no effect on the duration of expiration (demonstrates an "insensitive period" at the end of expiration).

suppress the intrinsic pacemaker-driven oscillations in the pre-BötC by the activation of post-I neurons which in turn inhibit the pre-I population of pre-BötC; (3) rPons provides inspiration-inhibitory and expiration-facilitatory PM reflexes that are independent of HB reflex and partly suppressed by pulmonary feedback; (4) Both HB and PM reflexes operate through the same medullary phase switching circuits.

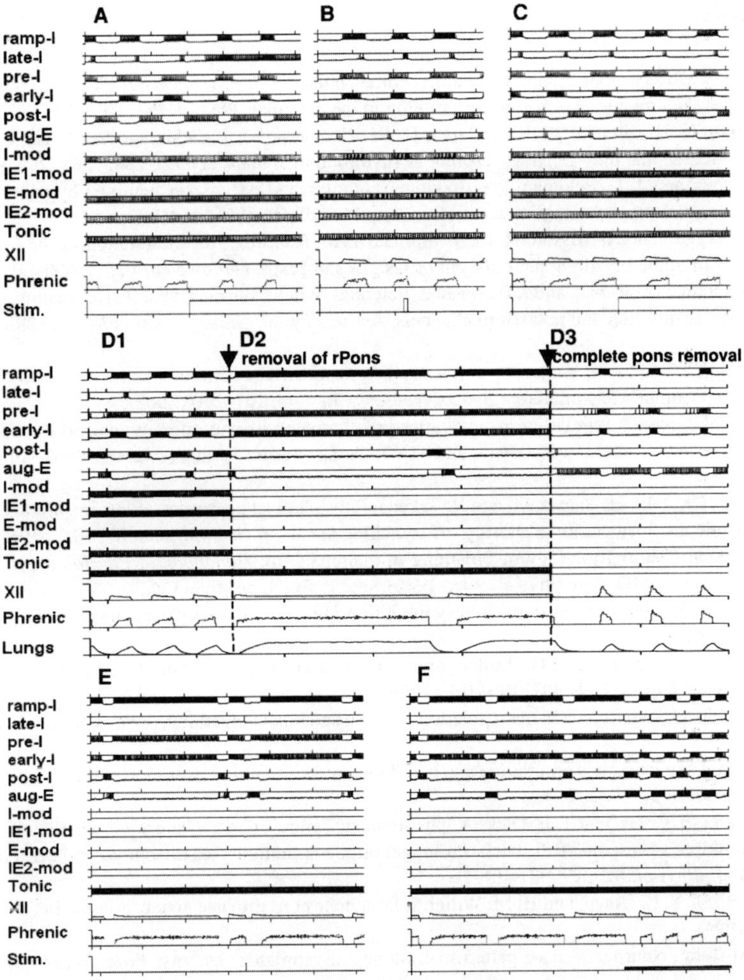

Figure 3. **Model performance: role of the pons** (vagal feedback is cut off). **A.** Continuous stimulation of the pontine IE_I population shortens inspiration and prolongs expiration. **B**. Short stimulation of IE_I population terminates inspiration. **C.** Continuous stimulation of the pontine E(-mod) population prolongs expiration. **D1–D3:** Simulation of pontine lesions. **D1.** The eupnoeic respiratory pattern (the pons is intact). **D2.** Removal of rPons (see Figure 1) converts the eupnoeic pattern to apneusis. **D3.** The following removal of cPons releases an intrinsic pacemaker-driven rhythm in the pre-I population of pre-BötC and converts apneusis to a gasping-like (*in vitro*-like) "decrementing" discharges. **E** and **F**: Apneusis (rPons is removed). **E.** Short vagal stimulation terminates apneustic inspiration. **F.** Continuous vagal stimulation shortens the apneustic inspiratory discharges.

4. Acknowledgements

This study was supported by NSF (0091942) and NIH (NS046062-02 and HL072415-01) grants to I. A. Rybak.

References

1. T. Lumsden, Observations on the respiratory centres in the cat, *J. Physiol. Lond.* **57**, 153–160 (1923).
2. M. I. Cohen, Neurogenesis of respiratory rhythm in the mammal, *Physiol. Rev.* **59**, 1105–1173 (1979).
3. J. C. Smith, H. H. Ellenberger, K. Ballanyi, D. W. Richter, and J. L. Feldman, Pre-Bötzinger complex: A brainstem region that may generate respiratory rhythm in mammals, *Science* **254**, 726–729 (1991).
4. J. C. Rekling and J. L. Feldman, Pre-Bötzinger complex and pacemaker neurons: hypothesized site and kernel for respiratory rhythm generation, *Ann. Rev. Physiol.* **60**, 385–405 (1998).
5. S. P. Lieske, M. Thoby-Brisson, P. Telgkamp, and J. M. Ramirez, Reconfiguration of the neural network controlling multiple breathing patterns: eupnea, sighs and gasps, *Nature Neurosci.* **3**, 600–607 (2000).
6. W. M. St.-John, I. A. Rybak, and J. F. R. Paton, Potential switch from eupnea to fictive gasping after blockade of glycine transmission and potassium channels, *Am. J. Physiol. (Integr. Comp. Physiol.)* **283**, R721–R731 (2002).
7. M. Okazaki, R. Takeda, H. Yamazaki, and A. Haji, Synaptic mechanisms of inspiratory off-switching evoked by pontine pneumotaxic stimulation in cats, *Neurosci. Res.* **44**, 101–110 (2002).
8. D. W. Richter, Neural regulation of respiration: rhythmogenesis and afferent control. In: *Comprehensive Human Physiology*, edited by R. Gregor and U. Windhorst (Berlin: Springer-Verlag, 1996), vol. II, pp. 2079–2095.
9. A. Haji, M. Okazaki, H. Yamazaki, and R. Takeda, Physiological properties of late inspiratory neurons and their possible involvement in inspiratory off-switching in cats, *J. Neurophysiol.* **87**, 1057–1067, (2001).
10. J. L. Feldman, Neurophysiology of breathing in mammals. In: *Handbook of Physiology*, edited by F. E. Bloom (Bethesda, MD: Am. Physiol. Soc., 1986), sec. 1, vol. 4, pp. 463–524.
11. E. E. Lawson, Prolonged central respiratory inhibition following reflex-induced apnea, *J. Appl. Physiol.* **50**, 844–879 (1981).
12. J. E. Remmers, D. W. Richter, D. Ballantyne, C. R. Bainton, and J. P. Klein, Reflex prolongation of stage I of expiration, *Pflügers Arch.* **407**, 190–198 (1986).
13. F. J. Clark and C. von. Euler, On the regulation of depth and rate of breathing, *J. Physiol. Lond.* **222**, 267–295 (1972).
14. C. K. Knox, Characteristics of inflation and deinflation reflexes during expiration in the cat, *J. Neurophysiol.* **36**, 284–295 (1973).
15. J. Jodkowski, S. Coles, and T. E. Dick, A 'pneumotaxic centre' in rats, *Neurosci. Lett*, **172**, 67–72 (1994).
16. J. Jodkowski, S. Coles, and T. E. Dick, Prolongation in expiration evoked from ventrolateral pons of adult rats, *J. Appl. Physiol.* **82**, 377–381 (1997).
17. S. F. Morrison, S. L. Cravo, and H. M. Wilfehrt, Pontine lesions produce apneusis in the rat, *Brain Res.* **652**, 83–86 (1994).
18. W. M. St.-John, Neurogenesis of patterns of automatic ventilatory activity, *Prog. Neurobiol.* **56**, 97–117 (1998).
19. J. H. Peever, J. H. Mateika, and J. Duffin, Respiratory control of hypoglossal motoneurones in the rat, *Pflügers Arch.* **442**, 78–86 (2001).
20. I. A. Rybak, K. Ptak, N. A. Shevtsova, and D.R. McCrimmon, Sodium currents in neurons from the rostroventrolateral medulla of the rat, *J. Neurophysiol.* **90**, 1635–1642 (2003).
21. F. P. Elsen and J. Ramirez, Calcium currents of rhythmic neurons recorded in the isolated respiratory network of neonatal mice, *J. Neurosci.* **18**, 10652–10662 (1998).
22. D. Frermann, B. U. D. Keller, and D. W. Richter, Calcium oscillations in rhythmically active respiratory neurones in the brainstem of the mouse, *J. Physiol. Lond.* **515**, 119–131 (1999).
23. I. A. Rybak, J. F. R. Paton, and J. S. Schwaber, Modeling neural mechanisms for genesis of respiratory rhythm and pattern: I. Models of respiratory neurons, *J. Neurophysiol.* **77**, 1994–2006 (1997).
24. I. A. Rybak, J. F. R. Paton, and J. S. Schwaber, Modeling neural mechanisms for genesis of respiratory rhythm and pattern: II. Network models of the central respiratory pattern generator, *J. Neurophysiol.* **77**, 2007–2026 (1997).
25. I. A. Rybak, N. A. Shevtsova, W. M. St.-John, J. F. R. Paton, and O. Pierrefiche, Endogenous rhythm generation in the pre-Bötzinger complex and ionic currents: Modelling and *in vitro* studies, *Eur. J. Neurosci.* **18**, 239–257 (2003).

5

Respiratory Activity During Sleep and Anesthesia

Ventilatory Instability Induced by Selective Carotid Body Inhibition in the Sleeping Dog

B.J. Chenuel, C.A. Smith, K.S. Henderson and J.A. Dempsey

1. Introduction

There is a considerable evidence that hypocapnia is a major contributor to the genesis of central apnea and periodic breathing (PB) in humans during sleep. Studies using mechanical ventilation to lower the arterial carbon dioxide pressure ($PaCO_2$) have shown that during non-rapid eye movement sleep (NREM), when respiration is under predominantly metabolic control, there is a highly sensitive apneic threshold (AT) induced by reduction in $PaCO_2$ that were only 2 to 4 mm Hg less than eupneic $PaCO_2$.[1–4] If there is a small difference between the AT for CO_2 ($P_{AT}CO_2$) and the eupneic $PaCO_2$ (a narrow "CO_2 reserve"), then a relatively small increase in ventilation (ventilatory overshoot), from whatever cause, could result in apnea. Conversely, if there is a large difference between eupneic $PaCO_2$ and $P_{AT}CO_2$ (a wide "CO_2 reserve") then a larger increment of ventilation is required to produce apnea.

Recently, Nakayama *et al.* have shown that the "CO_2 reserve" is plastic and depends on the background ventilatory drive.[5] More specifically, the "CO_2 reserve" is proportional to the ventilatory drive (*Fig. 1*); the greater the drive the wider the "CO_2 reserve" (*i.e.*, susceptible to apnea). For example, *Figure 1* shows that, for a given metabolic rate, steady-state metabolic acidosis resulted in hyperventilation and increased the "CO_2 reserve" from 5.1 to 6.7 Torr. Thus, an increase in alveolar ventilation of 1.4 l/mn was required to reach the new AT, 6.7 Torr below eupnea. Steady-state metabolic alkalosis resulted in hypoventilation. At this level, an increase in alveolar ventilation of only 0.3 l/min was required to reach the new AT, 3.7 Torr below eupnea. Current theory holds that specifically depressing carotid body (CB) chemoreceptor input and gain would tend to stabilize the ventilatory control system. So, the question we asked was would a CB-specific depressant be different from the apnea-enhancing effects of metabolic alkalosis, which depend on both peripheral and central chemoreceptors.

B.J. Chenuel • Laboratoire de Physiologie, EA 3450, Faculté de Médecine de Nancy, Université Henri Poincaré, France. **C.A. Smith, K.S. Henderson & J.A. Dempsey** • John Rankin Laboratory of Pulmonary Medicine, University of Wisconsin, Madison, U.S.A.

Post-Genomic Perspectives in Modeling and Control of Breathing, edited by Jean Champagnat, Monique Denavit-Saubié, Gilles Fortin, Arthur S. Foutz, Muriel Thoby-Brisson. Kluwer Academic/Plenum Publishers, 2004.

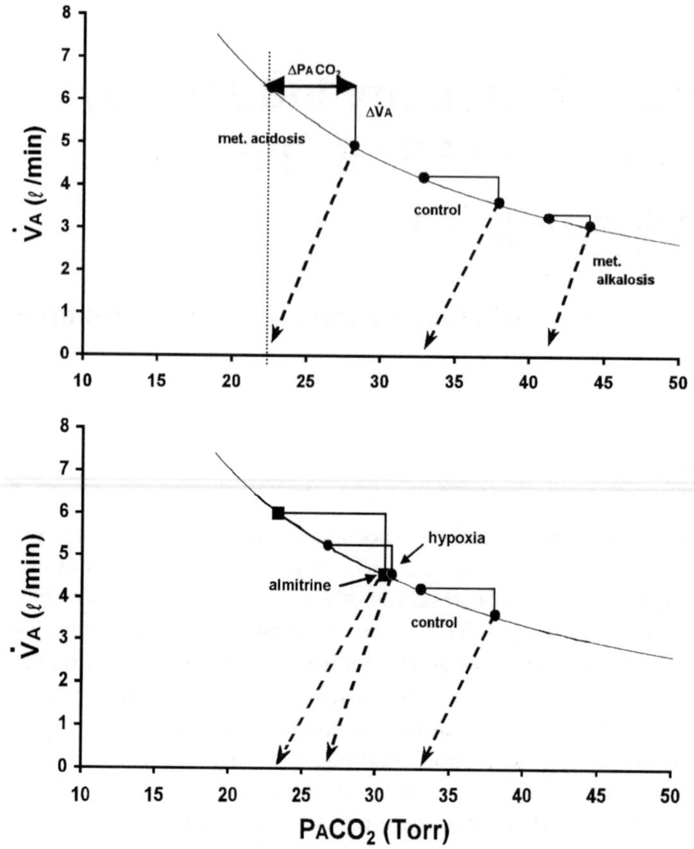

Figure 1. Effects of changes in background ventilatory drive on apnea susceptibility. Two principal determinants of apnea: the CO_2 reserve from spontaneous breathing to the apneic threshold and the change in ventilation above eupnea required to lower $P_{ET}CO_2$ sufficiently to reach the apneic threshold are shown during conditions of control, metabolic acidosis and alkalosis, non-hypoxic peripheral chemoreceptor stimulation via almitrine, and hypoxia in the sleeping dog. (adapted from Nakayama *et al.* with permission[5]). The isometabolic line describing the relationship between $PaCO_2$ and V_A is theoretical and was constructed using an assumed constant VCO_2 of 150 ml/min and the measured $P_{ET}CO_2$ during spontaneous eupneic breathing and at apneic thresholds. The diagonal dashed lines join eupneic and apneic points and their slopes indicate the gain below eupnea of the ventilatory response to CO_2 in each condition. Note the reduced increase in ventilation from spontaneous eupnea required to reach the apneic threshold during metabolic alkalosis whereas the slope of the ventilatory response to CO_2 remained unchanged from control, unlike hypoxia with a clear increased slope.

Hypoxia is the exception to this generalization; despite a marked increase in ventilatory drive and hyperventilation, the "CO_2 reserve" was narrowed and the slope of the CO_2 ventilatory response to CO_2 below eupnea was steepened. Specific pharmacological stimulation of the CB with almitrine resulted in a hyperventilation comparable to that seen with hypoxia, but the "CO_2 reserve" was widened proportional to the increase in ventilation just as it was with metabolic acidosis.

Another paper by Nakayama and colleagues demonstrated that the CBs are required to produce apneas within the time frame that they observed in during naturally occurring sleep apnea.[6] Taken together, these two papers by Nakayama *et al.* are strong evidence in support of a major role for carotid body chemoreceptors in the genesis of apnea/PB during sleep. However, Smith *et al.* using a vascularly isolated and perfused CB model, have shown that severe CB hypocapnia alone, or CB hyperoxia alone, will not significantly prolong TE.[7] However, there is evidence that hyperoxia at least does not completely silence the CBs.[8] These findings raise two questions: 1) if CBs are completely silenced, will apnea result? 2) if CBs are inhibited specifically, will the "CO_2 reserve" narrow, *i.e.*, will the susceptibility to apnea increase?

We addressed these questions by using dopamine, a relatively specific inhibitor of the carotid body ventilatory chemoreflex. We employed a high dose (bolus) of dopamine to completely silence the CB chemoreceptor. We also used a lower dose (infusion) of dopamine to test the effect of specific inhibition of the CB on the "CO_2 reserve". Preliminary findings to date suggest that a) complete silencing of the CB chemoreceptor will cause apnea and, b) specific CB chemoreceptor inhibition will narrow the "CO_2 reserve" and increase susceptibility to apnea.

2. Methods

Studies were performed over several days during non-rapid eye movement sleep on four unanesthetized female mixed-breed dogs (from 20 to 24 kg). They were trained to sleep in an air-conditioned, sound-attenuated chamber and chronically instrumented. This chronically instrumented model is described in detail elsewhere.[6] Briefly, the dogs were prepared with a chronic tracheostomy, a five-lead electroencephalogram montage and an arterial catheter (only two dogs).

Dogs breathed via a cuffed endotracheal tube, inserted into the chronic tracheostomy. Airflow, tracheal pressure, airway PO_2 and PCO_2 and systemic blood pressure were monitored, digitized and stored for subsequent analysis. All ventilatory data were analysed on a breath-by-breath basis.

Hypocapnia was created by pressure support ventilation (PSV). Using a silent balloon valve placed on the inspiratory way, the dog could breathe spontaneously from room air or be switched to PSV by inflation of the balloon. In PSV mode, the dog triggered each breath, once triggered the ventilator delivered a pre-set levels of inspiratory pressure support thereby increasing tidal volume (VT) and decreasing $P_{ET}CO_2$. A wide range of pressures was used over multiple trials on each dog. A threshold value for hypocapnia was determined at the point where apnea and PB were initiated. PB was identified visually by the presence of at least three cycles of hyperpnea and apnea with a consistent periodicity. Further, the apnea lengths had to be at least three standard deviations greater than baseline TE. The apneic threshold was taken to be the $P_{ET}CO_2$ observed in the breath immediately preceding the start of the PB.

Our index of propensity toward breathing instability is the "CO_2 reserve" defined as the difference in $P_{ET}CO_2$ between baseline ventilation (as the mean $P_{ET}CO_2$ during one minute of spontaneous breathing just before the start of PSV) and the apneic threshold.

The preliminary experimental protocol consisted of determination of the AT and the "CO_2 reserve" by means of PSV under control conditions and during low dose dopamine intra-venous infusion.

- Control: PSV was initiated at 4 cm H_2O during NREM sleep, maintained for two minutes and followed by return to spontaneous breathing for at least two minutes. This was repeated with 2 cm H_2O increments until apnea and clear PB occurred.
- Intra-venous infusion (or bolus) of Dopamine: two ranges of doses were used. A low dose of dopamine (from 1.5 to 5 g/kg/min) was intravenously infused to achieve a stable hypoventilation prior to and for the duration of, the PSV trials, in order to determine the "CO_2 reserve". The high dose of dopamine (from 6 to 50 g/kg/min) was performed to assess the ventilatory dose-response, in the absence of pressure support.

3. Results

The ventilatory effects of PSV resulted in a progressive pattern depending on the level of pressure support. At low level of pressure support, T_E was prolonged consistently but only by ~30 % and the V_T of the first spontaneous breath right after PSV was systematically reduced in the same proportion. At higher level of PSV and increased V_T and with further reductions in $P_{ET}CO_2$, apneas ($T_E >$ Mean baseline $T_E + 3*SD$) appeared regularly and periodic, cluster-type breathing was observed. This abrupt transition from small change in breath timing to substantial T_E prolongation and periodic breathing pattern was a consistent feature of achieving the apneic threshold during progressive hypocapnia via PSV.

In intact conditions, an increase in inspiratory minute ventilation via increased V_T from PSV sufficient to decrease the $P_{ET}CO_2$ by about 4 mm Hg (i.e., "CO_2 reserve" = $P_{AT}CO_2$—$P_{ET}CO_{2SPONTANEOUS}$) was required to produce apnea/periodicity.

With slow intra-venous infusion of low dose of dopamine, the minute ventilation during NREM sleep decreased approximately by 1 l/min, leading to an increase in $P_{ET}CO_2$ about 4 mm Hg.

In this background of decreased ventilatory drive and hypoventilation, there was a narrowed "CO_2 reserve" (approximately—3 mm Hg compared to control). The gain of the slope of the ventilatory reduction in response to hypocapnia below hypocapnia was not different from normoxic eupnea.

Using a high dose of dopamine, spontaneous ventilation during NREM sleep was dramatically reduced, leading to an initial apnea, followed by a clear hypoventilation or unstable breathing. The apnea occurs rapidly after the injection of dopamine and its length depends on the dose (T_E from 3 to 16 times control T_E for the highest dose).

4. Discussion

Our preliminary findings suggest two conclusions: 1) Abrupt and complete silencing of the carotid body chemoreceptor afferent output will cause apnea in the sleeping dog. 2) Specific inhibition of the carotid body chemoreceptor will narrow the "CO_2 reserve" in proportion to the decrease in background ventilation.

Our first conclusion implies that if the CB chemoreceptors are silenced rapidly apnea can result. This is unlikely to occur naturally in sleep apnea as a result of a ventilatory overshoot. We think it is more likely that an interaction between CB hypocapnia and lung stretch secondary to the increased VTs during the ventilatory overshoot is required to produce apnea in naturally occurring conditions.[9]

Our second conclusion suggests that specific CB inhibition alone, has no unique effects on ventilatory drive or gain of the CB. Therefore, inhibition of the CBs, or central chemoreceptors, or both in combination appear to produce qualitatively equivalent effects. In other words, dopamine (CB inhibition alone) produced effects on "CO_2 reserve" and CO_2 response gain below eupnea that was identical to that produced by metabolic alkalosis (both CB and central inhibition) at a comparable level of hypoventilation.

5. Acknowledgements

These studies were supported in part by NIH/NHLBI.

References

1. Skatrud JB, Dempsey JA. Interaction of sleep state and chemical stimuli in sustaining rhythmic ventilation. *J Appl Physiol* 1983; 55(3): 813–822.
2. Henke KG, Arias A, Skatrud JB, Dempsey JA. Inhibition of inspiratory muscle activity during sleep. Chemical and non chemical influences. *Am Rev Respir Dis* 1988; 138(1): 8–15.
3. Datta AK, Shea SA, Homer RL, Guz A. The influence of induced hypocapnia and sleep on the endogenous respiratory rhythm in humans. *J Physiol* 1991; 440: 17–33.
4. Meza S, Mendez M, Ostrowski M, Younes M. Susceptibility to periodic breathing with assisted ventilation during sleep in normal subjects. *J Appl Physiol* 1998; 85(5): 1929–1940.
5. Nakayama H, Smith CA, Rodman JR, Skatrud JB, Dempsey JA. Effect of ventilatory drive on carbon dioxide sensitivity below eupnea. *Am J Respir Crit Care Med* 2002; 165(9): 1251–1259.
6. Nakayama H, Smith CA, Rodman JR, Skatrud JB, Dempsey JA. Carotid body denervation eliminates apnea in response to transient hypocapnia. *J Appl Physiol* 2003; 94(1): 155–164.
7. Smith CA, Saupe KW, Henderson KS, Dempsey JA. Ventilatory effects of specific carotid body hypocapnia in dogs during wakefulness and sleep. *J Appl Physiol* 1995; 79(3): 689–699.
8. Lahiri S, Delaney RG. Relationship between carotid chemoreceptor activity and ventilation in the cat. *Respir Physiol* 1975; 24: 267–286.
9. Bajic J, Zuperku EJ, Tonkovic-Capin M, Hopp FA. Interaction between chemoreceptor and stretch receptor inputs at medullary respiratory neurons. *Am J Physiol* 1994; 266: R1951–1961.

Stability Analysis of the Respiratory Control System During Sleep

Zbigniew L. Topor, Konstantinon Vasilakos, and John E. Remmers

1. Introduction

Chemoreflex control of breathing during sleep consists of two negative feedback loops with delays depending on the controlled variables (Pa_{CO_2} and Pa_{O_2}). The peripheral chemoreceptor (PCR) loop operates with short delay and the central chemoreceptor (CCR) loop operates with long delay (see Figure 1). Periodic breathing, indicative of instability in the control system, is commonly seen during sleep in premature infants, in patients with heart failure, and during exposure to high altitude. These periodicities appear to relate to the operation of the two chemoreflex loops.

Precisely how the gains and delays of the two loops interact to determine overall stability of the system is uncertain, however.

The mathematical analysis of the control system having one negative feedback loop indicates that the oscillatory behavior develops when feedback gain and/or delay exceed critical levels [7]. Similar analysis of systems having multiple feedback loops, each having state-dependent delay has proven more challenging with positive results restricted to few special cases.

We employed our realistic model of the respiratory system and its chemoreflex control during sleep [17, 18] to develop a descriptive graphical method for stability analysis. The multi-compartmental model accurately describes the interaction of circulating blood with CO_2 and O_2, and separately simulates reflexes initiated by the PCR and CCR loop using state dependent delays. The results from a series of computer simulations reveal a complex interaction of the two chemoreflexes resulting in stable or unstable behavior of the system. These results are displayed on a two dimensional plot similar in concept to the phase plane with the chemosensitivities of the two loops serving as coordinates of each point. A region of stability exists with the normal operating point for the system lying well inside its boundaries. Changes to the sensitivities of either loop caused by known pathologies displace the operating point toward the border of the stability region and the distance to the

---

Zbigniew L. Topor, Konstantinon Vasilakos, and John E. Remmers • Department of Physiology, University of Calgary, Calgary, AB, T2N 4N1, Canada.

Post-Genomic Perspectives in Modeling and Control of Breathing, edited by Jean Champagnat, Monique Denavit-Saubié, Gilles Fortin, Arthur S. Foutz, Muriel Thoby-Brisson. Kluwer Academic/Plenum Publishers, 2004.

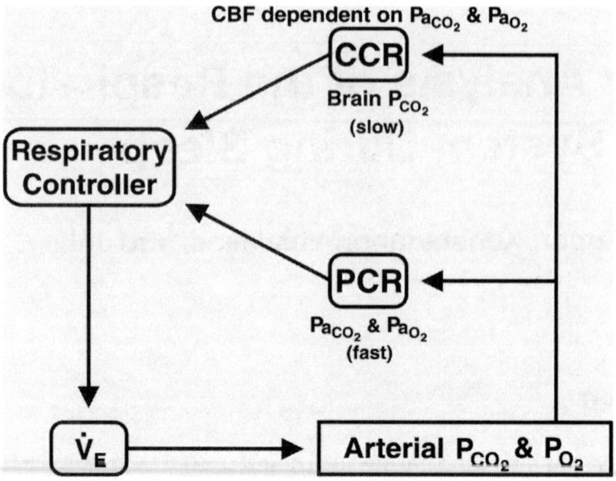

Figure 1. Two negative feedback loops responsible for chemoreflex control of breathing during sleep.

border depends upon the whether one or both gains are increased. The new position of the operating point determines if the system remains stable or develops sustained oscillations.

Our results indicate that stability of the respiratory control system depends critically on the relative gains and dynamic characteristics of the two chemoreflex loops.

2. Methods

2.1. General Structure of the Model

Our model, an extension of that described by Grodins *et al.* [5], is structured as a controller and a regulated system consisting of lungs, brain, and metabolizing tissue. The controller, by modulating the level of ventilation, determines the values of state variables (Pa_{O_2}, Pa_{CO_2}). The lungs have an alveolar compartment ventilated by a continuous unidirectional flow of respiratory gas and the alveolar gas partial pressures equal those in pulmonary venous blood for CO_2 and nitrogen but differ for O_2 by an alveolar-arterial P_{O_2} difference. Blood—respiratory gas interaction is described by a set of seven non-linear equations [11]. Arterial blood leaves the lungs and arrives at the brain and tissue compartments after circulatory transport delays which depend upon vascular volumes and blood flow rates. Variable circulatory delays are implemented as two delay lines that express the arterial blood gas values at the entrance to the brain and tissue compartments as equal to corresponding values in the arterial blood leaving the lungs at an appropriate time in the past. The gas tensions in each of these compartments are assumed to be uniform and equal to those in exiting venous blood. Three separate delay lines are implemented on the venous side of the circulation, two for circulatory delays of the venous blood leaving the brain and tissue compartments, and a third for the transport of mixed venous blood to the lungs. The plant equations comprise a set of 12 first-order time-delayed differential equations, with delays being a function of state variables, that describe the material balance relations for CO_2, O_2, and N_2 in each compartment. Cardiac output and cerebral blood flow are expressed as functions of arterial

CO_2 and oxygen tensions and incorporate steady state values reported recently [4, 8, 12, 15] with "on" and "off" dynamics described by Poulin *et al.* [13] and Poulin and Robbins [14].

The controller receives input from central and peripheral chemoreceptors and produces an output to respiratory muscles. The CCR transduces brain tissue P_{CO_2} (PB_{CO_2}). The PCR senses carotid body P_{CO_2} and P_{O_2}. The sum of these two inputs determines pulmonary ventilation based on steady state data describing ventilatory responses to P_{CO_2} as a family of straight lines with hypoxia-dependent slopes [3, 9, 10] for non-rapid eye movement sleep [16]. We determined the relationship between brain tissue P_{CO_2} and arterial P_{CO_2} at different levels of arterial P_{O_2} from the computational model of the plant and derived two controller equations. The first equation describes CCR component of ventilation as a function of PB_{CO_2}, and the second describes the contribution from PCR chemoreflex as a function of Pa_{CO_2} and Pa_{O_2}. The controller equations contain no dynamic terms so that all delays in the controller's response to an abrupt chemical stimulation derive from the characteristics of the plant. Thus, cardiac output and cerebral blood flow entirely determine the dynamics of each feedback loop.

2.2. Computer Simulations

The model was used to simulate experimental trials lasting 90 minutes and values of 70 critical variables describing the behavior of the system under particular simulated conditions were stored for later analysis. In addition, the progress of the simulation was observed on the computer monitor where four pre-selected variables were displayed in the real time mode. The model was embedded in an external routine, which step-wise changed the sensitivities of CCR and PCR loops from 0.5 to 5.0 times their normal values in increments of 0.1. The routine executed standard 90 min simulations for each pair of CCR and PCR values and an augmented breath was applied after 40 min. The data set from each 90 minutes simulation was analyzed off-line to evaluate stability and to determine the nature of the response to the transient perturbation.

3. Results

Each simulation was analyzed and Figure 2 illustrates the resulting stability display on a chemoreceptor sensitivity plane. The hatched area indicates region of stability. Outside of this region the behavior of the system is characterized by self-sustained bound oscillations. Note that a substantial increase in the overall loop gain of the system accomplished by proportional, 3.8 fold increase in both CCR and PCR chemosensitivities yields a stable system. Paradoxically, a smaller but disproportional increase in either chemoreflex loop (2.6 fold for CCR and 3.7 fold for PCR) leads to the unstable system. Figure 3 illustrates the position of the normal operating point of the system (point A with [1.0, 1.0] coordinates). A proportional increase in chemosensitivities of both loops translocates the operating point to B [3.75, 3.75], and a spontaneous deep breath results in a decaying transient oscillation, indicating that the system is stable (Panel B). Translocation to point C [1.0, 3.75], caused by a significant reduction in the PCR chemosensitivity, moves the operating point outside the boundary of the stability region and results in a self-sustained, bound oscillation consisting of alternating apnea and hyperpnea with a period length of 57 seconds (Panel C). The adjacent point C*[1.0, 3.0] reflects possible change in CCR chemosensitivity associated

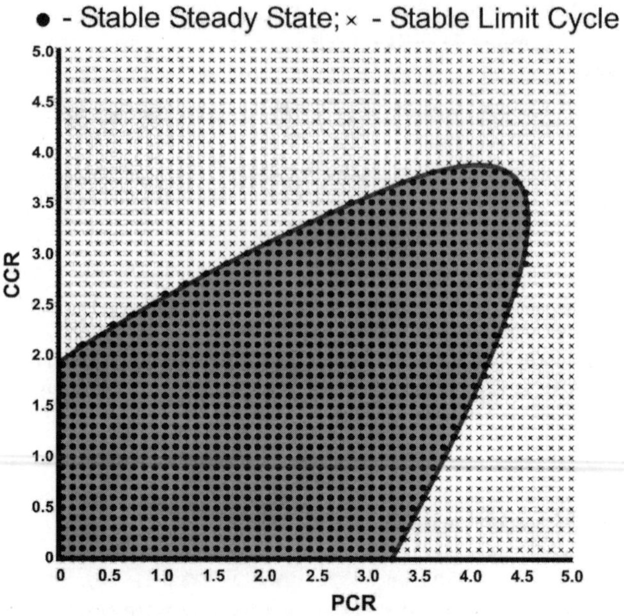

Figure 2. The region of stability (shaded) displayed on the chemoreceptor sensitivity plane.

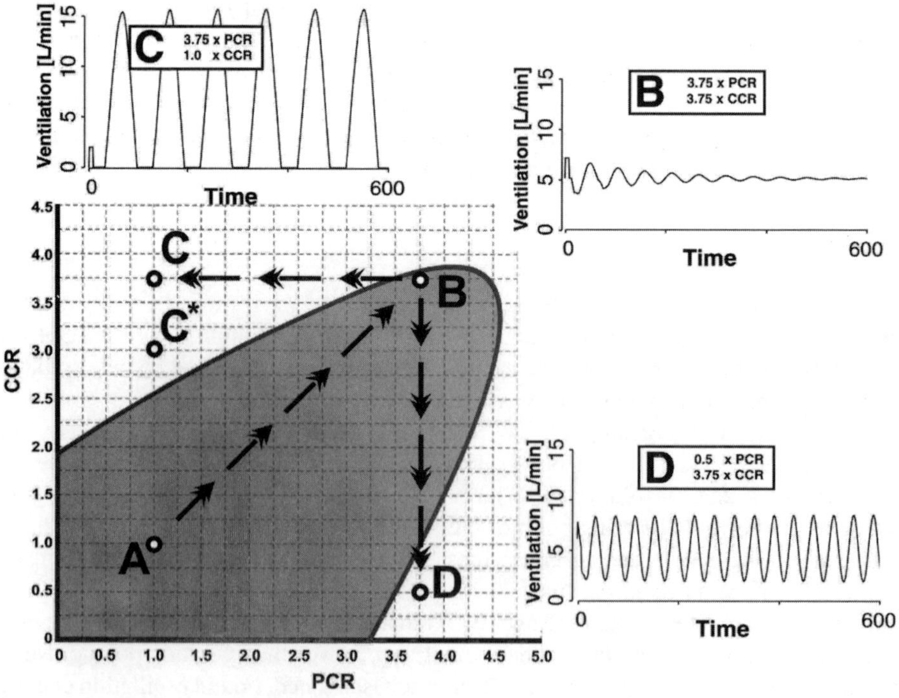

Figure 3. Translocation from stable locus (point A) to different loci on the chemoreceptor sensitivity plane (points B, C, C*, D) and behavior of the system in their vicinity.

with a congestive heart failure [19] and the behavior of the system here is similar to that depicted in panel C.

Translocation from point B to D [3.75, 0.5], associated with a substantial decrease in the chemosensitivity of CCR loop, also leads to the unstable behavior of the system (see Panel D). These oscillations are characterized by periods of hypopnea and hyperpnea and by an overall period length of 21 seconds. This behavior is reminiscent of a normal sleeping subject at high altitude.

4. Discussion

Our plot of component chemoresponsiveness in the two delayed feedback loop system of respiratory control during sleep reveals a stability field, similar to that described by Belair [1] and by Hale and Huang [6]. Using relative chemosensitivities of PCR and CCR loops as two variables defining our phase plane has particular utility since each determines component loop gain and each can be estimated experimentally in humans. Further, the graphical display allows visualization of a location of the operating point of the system in relation to the boundary of the stability region. Movement of this point may indicate a change in the system's stability. Changes in the size and shape of the stability field as a function of changes in system parameters, such as cerebral blood flow, cardiac output and/or their chemoresponsiveness, allow for evaluation of the role of these parameters in system stability.

Our results can provide plausible explanations for experimental observations and can suggest further experiments that may be informative. An example of the first use of our model is shown in Figure 3 beginning with stable locus (point B) with relatively high sensitivity for both chemoreceptors. The sensitivity of the PCR is held constant and the CCR sensitivity is progressively reduced until the unstable locus (point D) is reached. This provides a plausible explanation for the observation of Cherniack *et al.* [2] in which excluding the central chemoreflex from a high gain system led to ventilatory instability. An example of the second use of our model is illustrated by the translocation from B to locus C at constant CCR chemosensitivity. In this case, the system is predicted to become unstable by reduction of PCR chemosensitivity. To our knowledge, such has never been observed but could be revealed in an experiment such as Cherniack *et al.* [2] where the PCR could be excluded from a high gain system. Interestingly, the behavior of the system around locus C resembles sleep-disordered breathing in patients with congestive heart failure (Figure 3C) [19]. Their ventilation is characterized by periods of apnea and hyperpnea with a total length of a cycle between 50–60 seconds. Thus, possible 3-fold increase in chemosensitivity of both loops will move the operating point of the system well inside the stability field. Investigation into possible influence of cardiac output decrease on the size and shape of the stability field may give us some indication if proportional 3-fold increase in chemosensitivities of both loops could lead to ventilatory instability during sleep in congestive heart failure patients. Instability of the system around locus D (Figure 3D), which is characterized by much higher frequency of oscillations resembles closely the periodicity induced by an exposure to high altitude or periodic breathing in premature infants.

One potential limitation of our analysis is the fact that the model assumes ventilation as continuous unidirectional flow of air without considering its sub-components such as tidal volume (the depth of a breath) and respiratory timings (inspiratory time, expiratory time, and frequency of breathing). Consideration of these characteristics of ventilation may prove useful in light of the recent findings that the plasticity of the PCR chemoreflex has different effects on the various respiratory components [20]. On the other hand, our model simulates behavior of the human respiratory control system during non-REM sleep. In this condition spontaneous variations in the level of ventilation and possible changes in its sub-components values are minimized.

While generalization of our results may be limited by the specific nature of our computer simulation and the discrete analysis we employed, they have led us to novel insights regarding the stability of the respiratory control system. Specifically, we provide evidence that the relative gains and delays of the two loops play a key role in stability and that regions of stability can be usefully depicted on a chemoreceptor sensitivity plane.

References

1. Belair J., Stability of a differential-delay equation with two time lags, in: Proceedings of the Canadian Mathematical Society: 1986 Seminar on oscillation, bifurcation, and chaos, Langford W. and A. Mingarelli, ed., AMS, 1987.
2. Cherniack N. S., C. von Euler, I. Homma and F. F. Kao, Experimentally induced Cheyne-Stokes breathing, *Respir. Physiol.* **37** (1979), 185–200.
3. Cunningham D. J. C., Integrative aspects of the regulation of breathing: a personal view, Widdicombe J. G. Editor, "*Respiratory physiology*", UK: Butterworths, London, p 303–369, 1974.
4. Fortune J. B., D. Bock, A. M. Kupinski, H. H. Stratton, D. M. Shah and P. J. Feustel, Human cerebrovascular response to oxygen and carbon dioxide as determined by internal carotid artery duplex scanning, *The J. of Trauma* **32(5)** (1992), 618–628.
5. Grodins F. S., J. Buell and A. J. Bart, Mathematical analysis and digital simulation of the respiratory control system, *J. Appl. Physiol.* **22(2)** (1967), 260–276.
6. Hale J. K. and W. Huang, Global geometry of the stable regions for two delay differential equations, *J. Math. Anal. Appl.* **178** (1993), 344–362.
7. Khoo M. C. K., R. E. Kronauer, K. P. Strohl and A. S. Slutsky, Factors inducing periodic breathing in humans: a general model. *J. Appl. Physiol.* **53(3):** (1982), 644–659.
8. Kiely D. G., R. I. Cargill and B. J. Lipworth, Effects of hypercapnia on hemodynamic, inotropic, lusitropic, and electrophysiologic indices in humans, *Chest* **109(5)** (1996), 1215–1221.
9. Lloyd B. B., M. G. M. Jukes and D. J. C. Cunningham, The interactions between hypoxia and other ventilatory stimuli, *Quart. J. Exp. Physiol.* **43** (1958), 214–221.
10. Nielsen M. and H. Smith, Studies on the regulation of respiration in acute hypoxia, *Acta Physiol. Scand.* **24** (1952), 293–313.
11. Olszowka A. J. and L. E. Farhi, A system of digital computer subroutines for blood gas calculations, *Respir. Physiol.* **4** (1968), 270–280.
12. Phillips B. A., J. W. McConelly and M. D. Smith, The effects of hypoxemia on cardiac output. A dose-response curve, *Chest* **93(3)** (1988), 471–475.
13. Poulin M. J., P. J. Liang and P. A. Robbins, Dynamics of the cerebral blood flow response to step changes in end-tidal P_{CO_2} and P_{O_2} in humans, *J. Appl. Physiol.* **81(3)** (1996), 1084–95.
14. Poulin M. J. and P. A. Robbins, Indices of flow and cross-sectional area of the middle cerebral artery using Doppler ultrasound during hypoxia and hypercapnia in humans, *Stroke* **27** (1996), 2244–2250.
15. Serebrovskaya T. V., Comparison of respiratory and circulatory human responses to progressive hypoxia and hypercapnia, *Respiration* **59** (1992), 35–41.
16. Severinghaus J. W., Proposed standard determination of ventilatory responses to hypoxia and hypercapnia in man, *Chest* **70(1)** (1976), 129–131.

17. Topor Z. L., Investigation of the human respiratory control system by computer modeling and system identification techniques, *Ph.D. Thesis* University of Calgary, 1999.
18. Topor Z. L., M. Pawlicki and J. E. Remmers, A computational model of the human respiratory control system: responses to hypoxia and hypercapnia, *Ann. Biomed. Eng.* (in review).
19. Topor Z. L., L. Johannson, J. Kasprzyk and J. E. Remmers, Dynamic ventilatory response to CO_2 in congestive heart failure patients with and without central sleep apnea, *J. Appl. Physiol.* **91** (2001), 408–416.
20. Young D. L., F. L. Eldridge and C. S. Poon, Integration-differentiation and gating of carotid afferent traffic that shapes the respiratory pattern, *J. Appl. Physiol.* **94** (2002), 1213–1229.

A Physical Model of Inspiratory Flow Limitation in Awake Healthy Subjects

Abdelkebir Sabil, André Eberhard, Pierre Baconnier, Gila Benchetrit

1. Introduction

1.1. The Sleep Apnea Syndrome

The obstructive syndrome of sleep apnea/hypopnea is characterized by repeated partial or total obstructions of the upper airways (UA) [1] resulting in intermittent asphyxia and repeated micro-arousals. Sleep disordered breathing affects about 9% of males and 4.5% of females between the ages of 30 and 60 years [2], 15% to 20% of elderly people [3] and 5% to 10% of children [4]. Sleep apnea and hypopnea events are easily identified while more subtle events, such as inspiratory flow limitation episodes, are difficult to detect. Flow limitation is the result of a partial occlusion of the upper airway during sleep.

1.2. The Upper Airway

The upper airways (figure 1), located outside the thorax, are susceptible to collapse during inspiration in sleep. This behaviour does not concern the rigid segments of the nose and larynx. However, the pharynx, which is a soft muscular tube, is susceptible to collapse, for instance during swallowing or speech. Contraction of inspiratory muscles, particularly the diaphragm, creates a negative pressure in the upper airway that leads to the dynamic narrowing of the collapsible pharyngeal tube. To counteract the collapse of the pharynx, a protection mechanism (upper airway dilator reflex) is triggered prior to inspiration [5]. In fact, the application of a negative pressure to the upper airways will provoke a reflex contraction of the pharynx dilating muscles (genioglossus, geniohyoïd, styloglossus, masseters, pterygdoïdes) [6, 7]. This mechanism contributes to the stability of the upper airway and allows airflow to circulate through a nearly

Abdelkebir Sabil, André Eberhard, Pierre Baconnier, Gila Benchetrit • Laboratoire TIMC/IMAG (UMR CNRS 5525), Université Joseph Fourier (UJF), Faculté de Médecine de Grenoble, 38706 La Tronche Cedex, France.

Post-Genomic Perspectives in Modeling and Control of Breathing, edited by Jean Champagnat, Monique Denavit-Saubié, Gilles Fortin, Arthur S. Foutz, Muriel Thoby-Brisson. Kluwer Academic/Plenum Publishers, 2004.

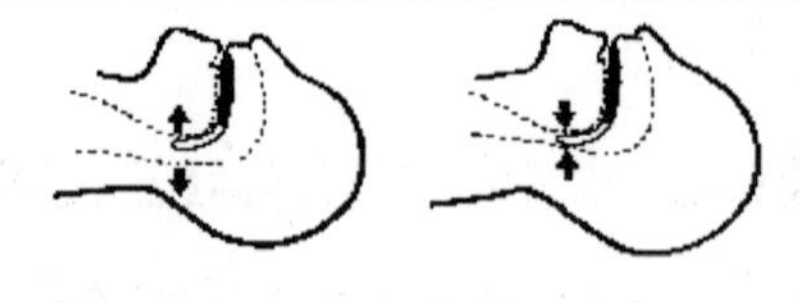

Figure 1. Collapsing of upper airway: open airway and normal airflow (a) and partially collapsed airway and limited airflow (b).

constant diameter during normal breathing. However, this protection is weakened during sleep and the diameter of the pharynx diminishes leading to a decrease in ventilation [6, 8]. This modification of the upper airway calibre during sleep does not have a substantial effect in normal subjects but becomes a problem in patients with a narrow pharynx.

1.3. Inspiratory Flow Limitation

Inspiratory flow limitation (IFL) is defined by an increase in negative intra-thoracic pressure that is not associated with an increase in airflow rate. IFL is characterised by the presence of a plateau in the flow-pressure relationship graph [9]. When this scenario lasts more than ten seconds or is present in more than two consecutive cycles, inspiratory flow is said to be limited [8]. Inspiratory flow limitation has been recently identified as a serious problem responsible for repeated micro-arousals [9]. To determine whether or not an inspiratory flow limitation has occurred, it is necessary to simultaneously measure inspiratory flow and the oesophageal pressure [10]. The pressure measurement allows a precise evaluation of the flow limitation; however, this method is invasive and difficult to perform as a routine measurement. Non invasive methods, such as measurement of pressure with a nasal canula or respiratory inductance plethysmography [9, 10, 11], have been proposed for the evaluation of inspiratory flow limitation. The shape of the inspiratory flow, obtained from a nasal canula signal or the derivative of inductance plethysmography signal, was analysed and proposed for the classification and identification of flow limitation events. Pollo et al. classified flow shapes into seven categories from a normal shape to the most limited shape [12]. Dempsey et al. classified flow shapes into four different categories [10] while Bloch et al. suggested a two category classification, non limited and limited flow shape [11]. This latter classification gave the best result in detection of flow limitation, with an 80% specifity and sensitivity [11]. This classification of the inspiratory flow based on the shape of the signal may not be generalized for all subjects. In fact, respiratory flow shapes vary between individuals [13] and an abnormal shape for an individual could be considered as a normal shape for another individual. It is therefore more relevant to evaluate the changes in shape for one individual than to rely on generalized preset shapes of inspiratory flow.

The detection of IFL, in the absence of oesophageal pressure, is based on changes in the flow shape. To investigate these changes, we designed a physical device to mimic upper airway collapsibility resulting in IFL.

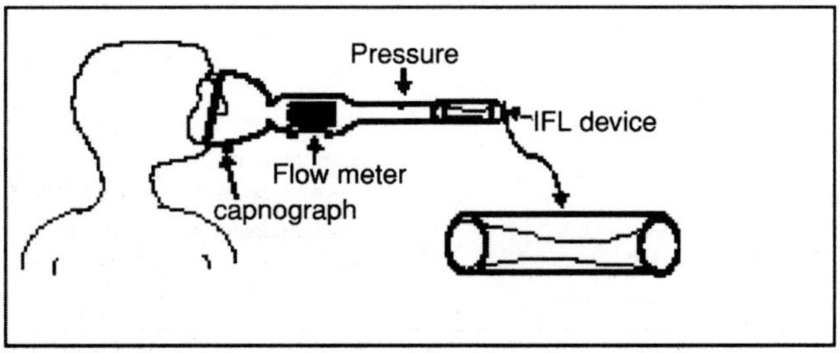

Figure 2. Apparatus for mimicking upper airway collapsibility.

2. Methods

2.1. Description of the Device

The purpose of the device is to allow a simple, non invasive and instantaneous simulation of inspiratory flow limitation in healthy awake subject. Based on the Starling resistance principle, we designed a device made of a plastic tube in which a flexible membrane is inserted and collapses if an aspiration is applied. When inserted in a breathing apparatus recording, the device mimics upper airway collapsibility and induces IFL. Lengthening the collapsible section of the device induces increasing IFL.

2.2. Experimental Protocol

Recordings were performed on healthy awake seated subjects (8 subjects, 5 males, age range 22 to 42). The subjects wore a facemask on which a pneumotachograph was mounted. The inspiratory flow limiting device was connected to the pneumotachograph and the pressure was measured at midpoint between the pneumotachograph and the device. The CO_2 level in the mask was also recorded during the experiment. Control breaths were recorded without the device in place, then increasing flow limitation was induced using the device. The limiting device was removed after 10 to 15 breaths and the subject returned to normal breathing.

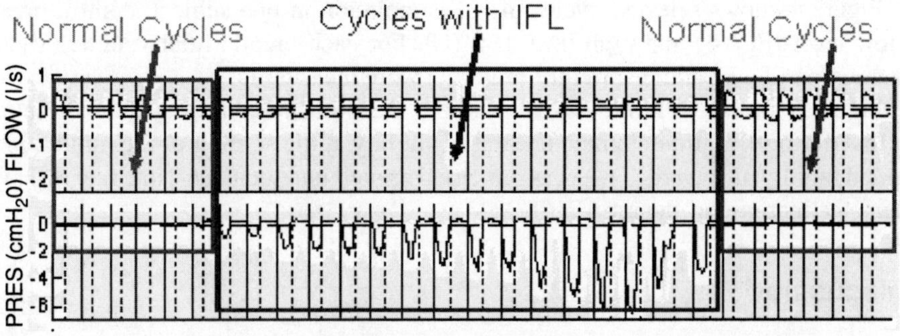

Figure 3. An example of recording before, during and after inducing an inspiratory flow limitation.

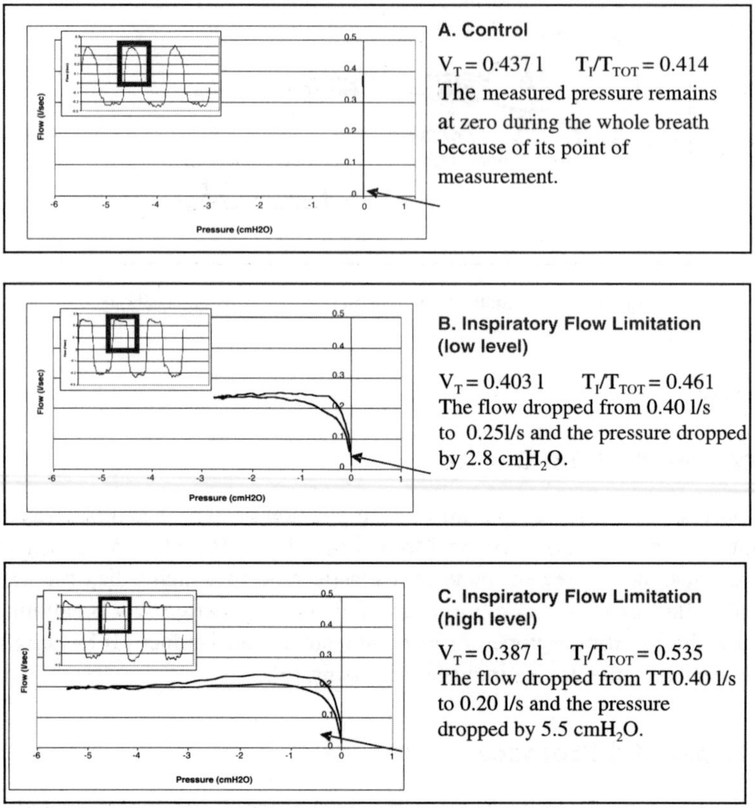

Figure 4. The windows show three respiratory flow cycles for a healthy awake subject. The **pressure/flow** curves of the middle cycle during inspiration are represented as well. A represents control cycles without the use of the device; we note that the pressure remained at zero during the entire breath because of its point of measurement. A low level IFL is seen in B, the pressure became progressively more negative while the inspiratory flow remained constant. A higher level IFL is seen in C, the pressure drop was more important than it was in B and the inspiratory flow remained constant.

3. Result

Figure 4 shows selected cycles of IFL simulation in one subject: control breathing (A), low level IFL (B) and high level IFL (C). For each breath, tidal volume (V_T), total cycle duration (T_{TOT}) and inspiratory time (T_I) were measured.

Figure 5 shows 14 limited cycles and the mean of the control cycles. These results show that during IFL, tidal volume decreased slightly at the starting of the limitation but it increased rapidly afterward. T_{TOT} was increased as well but not at the same rate as T_I which explains the increase of the ratio T_I/T_{TOT}.

4. Discussion

Our findings show that this simple device could be used to mimic upper airway collapsibility and inspiratory flow limitation. Indeed, with the device in place, pressure became

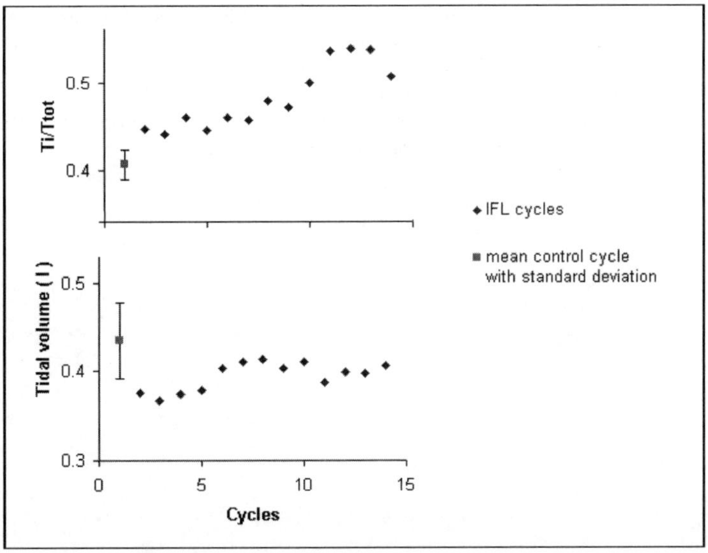

Figure 5. A comparison between the limited cycles and the mean of the control cycles. The ratio of T_I/T_{TOT} increased with the respiratory effort. The tidal volume decreased slightly when IFL was induced.

progressively more negative while inspiratory flow remained constant and respiratory parameters changed: V_T decreased slightly while T_{TOT} and T_I/T_{TOT} increased.

In conclusion, the device that we designed represents an appropriate tool to induce IFL and therefore characterize flow shape changes due to IFL. In addition, mimicking IFL in healthy awake subjects provides the means to induce controlled IFL and determine the relationship between the intensity of IFL and flow shape characteristics.

References

1. Farré R., Peslin R., Rotger M., and Navajas D. Inspiratory Dynamic Obstruction Detected by Forced Oscillation during CPAP, A Model Study. Am. J. Respi. Crit. Care Med., 1997, 155: 952–956.
2. Young T., Palta M., Dempsey J., Skatrud J., *et al.* The Occurrence of Sleep Disordered Breathing Among Middle-aged Adults. New Engl. J. Med., 1993, 328: 1230–1235.
3. Ancoli-Israel S. Epidemiology of Sleep Disorders. Clin. Geriat. Med., 1989, 5: 347–362.
4. Guilleminault C. and Pelayo R. Sleep-disordered Breathing in Children. Ann. Med., 1998, 30: 350–6.
5. Horner R. L., Innes J. A., Murphy K., Guz A. Evidence for Reflex Upper Airway Dilator Muscle Activation by Sudden Negative Airway Pressure in Man. J Physiol (London), 1991, 436: 15–29.
6. Remmers J. E., deGroot W. J., Sauerland E. K., and Anch A. M. Pathogenesis of Upper Airway Occlusion During Sleep. *J Appl Physiol*, 1978, 44: 931–938.
7. Tantucci C., Mehiri S., Duguet A., Similowski T., Arnulf I., Zelter M., Derenne J.P., Milic-Emili J. Application of Negative Expiratory Pressure During Expiration and Activity of Genioglossus in Humans. *J Appl Physiol*, 1998, 84: 1076–82.
8. Tangel D. J., Mezzanotte M. S., White D. P. Influence of Sleep on Tensor Palatini EMG and Upper Airway Resistance in Normal Men. *J Appl Physiol*, 1991, 70: 2574–81.
9. Hosselet J. J., Norman R. G., Ayappa I., and Papoport D. Detection of Flow Limitation with Nasal Cannula/Pressure Transducer System. Am J Respir Crit Care Med, 1998, 157: 1461–1467.
10. Clark S. A., Wilson C. R., Satoh M., Pegelow D., Dempsey J. A. Assessment of Inspiratory Flow Limitation Invasively and Noninvasively During Sleep. Am J Respir Crit Care Med. 1998, 158: 713–22.

11. Kaplan V., Zhang J. N., Russi E. W., Bloch K. E. Detection of inspiratory flow limitation during sleep by computer assisted respiratory inductive plethysmography. Eur Respir J, 2000, 15: 570–578.
12. Aittokallio T., Saaresranta T., Polo-Kantola P., Nevalainen O., and Pollo O. Analysis of Inspiratory Flow Shapes in Patients with Partial Upper-airway Obstruction During Sleep. Chest 2001; 119: 37–44.
13. Benchetrit G., Shea S. A., Pham Dinh T., Bodcco S., Baconnier P., Guz, A. Individuality of Breathing Patterns in Adults Assessed Over the Time. Respir. Physio, 1989, 75: 199–210.

Antioxidants Prevent Blunting of Hypoxic Ventilatory Response by Low-Dose Halothane

Albert Dahan, Raymonda Romberg, Elise Sarton and Luc Teppema

1. The Ventilatory Response to Acute Isocapnic Hypoxia

In adult humans, acute isocapnic hypoxia induces a brisk ventilatory response. This acute hypoxic response or AHR originates at the peripheral chemoreceptors of the carotid bodies.[1,2] The full mechanisms of oxygen sensing at the carotid bodies (CB) is still poorly understood. At present it is thought that membrane ion channels (e.g. potassium channels) are critically involved and that low oxygen inhibits various K^+-currents through the CB type I cell membrane (see reference #3 and references cited therein). This causes membrane depolarization and consequently the influx of calcium ions into the cell and the activation of a complex cascade of events within the type I cell. At the end of this cascade, the cell releases neurotransmitters (*e.g.*, acetylcholine and ATP) which activate postsynaptic receptors located on afferent endings of the carotid sinus nerve.

2. Influence of Inhalational Anesthetics on the Ventilatory Response to Acute Isocapnic Hypoxia

Volatile halogenated anesthetic agents are potent ventilatory depressants and at anesthetic concentrations affect ventilatory control at various sites within the body. For example, halothane causes ventilatory depression by abolishing the peripheral drive of the carotid bodies, by changing the balance between excitatory and inhibitory neuromodulators within the central nervous system toward a net inhibitory effect on the respiratory neuronal pool and by suppression of the respiratory muscles. At subanesthetic concentrations (0.05 to 0.2 minimum alveolar concentration [MAC]), the volatile anesthetics halothane, enflurane, isoflurane and sevoflurane depress the ventilatory response to acute (that is, 3 to 5 min) isocapnic hypoxia by 20 to 60%.[3–11] There is strong evidence that this effect is due to

---

Albert Dahan, Raymonda Romberg, Elise Sarton and Luc Teppema • Department of Anesthesiology, Leiden University Medical Center, 2300 RC Leiden, The Netherlands.

Post-Genomic Perspectives in Modeling and Control of Breathing, edited by Jean Champagnat, Monique Denavit-Saubié, Gilles Fortin, Arthur S. Foutz, Muriel Thoby-Brisson. Kluwer Academic/Plenum Publishers, 2004.

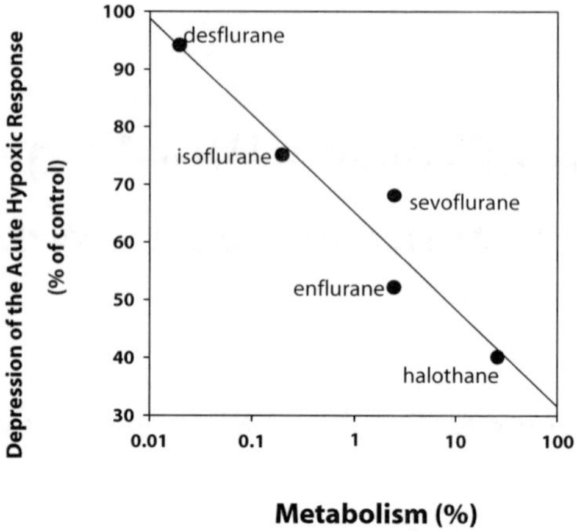

Figure 1. Relationship between depression of the acute hypoxic response at 0.1 MAC and the relative metabolism of the halogenated anesthetic agents.

a selective, potent and preferential effect at the peripheral chemoreceptors of the carotid bodies. For example, *i*) hypoxia-driven ventilation decreases by 25% within 30 s and by 40% within 1 min of exposure to 0.15–0.30% inspired halothane.[4] Within this time period brain concentrations of halothane cannot contribute to this effect taken the relatively slow blood-effect-site equilibration of halothane (at 30 s of halothane-wash-in brain concentrations will be 2% or less of end-tidal, while the CB concentration will be 70% of end-tidal; at 60 s of wash-in these values are 10% and 90%, respectively); *ii*) Using the 'dynamic-end tidal forcing' technique, the ventilatory response to a square-wave change in end-tidal PCO_2 may be separated into a fast component arising from the peripheral chemoreceptors and a slow component arising from the central chemoreceptors. Halothane, isoflurane and sevoflurane at 0.1 MAC reduced the peripheral CO_2 sensitivity without affecting central CO_2 sensitivity.[5,7,11]

There are several observations that may shed some light on the mechanism via which anesthetics affect the acute hypoxic response. 1) Volatile anesthetic agents with appreciable metabolism and the production of free radicals under hypoxic conditions such as halothane and enflurane induce severe depression of the AHR On the other hand, desflurane, an agent with low metabolism and little free radical production, shows little to no depression of the normocapnic AHR (see also figure 1). 2) Propofol, an intravenous anesthetic agent with anti-oxidant properties, has no direct effect at the carotid bodies during hypoxia.[12] 3) In animals, the degree of depression of the AHR may depend on their oxidant/anti-oxidant status.[3] For example, animals such as goats that produce large quantities of ascorbic acid show little depression of the AHR by halothane (0.5 MAC).[3,13,14] The reverse is true for species that produce little to no ascorbic acid such as cats and humans.[3,15] (4) Volatile anesthetics open oxygen and acid sensitive background potassium (TASK) channels in rat carotid body cells.[16]

Items 1–3 suggest an important role for free radicals in anesthesia induced depression of the carotid bodies. Possibly these free radical species play an important modulatory role via an effect at potassium channels located in the cell membrane of the type I glomus cells.

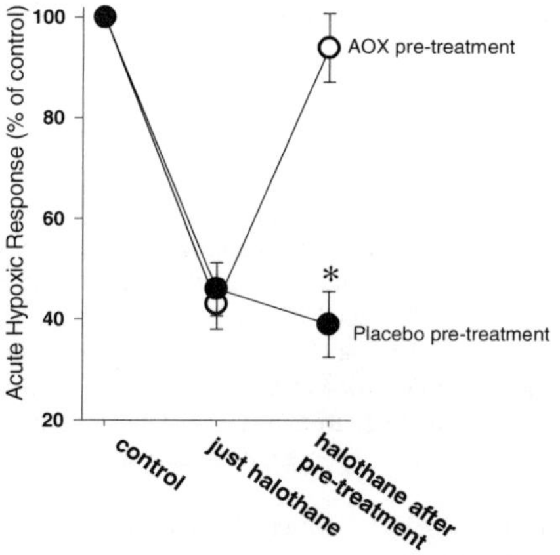

Figure 2. Pretreatment with antioxidants (AOX) but not with placebo prevents depression of the Acute Hypoxic Response (AHR) by low-dose halothane (0.13 MAC). 100% is the control value (*i.e.*, no drugs given). Values are mean $\pm$ SD. * $P < 0.05$ *versus* AOX-pretreatment and control (anova).

3. Influence of Antioxidant Pretreatment on Halothane-Induced Depression of the Acute Hypoxic Response

We studied the effect of the antioxidants (AOX) ascorbic acid (2 gram iv) and α-tocopherol (200 mg, p.o.) on the ventilatory response to isocapnic hypoxia (SpO_2 82 $\pm$ 2%; duration of hypoxia 3 min) during inhalation of 0.11% end-tidal halothane (~0.15 MAC) in healthy male volunteers. After performing a control hypoxic response and a response during the inhalation of halothane subjects either received AOX (n = 8) or placebo (n = 8). The results are shown in figure 2. Halothane reduced the AHR by more than 50%. Pretreatment with AOX but not with placebo prevented any depression of the AHR (AHR control $-$ 0.79 $\pm$ 0.35 L/min per % desaturation; AHR halothane = 0.36 $\pm$ 0.16 $L.min^{-1}.\%^{-1}$; AHR halothane AOX = 0.77 $\pm$ 0.32 $L.min^{-1}.\%^{-1}$; AHR halothane placebo = 0.36 $\pm$ 0.27 $L.min^{-1}.\%^{-1}$).

4. Concluding Remarks

Our data indicate that systemically administered antioxidants are able to prevent or reverse any depression induced by low-dose halothane in humans. We suggest that the depressant effect of halothane and possibly also by the other inhalational anesthetics (see figure 1) may be caused by reactive oxygen species produced by their reductive metabolism during hypoxia. The AOX cocktail effectively removed these radical species preventing their depressant effect on the carotid body oxygen sensing mechanism. An alternative explanation may be that the systemic AOX exposure change in redox state of the carotid body type I glomus cells prevented the binding of halothane to its effector-site (*e.g.*, an oxygen-sensitive background K^+-channel).

The finding that the depressant effect of anesthetics on the hypoxic chemoreflex may pharmacologically be prevented or reversed is of evident clinical importance as it may open the door to the treatment of respiratory events in the post-anesthesia carue unit using cheap but effective measures.

References

1. M. Fatemian, D. Nieuwenhuijs, L.J. Teppema, S. Meinesz, A. van der Mey, A. Dahan and P.A. Robbins. The respiratory response to carbon dioxide in humans with unilateral and bilateral resections of the carotid bodies. *J. Physiol. (Lond.)* **549**, 965–973 (2003).
2. M. Vizek, C.K. Picket and J.V. Weil. Biphasic ventilatory response of adult cats to sustained hypoxia has central origin. *J. Appl. Physiol.* **63**, 1659–1664 (1987).
3. L.J. Teppema, D. Nieuwenhuijs, E. Sarton, R. Romberg, C.N. Olievier, D.S. Ward and A. Dahan. Antioxidants prevent depression of the acute hypoxic response by subanaesthetic halothane in men. *J. Physiol. (Lond.)* **54**, 931–938 (2002).
4. R.L. Knill and J.L. Clement. Site of selective action of halothane on the peripheral chemoreflex pathway in humans. *Anesthesiology* **61**, 121–126 (1984).
5. A. Dahan, M. van den Elsen, A. Berkenbosch, J. DeGoede, I.C.W. Olievier, J.W. van Kleef and J.G. Bovill. Effects of subanesthetic halothane on the ventilatory responses to hypercapnia and acute hypoxia in healthy volunteers. *Anesthesiology* **80**, 727–738 (1994).
6. A. Dahan, M. van den Elsen, A. Berkenbosch, J. DeGoede, I. Olievier and J.W. van Kleef. Influence of a subanesthetic concentration of halothane on the ventilatory response to step changes into and out of sustained hypoxia in healthy volunteers. *Anesthesiology* **81**, 850–859 (1994).
7. M. van den Elsen, A. Dahan, J. DeGoede, A. Berkenbosch and J. van Kleef. Influences of subanesthetic isoflurane on ventilatory control in humans. *Anesthesiology* **83**, 478–490 (1995).
8. A. Dahan, E. Sarton, M. van den Elsen, J. van Kleef, L. Teppema and A. Berkenbosch. Ventilatory responses to hypoxia in humans: influences of subanesthetic desflurane. *Anesthesiology* **85**, 60–68 (1996).
9. E. Sarton, A. Dahan, L. Teppema, M. van den Elsen, E. Olofsen, A. Berkenbosch and J. van Kleef. Acute pain and central nervous system arousal do not restore impaired hypoxic ventilatory response during sevoflurane sedation. *Anesthesiology* **85**, 295–303 (1996).
10. B. Nagyova, K.L. Dorrington, M.J. Poulin and P.A. Robbins. Influence of 0.2 minimum alveolar concentration of enflurane on the ventilatory response to sustained hypoxia in humans. *Br. J. Anaesth.* **78**, 707–713 (1997).
11. M. van den Elsen, E. Sarton, L. Teppema, A. Berkenbosch and A. Dahan. Influence of 0.1 minimum alveolar concentration of sevoflurane, desflurane and isoflurane on dynamic ventilatory response to hypercapnia in humans. *Br. J. Anaesth.* **80**, 174–82 (1998).
12. D. Nieuwenhuijs, E. Sarton, L.J. Teppema, E. Kruyt, I. Olievier, J. van Kleef and A. Dahan. Respiratory sites of action of propofol – absence of depression of peripheral chemoreflex loop by low-dose propofol. *Anesthesiology* **95**, 889–895 (2001).
13. S.O. Koh and J.W. Severinghaus. Effect of halothane on hypoxic and hypercapnic ventilatory responses of goats. *Br. J. Anaesth.* **65**, 713–717 (1990).
14. J. Ponte and C.L. Sadler. Effect of halothane, enflurane and isoflurane on carotid body chemoreceptor activity in the rabbit and the cat. *Br. J. Anaesth.* **62**, 33–40 (1989).
15. R.O. Davies, M.W. Edwards and S.L. Lahiri. Halothane depresses the response of carotid body chemoreceptors to hypoxia and hypercapnia in the cat. *Anesthesiology* **57**, 153–159 (1982).
16. K.J. Buckler, B.A. Williams and E. Honore. An oxygen-, and acid and anaesthetic-sensitive TASK-like background potassium channel in rat arterial chemoreceptor cells. *J. Physiol. (Lond.)* **525**, 135–142 (2000).

Mechanism of Propofol-Induced Central Respiratory Depression in Neonatal Rats

Anatomical Sites and Receptor Types of Action

Masanori Kashiwagi, Yasumasa Okada, Shun-ichi Kuwana, Shigeki Sakuraba, Ryoichi Ochiai, and Junzo Takeda

1. Introduction

Propofol (2, 6-diisopropylphenol) is an intravenous anesthetic, which has been increasingly used for both the induction and maintenance of general anesthesia[1,2] as well as in critical care medicine.[3] When propofol is used, one of the most important adverse effects is respiratory depression,[4–8] which is caused by suppression of the central respiratory neuronal network. It is unclear, however, whether the main site of propofol action is the brainstem or the spinal cord.[4–8] In most previous studies, the response of respiratory neurons to propofol has been largely neglected. We address this lacuna here.

The brainstem-spinal cord preparation of the neonatal rat is an established *in vitro* model for physiological studies of the mammalian respiratory neuronal network.[9–11] Because pharmacological agents can be applied to the preparation in a precise concentration in an anesthetic- and muscle relaxant-free condition, this preparation is also useful for pharmacological studies.[12,13]

The purposes of the present study are first, to separately evaluate the propofol effect on the medulla oblongata and the spinal cord in order to identify the anatomical site of propofol action, and second, to identify the receptor type on which propofol exerts its effect in brainstem respiratory neurons. For these purposes, we used the brainstem-spinal cord preparation of the neonatal rat.

Masanori Kashiwagi, Shigeki Sakuraba, Ryoichi Ochiai, and Junzo Takeda • Department of Anesthesiology, School of Medicine, Keio University, Tokyo 160-8582 Japan. **Yasumasa Okada** • Department of Medicine, Keio University Tsukigase Rehabilitation Center, Shizuoka-ken 410-3215 Japan. **Shun-ichi Kuwana** • Department of Physiology, Teikyo University, Itabashi-ku, Tokyo 173-8605 Japan.

Post-Genomic Perspectives in Modeling and Control of Breathing, edited by Jean Champagnat, Monique Denavit-Saubié, Gilles Fortin, Arthur S. Foutz, Muriel Thoby-Brisson. Kluwer Academic/Plenum Publishers, 2004.

2. Methods

2.1. Brainstem-Spinal Cord Preparation

The surgical procedure for this preparation has been described elsewhere.[10–14] Briefly, the brainstem and the cervical spinal cord of 1- to 4-day old Wistar rats of either sex were isolated en bloc under deep ether anesthesia. The brainstem was transected at the level of the VIth cranial nerve root. The preparation was superfused continuously at 25–26°C with artificial cerebrospinal fluid of the following composition (in mM): NaCl 124, KCl 5.0, KH_2PO_4 1.2, $CaCl_2$ 2.4, $MgSO_4$ 1.3, $NaHCO_3$ 26, glucose 30, equilibrated with a gas mixture of 95% O_2-5% CO_2. We monitored rhythmic inspiratory activity from the C4 ventral root.

2.2. Neuronal Recording

The activity of neurons in the rostral ventrolateral medulla was recorded intracellularly with a perforated patch configuration. The method of neuronal recording was almost the same as that described previously.[13–16] Briefly, a glass pipette (GC100-TF-10; Clark; Reading, UK) was pulled with a horizontal puller to a tip size of approximately 2 m. Electrode resistance was between 10 and 14 MΩ when filled with a solution containing (in mM) K-gluconate 130, KCl 3, EGTA 10, HEPES 10, $CaCl_2$ 1, $MgCl_2$ 1 and nystatin (100 g/mL). The pH of the filling solution was adjusted between 7.2 and 7.3 with KOH. Membrane potentials were recorded with a whole cell patch amplifier (CEZ 3100, Nihon Kohden, Tokyo, Japan). The membrane potential was presented without correcting the liquid junction potential. Recorded medullary neurons were classified into four types according to their firing patterns: pre-inspiratory, inspiratory, expiratory, and non-respiratory neurons.[9,14,16]

2.3. Experimental Protocol

2.3.1. Experiment 1: Selective Application of Propofol to the Medulla or Spinal Cord

We used a chamber partitioned at the spino-medullary junction with two thin polyvinyl chloride plates placed parallel to each other (with woven nylon packed between them for seal and drainage) for discrete application of pharmacological agents to the brainstem and to the spinal cord. After baseline recording, the superfusate bathing either the medulla or the spinal cord was replaced by a solution containing 20 M propofol (whereas the other was continuously superfused with the control solution), and the recording was continued for 45 min. In the control group, a recording was made with both parts superfused with the control solution. Each group contained ten preparations.

2.3.2. Experiment 2: Responses of Bicuculline to Propofol Action on Medullary Neurons

We added 20–100 M propofol with or without 4 M bicuculline to the superfusate. After baseline recording with the control solution, the superfusate was replaced as follows:

propofol, propofol with 4 M bicuculline, and propofol again. Duration of superfusion with each solution was within 10 min.

2.4. Data Analysis

Data were presented as mean ± standard deviation (SD). The significance of difference in values was evaluated using an analysis of variance with Dunnett's test (in Experiment 1) and a paired t-test (in Experiment 2). We considered a difference to be statistically significant when $P < 0.05$.

3. Results

3.1. Selective Application of Propofol to the Medulla or Spinal Cord

Superfusion of the medulla with 20 M propofol for 15 min reduced the C4 burst rate by 73% ($P < 0.05$). By contrast, application of propofol to the spinal cord exerted little influence on either C4 burst rate or integrated C4 amplitude (Figure 1).

3.2. Responses of Bicuculline to Propofol Action on Medullary Neurons

We successively recorded one pre-inspiratory neuron, three expiratory neurons and four inspiratory neurons from the rostral ventrolateral medulla. Action potential firing of pre-inspiratory and expiratory neurons disappeared during propofol superfusion. In pre-inspiratory and expiratory neurons, bicuculline provided an antagonistic action against propofol-induced decreases in burst rate and arrest of action potential firing, and antagonized propofol-induced hyperpolarization of resting membrane potentials (from −62.5 ± 14.6 mV to −55.8 ± 18.4 mV, $P > 0.05$). A representative recording of the membrane potential of an expiratory neuron is shown in Figure 2. In inspiratory neurons, on the other hand, neither propofol nor bicuculline influenced membrane potential trajectories except for synchronous changes in burst rate. Resting membrane potentials were −63.4 ± 9.1 mV during bath application of propofol and −64.6 ± 10.6 mV during bath-application of propofol and bicuculline.

4. Discussion

Bath application of propofol to the spinal cord did not affect C4 inspiratory output; we did not observe any decrement in C4 inspiratory burst activity (Figure 1). The descending transmission of the inspiratory neural drive to phrenic motoneurons has been reported to depend on the activation of non-N-methyl-D-aspartate (non-NMDA) receptors, especially 2-amino-4-phosphonobutyric acid (AP-4)-sensitive presynaptic glutamate receptors.[17–19] Our results are consistent with these previous studies of propofol action on glutamate receptors; it has been shown that propofol has a minimal effect on the non-NMDA subtype of glutamate receptors.[20,21] It also is known that propofol acts on the sensory neurons of the spinal dorsal horn.[22] However, our finding clearly indicates that propofol does not exert

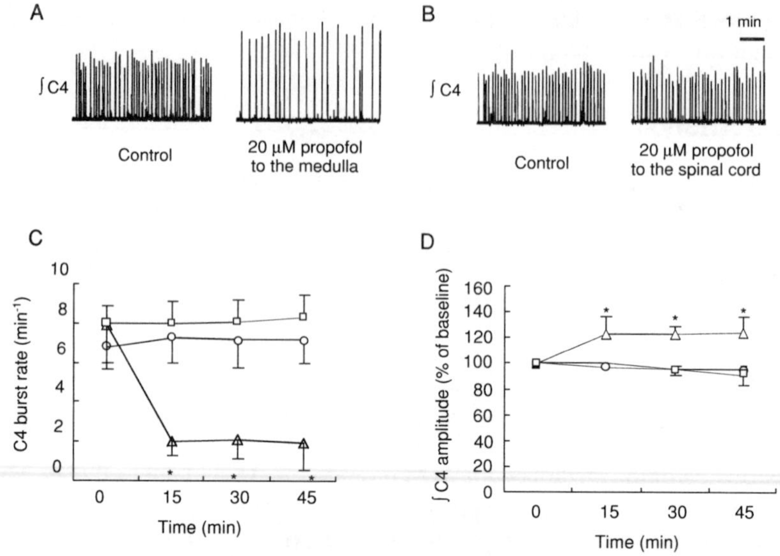

Figure 1. Representative recording of integrated C4 activity (∫C4) before and during discrete bath application of propofol. (**A**) The C4 inspiratory burst rate decreased when propofol was selectively applied to the medulla. (**B**) The C4 inspiratory burst rate was not affected when propofol was selectively applied to the spinal cord. (**C**) Effect of selective bath-application of propofol to either the medulla or the spinal cord on C4 inspiratory burst rates and (**D**) ∫C4 amplitude. (○) Control, (△) medulla, (□) spinal cord (n = 10 in each group). Data are shown as mean ± SD. $*P < 0.05$ versus control group.

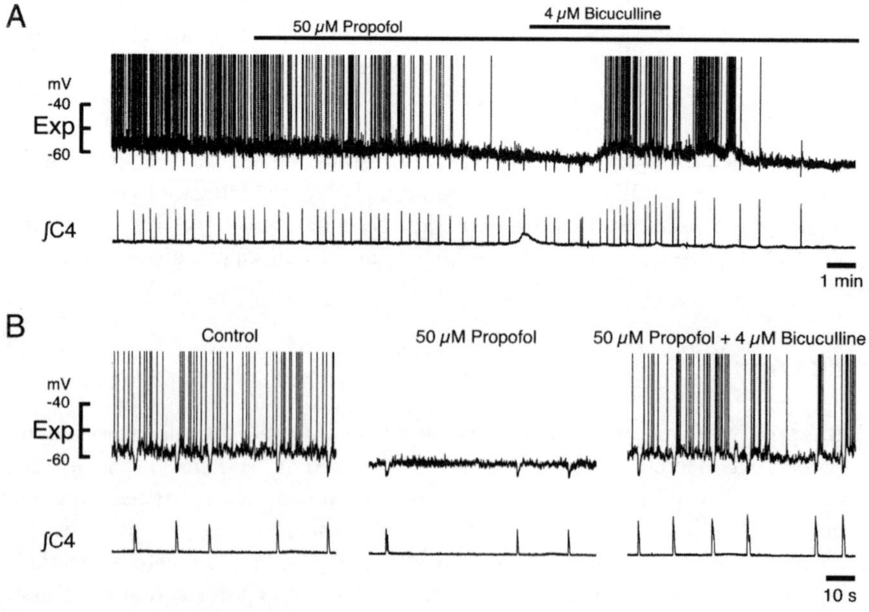

Figure 2. Effects of propofol and bicuculline on the membrane potential of a representative expiratory neuron (Exp). (**A**) Continuous recording of a membrane potential. Propofol hyperpolarized the expiratory neuron and suppressed action potential firing. Bicuculline reversed the effect of propofol. (**B**) Recording in (**A**) on an extended time scale. ∫C4 = integrated C4.

any direct effect on descending transmission of respiratory neuronal activity at the level of the spinal cord. Therefore, propofol-induced respiratory depression can be explained by inhibition of the medullary respiratory neuronal network, at least in this brainstem-spinal cord preparation of the neonatal rat.

In Experiment 2, we analyzed the effect of the $GABA_A$ receptor-antagonist bicuculline on propofol-induced suppression of respiratory neurons. An account of the effects of propofol itself on respiratory neuron activity is currently in preparation for publication, independent of the present paper. In the present paper, we demonstrated that bicuculline exerts an antagonistic effect on propofol-induced neuronal suppression in pre-inspiratory and expiratory neurons in the rostral ventral medulla. It has been reported that propofol action on neurons are $GABA_A$ receptor-mediated in various parts of the brain, e.g. the neocortex.[23] However, the receptor on respiratory neurons on which propofol exerts its action has not been identified, and this is the first report to demonstrate that propofol-induced hyperpolarization of brainstem respiratory neurons is mediated by the activation of a $GABA_A$ receptor. We recently reported that GABA in the brainstem plays an essential role in the maintenance of normal inspiratory burst activity using GABA synthesizing enzyme-deficient mice.[16] Also, pre-inspiratory neurons have been assumed to act as a pacemaker cell in respiratory rhythm generation, and the burst rates of inspiratory neurons are assumed to depend on the activity of pre-inspiratory neurons.[24] Therefore, we propose that the observed suppression of respiratory activity induced by propofol is very likely mediated, at least partly, by the activation of $GABA_A$ receptors on pre-inspiratory neurons. It has recently been reported in a study employing single-unit neuronal recording of the rat substantia nigra, however, that $GABA_B$ receptors also are involved in the mechanism of propofol-induced anesthesia.[25] Further studies should be conducted to also clarify the involvement of $GABA_B$ receptors in propofol-induced suppression of respiratory neurons.

In summary, we have demonstrated that propofol inhibits neural respiratory activity in the medulla, but does not directly suppress descending neural transmission in the spinal cord. Propofol-induced suppression of pre-inspiratory/expiratory neurons in the rostral ventrolateral medulla was mediated, at least partly, through $GABA_A$ receptor activation.

5. Acknowledgements

This study was supported by Keio Gijuku Academic Development Funds (MK), the Keio University Medical Science Fund (MK), a Grant-in-Aid from the Japanese Ministry of Education, Science, Sports and Culture (MK, YO and SK) and a Research Grant for Specific Diseases from the Japanese Ministry of Health and Welfare (YO).

References

1. P. S. Sebel, and J. D. Lowdon, Propofol: a new intravenous anesthetic, *Anesthesiology* **71**, 260–277 (1989).
2. J. Tang, L. Chen, P. F. White, M. F. Watcha, R. H. Wender, R. Naruse, R. Kariger, and A. Sloninsky, Recovery profile, costs, and patient satisfaction with propofol and sevoflurane for fast-track office-based anesthesia, *Anesthesiology* **91**, 253–261 (1999).
3. G. Angelini, J. T. Ketzler, and D. B. Coursin, Use of propofol and other nonbenzodiazepine sedatives in the intensive care unit, *Crit. Care Clin.* **17**, 863–880 (2001).

4. R. M. Grounds, D. L. Maxwell, M. B. Taylor, V. Aber, and D. Royston, Acute ventilatory changes during i.v. induction of anaesthesia with thiopentone or propofol in man. Studies using inductance plethysmography, *Br. J. Anaesth.* **59**, 1098–1102 (1987).
5. N. W. Goodman, A. M. Black, and J. A. Carter, Some ventilatory effects of propofol as sole anaesthetic agent, *Br. J. Anaesth.* **59**, 1497–1503 (1987).
6. P. Kulkarni, and K. A. Brown, Ventilatory parameters in children during propofol anaesthesia: a comparison with halothane, *Can. J. Anaesth.* **43**, 653–659 (1996).
7. J. E. Sternlo, and R. H. Sandin, Recurrent respiratory depression after total intravenous anaesthesia with propofol and alfentanil. *Anaesthesia* **53**, 378–381 (1998).
8. J. E. Quandt, E. P. Robinson, W. J. Rivers, and M. R. Raffe, Cardiorespiratory and anesthetic effects of propofol and thiopental in dogs, *Am. J. Vet. Res.* **59**, 1137–1143 (1998).
9. H. Onimaru, and I. Homma, Respiratory rhythm generator neurons in medulla of brainstem-spinal cord preparation from newborn rat, *Brain Res.* **403**, 380–384 (1987).
10. Y. Okada, K. Mückenhoff, G. Holtermann, H. Acker, and P. Scheid, Depth profiles of pH and PO_2 in the isolated brain stem-spinal cord of the neonatal rat, *Respir. Physiol.* **93**, 315–326 (1993).
11. Y. Okada, A. Kawai, K. Mückenhoff, and P. Scheid, Role of the pons in hypoxic respiratory depression in the neonatal rat, *Respir. Physiol.* **111**, 55–63 (1998).
12. A. Kawai, Y. Okada, K. Mückenhoff, and P. Scheid, Theophylline and hypoxic ventilatory response in the rat isolated brainstem-spinal cord, *Respir. Physiol.* **100**, 25–32 (1995).
13. Y. Okada, S. Kuwana, and M. Iwanami, Respiratory suppression induced by nicotine withdrawal in the neonatal rat brainstem: implications in the SIDS risk factor, *Adv. Exp. Med. Biol.* **499**, 187–194 (2001).
14. S. Kuwana, Y. Okada, and T. Natsui, Effects of extracellular calcium and magnesium on central respiratory control in the brainstem-spinal cord of neonatal rat, *Brain Res.* **786**, 194–204 (1998).
15. Y. Okada, Z. Chen, W. Jiang, S. Kuwana, and F. L. Eldridge, Anatomical arrangement of hypercapnia-activated cells in the superficial ventral medulla of rats, *J. Appl. Physiol.* **93**, 427–439 (2002).
16. S. Kuwana, Y. Okada, Y. Sugawara, N. Tsunekawa, and K. Obata, Disturbance of neural respiratory control in neonatal mice lacking GABA synthesizing enzyme 67-kDa isoform of glutamic acid decarboxylase, *Neurosocience* **120**, 861–870 (2003).
17. G. Liu, J. L. Feldman, and J. C. Smith, Excitatory amino acid-mediated transmission of inspiratory drive to phrenic motoneurons, *J. Neurophysiol.* **64**, 423–436 (1990).
18. J. J. Greer, J. C. Smith, and J. L. Feldman, Role of excitatory amino acids in the generation and transmission of respiratory drive in neonatal rat, *J. Physiol. Lond.* **437**, 727–749 (1991).
19. J. J. Greer, J. C. Smith, and J. L. Feldman, Glutamate release and presynaptic action of AP4 during inspiratory drive to phrenic motoneurons, *Brain Res.* **576**, 355–357 (1992).
20. B. A. Orser, M. Bertlik, L. Y. Wang, and J. F. MacDonald, Inhibition by propofol (2, 6 di-isopropylphenol) of the N-methyl-D-aspartate subtype of glutamate receptor in cultured hippocampal neurones, *Br. J. Pharmacol.* **116**, 1761–1768 (1995).
21. A. Kitamura, W. Marszalec, J. Z. Yeh, and T. Narahashi, Effects of halothane and propofol on excitatory and inhibitory synaptic transmission in rat cortical neurons, *J. Pharmacol. Exp. Ther.* **304**, 162–171 (2003).
22. X. P. Dong, and T. L. Xu, The actions of propofol on gamma-aminobutyric acid-A and glycine receptors in acutely dissociated spinal dorsal horn neurons of the rat, *Anesth. Analg.* **95**, 907–914 (2002).
23. B. Antkowiak, Different actions of general anesthetics on the firing patterns of neocortical neurons mediated by the GABA(A) receptor, *Anesthesiology* **91**, 500–511 (1999).
24. K. Ballanyi, H. Onimaru H, and I. Homma, Respiratory network function in the isolated brainstem-spinal cord of newborn rats, *Prog. Neurobiol.* **59**, 583–634 (1999).
25. L. Schwieler, D. S. Delbro, G. Engberg, and S. Erhardt, The anaesthetic agent propofol interacts with GABA(B)-receptors: an electrophysiological study in rat, *Life Sci.* **72**, 2793–2801 (2003).

Interaction of Arousal States with Depression of Acute Hypoxic Ventilatory Response by 0.1 MAC Halothane

Jaideep J. Pandit, Ben Moreau, Peter A. Robbins

1. Introduction

It is well-established that volatile anesthetic agents at doses of < 0.2 MAC depress the acute hypoxic ventilatory response (AHVR) by ~50–70%[1,2]. However, the effect of anesthetic agents can be very variable[3,4].

The source of this variability of effect is unclear, but van den Elsen *et al.* have suggested that the state of arousal of subjects (and especially the effect of audiovisual (AV) stimulation in the form of watching television) is important[5,6].

Additionally, Sarton *et al.* found that acute pain (which might seem at first to be more arousing than AV stimulation) did not reverse the depression of AHVR by sevoflurane[7]. They therefore concluded that AV stimulation was a "specific" stimulus which interacted with the hypoxic chemoreflex in a unique manner. Pain, they argued, was a "non-specific" stimulus which, while arousing the central nervous system in general, did not do so in a manner which specifically interacted with the hypoxic chemoreflex. In summary, the hypothesis was that background study conditions (ie, arousal states) explained the variability of results seen in different studies for the effect of anesthetics on AHVR[8].

Pandit has offered an alternative interpretation[9]. Using the technique of meta-analysis he concluded that the largest source of variability in published, pooled results was not due to the type of arousal stimulus, but simply due to the type of anesthetic agent used[9]. In other words, different anesthetic agents have intrinsically different effects on the hypoxic chemoreflex, regardless of background arousal state.

We planned to assess whether AV stimulation would prevent the depression of AHVR by halothane in the manner predicted by Dahan and colleagues[5–8]; or whether AV stimulation would have no effect, as predicted by Pandit[9]. We also wished to investigate the effects of acute pain on the ability of 0.1 MAC halothane to depress AHVR.

Jaideep J. Pandit • Nuffield Department of Anaesthetics, John Radcliffe Hospital, Oxford OX3 9DU, UK. **Ben Moreau and Peter A. Robbins** • University Laboratory of Physiology, Parks Road, Oxford OX1 3PT, UK.

Post-Genomic Perspectives in Modeling and Control of Breathing, edited by Jean Champagnat, Monique Denavit-Saubié, Gilles Fortin, Arthur S. Foutz, Muriel Thoby-Brisson. Kluwer Academic/Plenum Publishers, 2004.

2. Methods

2.1. Subjects

With ethical approval, we studied 10 healthy volunteers (9 men and 1 woman; mean age 20.6 years [range 20–22 years]; mean height 1.78 m [1.65–1.85 m]; mean weight 72.1 kg [55–84 kg]).

2.2. Control of End-Ridal Gases

During experiments subjects were seated in a chair, wore a noseclip and breathed through a mouthpiece. Respiratory volumes were measured by a turbine volume measuring device and flows by a pneumotachograph in series with the mouthpiece. Expired gas at the mouth was sampled continuously by a mass spectrometer and analysed for PCO_2 and PO_2. The volumes and flows and the PCO_2 and PO_2 at the mouth were recorded in real time with a 50Hz sampling frequency by a computer, which also executed a peak-picking program to locate end-tidal PCO_2 (PET_{CO2}) and end-tidal PO_2 (PET_{O2}). End-tidal gases were controlled by dynamic end-tidal forcing, to maintain desired end-tidal values independently of changes in ventilation[10,11].

2.3. Protocols

Before each experimental period, subjects underwent a period of quiet breathing to establish their ambient end-tidal PCO_2 (PET_{CO2}). In those experimental periods involving administration of halothane, this time interval was also used to reach the target end-tidal value (0.1 MAC) of halothane. Dynamic end-tidal forcing was then used to hold the PET_{CO2} 1–2 mmHg above this ambient value throughout each protocol. The end-tidal PO_2 (PET_{O2}) was controlled in the following manner: an initial 4 min period of 100 mmHg, followed by three steps of hypoxia (PET_{O2} 50 mmHg), each of 4 min duration, and each separated by 4 min of euoxia (PET_{O2} 50 mmHg).

These end-tidal gas profiles were undertaken in six protocols (in random order on different days), each protocol characterized by a distinct background arousal state (separately with and then without 0.1 MAC halothane):

1. *Control*: subjects were in a darkened, quiet room with their eyes closed;
2. *AV stimulation*: subjects watched television with sound in a bright, noisy environment;
3. *Pain*: subjects were exposed to experimentally-induced acute pain, while in a darkened, quiet room with their eyes closed.

2.4. Administration of Arousal Stimuli

For audiovisual stimulation protocols, subjects watched the television program or video of their choice. The sound was high, the room brightly-lit and all experimenters freely talked and entered and left the room at will.

For those control protocols in which audiovisual stimulation was absent, and for those in which experimental pain was used, the room was darkened by the use of window blinds, all noise was reduced to bare minimum whispers between experimenters required for the conduct of the study, the subject's eyes were closed, and the subject wore headphones further to eliminate noise.

For the pain protocols, two electrodes were placed on the skin overlying the tibial bone. The electrodes were attached to an electrostimulator which delivered noxious electrical stimuli of 0.2 ms duration at 1 Hz. The current strength could vary from 0–80 mA and was adjusted to achieve a target visual analog pain score (VAS) of no less than 5/10 and no higher than 6/10 (on a scale 0 = no pain and 10 = worst pain possible). The painful stimulus was administered for fixed durations of 6 min, spanning the hypoxic step and the 2 min of euoxia before hypoxia. The absence of pain for 2 min during euoxic periods was designed to minimize the possibility of adaptation to the pain. The VAS was tested at the end of each protocol to assess stability of the stimulus.

2.5. Data Analysis

Data were averaged into 1 min periods. The ventilation in the last min of euoxia, before any hypoxia was administered was used as the baseline, euoxic ventilation for each study condition. The AHVR for each hypoxic step was then calculated as the difference between the peak ventilation reached in the 4 min of hypoxia, and the ventilation in the last min before the hypoxic exposure. Thus, for each experimental period, there were three values of AHVR obtained, and these were averaged to yield the average value of AHVR for the protocol. These individual subject values were then averaged to obtain the mean for the group.

2.6. Statistical Analysis

The values for baseline euoxic ventilation, AHVR and BIS were first subjected to analysis of variance (ANOVA, SPSS for Windows). The "response" was "ventilation" or "AHVR", and there were three "factors": "arousal" (fixed factor, three levels); "halothane" (fixed factor, two levels); and "subject (random factor, ten levels). If ANOVA indicated a statistically significant effect, then post-hoc paired Student's t-tests were undertaken to locate the precise source of the significant effect. A value of $P < 0.05$ was taken as statistically significant. For those comparisons involving multiple post-hoc tests, the Bonferroni correction was applied and statistical significance was taken at a value of $P < 0.05/n$, where n was the number of comparisons made for the hypothesis tested.

3. Results

Figure 1 shows the results for baseline euoxic ventilation. ANOVA did not show any significant effect of halothane, but did suggest that the interactive term of halothane and arousal was significant ($P < 0.04$), indicating that halothane might influence baseline ventilation depending on the prevailing arousal state. However, post-hoc t-tests did not confirm this suggestion, and within each of the arousal states, there was no significant effect of halothane. There was a significant influence of arousal alone (ANOVA, $P < 0.008$). Post hoc t-tests showed that the mean euoxic ventilation in the control protocol differed

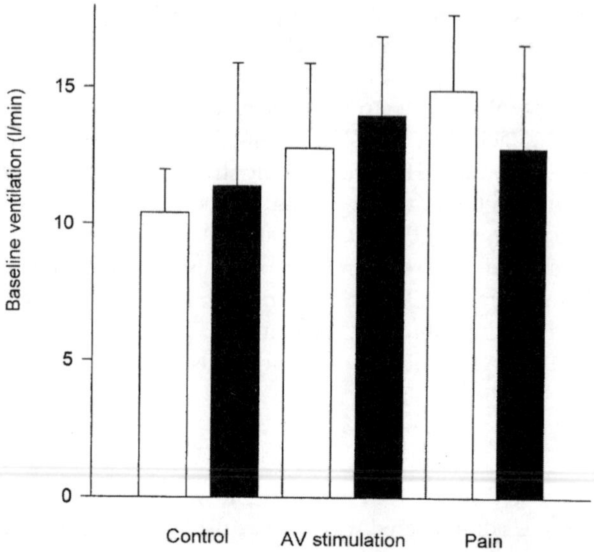

Figure 1. Mean values (± SD) of euoxic baseline ventilation for the without halothane (hollow bars) and with halothane (filled bars).

from ventilation in both AV and pain protocols ($P < 0.002$ and $P < 0.001$ respectively). The mean ventilations in AV and pain protocols were not different. Thus, both types of arousal significantly and modestly increased baseline euoxic ventilation as compared with control.

Figure 2 shows the values for AHVR in each of the six protocols. ANOVA showed significant effects of halothane ($P < 0.001$), but not of the interactive term of halothane and arousal, which suggested that halothane significantly reduced AHVR, regardless of the background arousal state. The degree of depression of AHVR by halothane was 49% in the control protocol; 51% in the AV stimulation protocol; and 53% in the pain protocol (Figure 2). ANOVA also indicated that arousal alone had a significant influence on AHVR ($P < 0.004$). Post-hoc t-tests showed that AHVR in the control protocol differed from AHVR in both AV and pain protocols ($P < 0.003$ and $P < 0.001$ respectively), but AHVR did not differ between pain and AV protocols. Thus, both types of arousal increased AHVR modestly as compared with control.

4. Discussion

The striking result of this study is that AV stimulation does not prevent the blunting of AHVR by 0.1 MAC halothane. Previous reports that AV stimulation prevents the blunting of AHVR by 0.1 MAC isoflurane[5,6] might lead to the prediction that the same is true for all agents, but this is not the case. A second result is that pain does not prevent the blunting of AHVR by 0.1 MAC halothane. A third result is that, regardless of the effect of halothane, arousal (both AV and pain) can modestly augment euoxic ventilation and AHVR.

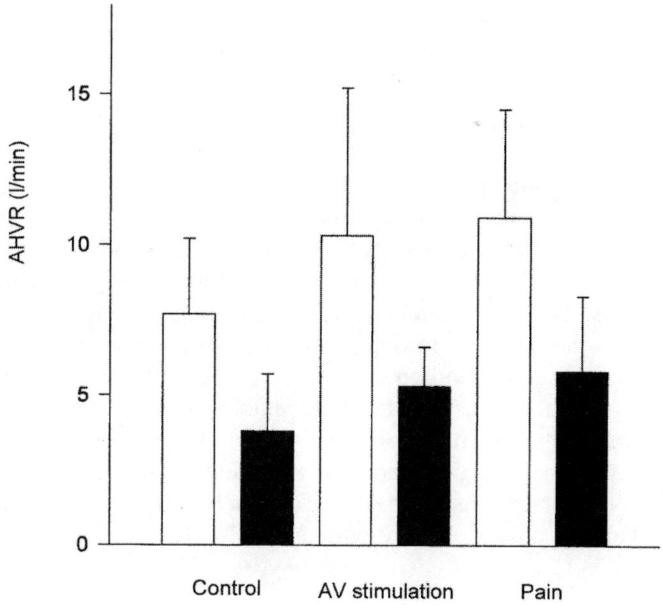

Figure 2. Mean values (± SD) of AHVR for the protocols without halothane (hollow bars) and with halothane (filled bars).

4.1. Effects of Arousal on Baseline Ventilation and AHVR

Our result that we did not observe any effects of 0.1 MAC halothane on euoxic baseline ventilation is consistent with previous results for halothane[12,13], isoflurane[3–6] and sevoflurane[7,14]. Thus, volatile anesthetics have a more profound effect on the hypoxic chemoreflex than they do on the mechanisms controlling basal ventilation.

Like Sarton *et al.*[7,15], we also found that pain *per se* increased baseline ventilation, suggesting a degree of interaction between pain and the neural factors which control basal ventilation.

In addition we found that pain increased AHVR modestly as compared with control. This result differs from previous work which suggests that pain does not augment the peripheral hypoxic chemoreflex[7,15–17]. Possible explanations are (a) that we used a slightly higher stimulus level; (b) that at least a small part of the increase in AHVR we observed was due to a "startle" response in ventilation at the onset of acute pain (c) that an augmented AHVR is specific to the quality (and not just intensity) of pain stimulus we used: other groups have reported different patterns of ventilatory response with heat[16] and pressure-induced pain[17].

We found that the effect of AV stimulation on baseline ventilation was significant (mean increase 2.4 l/min in the control protocol; Figure 1). This is consistent with a recent report of Karan *et al.*[16], and previously van den Elsen *et al.*[5,6] reported a larger increase of 3.9 l/min, but this did not reach statistical significance in their study.

We also found that AV stimulation increased AHVR significantly (by 2.6 l/min in the control protocol; Figure 2). This result is similar to that of Karan *et al.*[16], but van den Elsen *et al.* observed a rise of only 0.4 l/min (NS)[5,6]. Despite modestly increased AHVR with AV stimulation, halothane still reduced AHVR by 50% (Figure 2).

4.2. Implications of the Results for Respiratory Control

If we accept the widely-held axiom that all volatile anesthetics act by the same mechanism, why is our result for halothane so different from the previous result by another group for isoflurane? One possibility is that one research group has erred in its methodology (though this seems unlikely). The other possibility is that the original axiom is incorrect, and that in fact, anesthetics are different. Although this suggestion is speculative, functional MRI (fMRI) offers one means of testing this hypothesis. Heinke and Schwarzbauer have reported that isoflurane significantly reduces task-induced activation in only three specific brain areas, and not homogenously in the entire brain[18]. Willis *et al.* have found significant differences between isoflurane and propofol anesthesia when visually-induced fMRI signals in dogs were analysed[19]. Taken together, these early findings are consistent with the notion of specific target pools of neurons for anesthetic action. It would therefore seem important to assess hypoxia-induced and AV stimulation-induced fMRI signals with different anesthetic agents in humans.

References

1. R.L. Knill and A.W. Gelb. Ventilatory responses to hypoxia and hypercapnia during halothane sedation and anesthesia in man. *Anesthesiology* **49:** 244–251 (1978).
2. R.L. Knill, P.H. Manninen and J.L. Clement. Ventilation and chemoreflexes during enflurane sedation and anaesthesia in man. *Can Anaesth Soc J* **26:** 353–360 (1979).
3. J.A. Temp, L.C. Henson and D.S. Ward. Does a subanesthetic concentration of isoflurane blunt the ventilatory response to hypoxia? *Anesthesiology* **77:** 1116–1124 (1992).
4. J.A. Temp, L.C. Henson and D.S. Ward. Effect of a subanesthetic minimum alveolar concentration of isoflurane on two tests of the hypoxic ventilatory response. *Anesthesiology* **80:** 739–750 (1994).
5. M.J.L.J. van den Elsen, A. Dahan, A. Berkenbosch, J. DeGoede, J.W. van Kleef and I.C.W. Olievier. Does subanesthetic isoflurane affect the ventilatory response to acute isocapnic hypoxia in healthy volunteers? *Anesthesiology* **81:** 860–867 (1994).
6. M. van den Elsen, A. Dahan, J. DeGoede, A. Berkenbosch and J. van Kleef. Influence of subanesthetic isoflurane on ventilatory control in humans. *Anesthesiology* **83:** 478–490 (1995).
7. E. Sarton, A. Dahan, L. Teppema, M. van den Elsen, E. Olofsen, A. Berkenbosch A. and J. van Kleef. Acute pain and central nervous system arousal do not restore impaired hypoxic ventilatory response during sevoflurane sedation. *Anesthesiology* **85:** 295–303 (1996).
8. A. Dahan and L. Teppema. Editorial 1. Influence of low-dose anaesthetic agents on ventilatory control: where do we stand? *Br J Anaesth* **83:** 199–201 (1999).
9. J.J. Pandit. The variable effect of low-dose volatile anaesthetics on the acute ventilatory response to hypoxia in humans: a quantitative review. *Anaesthesia* **57:** 632–43 (2002).
10. P.A. Robbins, G.D. Swanson and M.G. Howson. A prediction-correction scheme for forcing alveolar gases along certain time-courses. *J Appl Physiol* **52:** 1353–7 (1982).
11. J.J. Pandit and P.A. Robbins. The ventilatory effects of sustained isocapnic hypoxia during exercise in humans. *Respir Physiol* **86:** 393–404 (1991).
12. A. Dahan, M.J.L.J. van den Elsen, A. Berkenbosch, J. DeGoede, I.C.W. Olievier, J.W. van Kleef and J. Bovill. Effects of subanesthetic halothane on the ventilatory responses to hypercapnia and hypoxia in healthy volunteers. *Anesthesiology* **80:** 727–738 (1994).

13. B. Nagyova, K.L. Dorrington, E.W. Gill and P.A. Robbins. Comparison of the effects of sub-hypnotic concentrations of propofol and halothane on the acute ventilatory response to hypoxia. *Br J Anaesth* **75:** 713–718 (1995).
14. J.J. Pandit, J. Manning-Fox, K.L. Dorrington and P.A. Robbins. Effects of subanaesthetic sevoflurane on ventilation. 2: Response to acute and sustained hypoxia in humans. *Br J Anaesth* **83:** 210–6 (1999).
15. E. Sarton, A. Dahan, L. Teppema, A. Berkenbosch, M. van den Elsen and J. van Kleef J: Influence of acute pain induced by activation of cutaneous nociceptors on ventilatory control. *Anesthesiology* **87:** 289–96 (1997).
16. R. Duranti, T. Pantaleo, F. Bellini, F. Bongianni and G. Scanno. Respiratory responses induced by activation of somatic nociceptive afferents in humans. *J Appl Physiol* **71:** 2440–8 (1991).
17. S.B. Karan, D.S. Ward, P. Bailey, R. Norton and W. Voter. Effects of pain and audiovisual stimulation on the hypoxic ventilatory response. *Anesthesiology* **99:** A-1532 (2003).
18. W. Heinke and C. Schwarzbauer. Subanesthetic isoflurane affects task-induced brain activation in a highly specific manner. A functional magnetic resonance imaging study. *Anesthesiology* **94:** 973–81 (2001).
19. C.K. Willis, R.P. Quinn, W.M. McDonell, J. Gati, J. Parent and D. Nicolle. Functional MRI as a tool to assess vision in dogs: the optimal anesthetic. *Vet Opthalmol* **4:** 243–53 (2001).

6

CARDIO-RESPIRATORY REGULATIONS AND CEREBRAL BLOOD FLOW

Relationship Between Ventilatory and Circulatory Responses to Sustained Mild Hypoxia in Humans

Toshio Kobayashi, Atsuko Masuda, Yoshikazu Sakakibara, Michiko Tanaka, Shigeru Masuyama and Yoshiyuki Honda

Key words: sustained hypoxia, hypoxic ventilatory depression, hypoxic circulatory response, pulse dye densitometry.

1. Introduction

The ventilatory response to isocapnic moderate hypoxia in humans is biphasic, consisting of an initial brisk increase followed by a gradual decrease in ventilation, namely hypoxic ventilatory depression (HVD)[1]. A similar biphasic response in the heart rate (HR) during sustained hypoxia was also observed in our previous study[2]. Although some mechanisms of the HVD have been proposed (the increase in inhibitory neuromodulators in the central nervous systems[3], adaptation of the peripheral chemoreceptors[4], increase in brain blood flow[5] etc.), little has been known about the circulatory parameters, especially about stroke volume (SV) and cardiac output (CO) behaviors during sustained hypoxia in humans. Accurate measurement of SV (and CO) is difficult noninvasively, during sustained hypoxic exposure. In order to compare the effect of sustained mild hypoxia ($SpO_2 = 80\%$) on ventilatory vs. circulatory responses, healthy young humans were exposed to sustained hypoxia (end-tidal $Po_2 = 55 \sim 60$ mmHg) for 20 min, applying a newly developed apparatus for measuring CO (pulse dye densitometry).

2. Methods

Twelve healthy young volunteers (8 males and 4 females) ranging in age from 18 to 33 years participated in this study. All subjects gave their informed consent, and the study protocol was approved by the local ethics committee.

Toshio Kobayashi • Hiroshima Univ, Hiroshima, 734-8551, Japan. **Atsuko Masuda** • Tokyo Med & Dent Univ, Tokyo. **Yoshikazu Sakakibara** • Kanazawa Inst Technol, Kanazawa. **Michiko Tanaka** • Miyazaki Pref Nurs Univ, Miyazaki. **Yoshiyuki Honda** • Chiba Univ, Chiba.

Post-Genomic Perspectives in Modeling and Control of Breathing, edited by Jean Champagnat, Monique Denavit-Saubié, Gilles Fortin, Arthur S. Foutz, Muriel Thoby-Brisson. Kluwer Academic/Plenum Publishers, 2004.

The subjects were seated and breathed in a closed circuit containing a rubber bag. End–tidal Po_2 was maintained at 55 ~ 60 mmHg ($SpO_2 = 80\%$) by adjusting the inflow of N_2 and O_2 into the circuit. End-tidal Pco_2 was maintained at room-air breathing level by adjusting the by-pass flow to a CO_2 absorber.

A hot-wire flow meter, which was connected to rapid responding O_2 (zirconia) and CO_2 (infra-red) analyzers was inserted between a mouth piece and a one-way valve. These devices were incorporated into a data acquisition and processing system [RM 280, MINATO]. The respiratory flow, O_2 and CO_2 signals were real time treated in order to obtain tidal volume, respiratory frequency, minute ventilation (V_E), end-tidal O_2 and CO_2 concentration, inspiratory O_2 and CO_2 concentrations on a breath-by-breath basis.

Stroke volume (SV), heart rate (HR) and cardiac output (CO) were measured by a newly developed pulse dye densitometry which based on the accurate determination of intra-vascular indocyanine green (ICG) concentration [DDG–2001: Nihon Kohden, Tokyo, Japan]. Pulse dye densitometry is based on the principle of 'pulse spectrophotometory'. It uses wavelengths of 805 nm and 940 nm for calculation of blood ICG concentration. The peak optical absorption of ICG occurs at 805 nm and no significant optical absorption occurs at 940 nm. The absorption of oxyhemoglobin and deoxyhemoglobin are equal at 805 nm, and the influence of light absorption of hemoglobin on oxygen saturation is quite small at 940 nm. Details of the basic principle are described in the preceding publication[6]. Injected ICG was detected by the optical probe from the fractional pulsatile change in optical absorption at 805 nm from which the ICG concentration was continuously computed.

After sufficient rest, the subject was connected to a breathing circuit through a mouth-piece with a one-way valve and breathed room air. When stable breathing and HR had been achieved, the valves in the circuit were turned at the end of expiration to cause re-breathing from the rubber bag initially filled with gas with Po2 at about 55 mmHg. $PETCO_2$ was maintained at the air-breathing level during hypoxia. Isocapnic hypoxia was sustained for 20 min with continuous measurement of arterial O_2 saturation (Spo_2) by pulse oximeter [Ohmeda, BIOX, USA]. A bolus of 5 mg ICG in 1 ml distilled water was injected into an antecubital vein just before and at 5, 7, 10, 15 and 20 min during hypoxic exposure. The optical probe was attached to the ear lobe and blood sample was obtained for Hb measurement. Ventilatory and circulatory variables which were normalized by body surface area were compared at base line and at 5, 7, 10, 15 and 20 min during hypoxic challenge.

3. Results

The average ventilatory and circulatory responses to sustained isocapnic hypoxia are presented in Table 1. During hypoxic exposure, V_E showed a biphasic profile. Minute ventilation initially increased to about 146% of control at 5 min ($p < 0.01$ vs. control) and then gradually decreased to about 125% at 15 min ($p < 0.05$ vs. the value at 5 min). As for the cardiac parameters, HR initially increased to about 117% of control at 5 min ($p < 0.01$ vs. control) as $PETo_2$ fell, and then gradually decreased to about 112% at 15 min ($p < 0.05$ vs. the value at 5 min), exhibiting a similar change to V_E in the time course. In contrast, CO attained delayed peak at 7 min ($p < 0.01$ vs. control) and followed by slight depression, but not significant. Peak response in SV was further delayed to 10 min ($p < 0.05$ vs. 5 min), but it was not a significant increase from the control level.

Table 1. Time course of ventilatory and circulatory parameters during isocapnic hypoxia.

		Hypoxia duration				
	room air	5 min	7 min	10 min	15 min	20 min
VE (l/min/BSA)	6.71 ± 1.61**	9.77 ± 4.26	9.40 ± 3.61	8.47 ± 3.28**	8.37 ± 2.78*	8.52 ± 2.52
HR (beats/min)	74.4 ± 6.3**	86.7 ± 8.4	85.9 ± 7.7	85.0 ± 6.5	83.6 ± 7.0#,*	83.7 ± 5.9##,*
CO (l/min/BSA)	3.06 ± 0.58##	3.58 ± 0.63	3.77 ± 0.73*	3.77 ± 0.62	3.62 ± 0.67	3.64 ± 0.65
SV (ml/BSA)	41.6 ± 9.4	41.8 ± 9.3	44.4 ± 10.1*	44.7 ± 9.4*	43.7 ± 9.0*	43.8 ± 9.3

** $p < 0.01$, * $p < 0.05$ vs. 5 min, # $p < 0.05$, ## $p < 0.01$ vs. 7 min

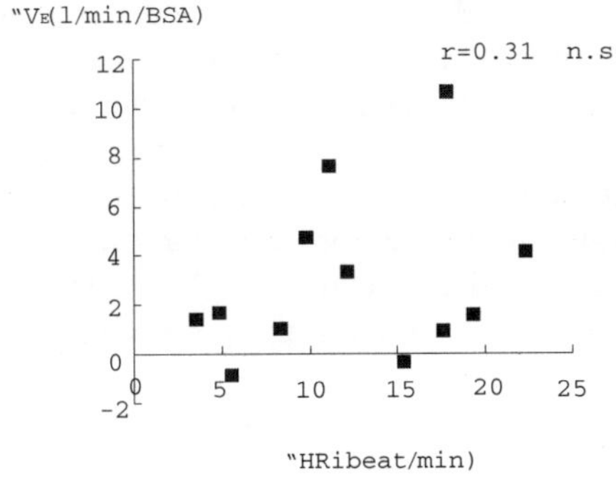

Figure 1. Relation between the magnitude of initial increase in HR and V_E during hypoxia.

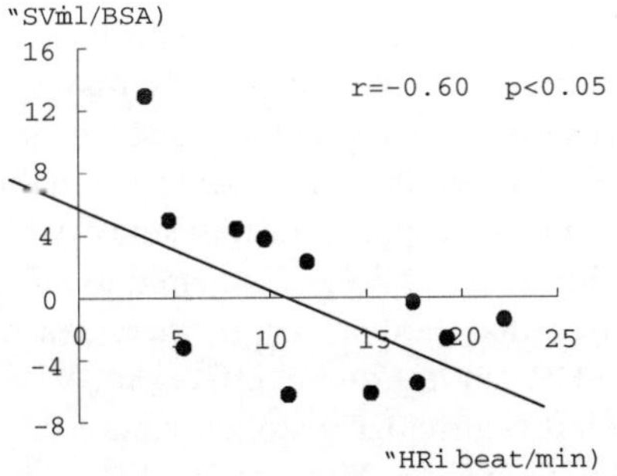

Figure 2. Relation between the magnitude of initial increase in HR and SV during hypoxia.

During initial hypoxic exposure, the rate of initial increase in HR showed weak positive correlation with that in VE, but not significant ($r = 0.31$, $p = .0.33$, Fig. 1). The rate of subsequent decline in HR was not correlated with that in V_E ($r = 0.22$, $p = 0.49$). Neither the rate of initial increase nor subsequent decline in SV and CO showed any correlation with that in V_E.

On the other hand, both the rate of initial increase (Fig. 2) and subsequent decline in HR showed significant negative correlation with those in SV ($r = -0.60$, $p < 0.05$ and $r = -0.63$, $p < 0.05$, respectively).

4. Discussion

In the present study, although the similar response profile in both V_E and HR activities was ascertained, there was little correlation between the rate of changes in V_E and HR recognized either during initial increase or subsequent decline phase. This suggested that the effect of pulmonary stretch reflex on heart rate modulation was not evident during sustained mild hypoxia in humans. Slutsky et al.[7] also found in healthy humans that HR was best fitted to an inverse linear relation to arterial oxygen saturation and that there was no relationship between HR and ventilatory responses to hypoxia. Yamamoto et al.[8] also showed that there was little effect of pulmonary stretch reflex on heart rate modulation during mild hypoxia in humans. It was also suggested that the initial tachycardia was sympathetically mediated and the secondary bradycardia indicated the direct effect of hypoxia on cardiac pacemaker tissue[9]. On the other hand, it was suggested that the HR response to hypoxemia is strongly related to the strength of the Hering-Breuer reflex in dogs[10]. Our previous study[2] in humans also demonstrated positive correlation between V_E and HR during sustained hypoxia, which suggested the important role of pulmonary stretch reflex in the positive V_E-HR relationship. One of the reasons for these controversial results might be due to species difference. Respiratory reflex changes of HR in man were reported to be weaker than in other mammalian species[11]. From our present results, we conclude that there may be little relationship between V_E and HR profiles during sustained hypoxia, or, there might be only a weak relationship between V_E and HR under some specific conditions in awake humans.

In this study, SV and CO responses were dissimilar to ventilatory response. Their profiles showed no correlation with that in V_E during sustained hypoxia. Stroke volume response revealed only a slight increase during hypoxic exposure and reached peak at 10 min, but no significant difference was seen between the control and the hypoxic exposure phase. The tendency of gradual increase in SV in this study may be partly due to sympathetic activation by sustained hypoxia. In our previous study in humans[2], we used the same experimental protocol and found a gradual increase of adrenaline during 15 min sustained hypoxic exposure. Cardiac output response in this study, which is the multiplication of HR by SV, attained delayed peak at 7 min as compared to HR and V_E, followed by slight depression. Cargill[12] reported that cardiac output increases during hypoxia, and this is entirely due to increases in HR but not to any effect on SV in awake humans. They also reported that parameters of left ventricular systolic function and myocardial inotropic state were not affected by severe hypoxemia. Marshall et al.[13] also reported that cardiac index increased, and vascular conductance of most body tissues was increased by peripheral vasodilatation in unanaesthetized rats during hypoxia. So, delayed response of CO increase in this study may be partly due to the decreased venous return by peripheral venous pooling during hypoxia. Interestingly, we found that HR and SV revealed clear negative correlation during hypoxic exposure. This may imply that rapid increase in HR in initial hypoxic phase was compensated by decreased SV because of limited venous return, which resulted in stable CO.

In conclusion, interaction of ventilatory and circulatory responses during hypoxia was not evident in this study, suggesting that mechanisms for cardiovascular controls were not closely linked with those for ventilatory controls during sustained mild hypoxia in awake humans.

References

1. Easton, P.A., Slykerman, L.J., Anthonisen, N.R., Ventilatory response to sustained hypoxia in normal adults, *J. Appl. Physiol.* **61**(3), 906–911 (1986).
2. Tanaka, M., Takaishi, S., Ohdaira, T., Kobayashi, T., Maruyama, R., Ahn, B., Masuda, A., Masuyama, S., Honda, Y., Dependence of biphasic heart rate response to sustained hypoxia on magnitude of ventilation in man, *Jpn. J. Physiol.* **42**(6), 865–75 (1992).
3. Yamamoto, M., Nishimura, M., Kobayashi, S., Akiyama, Y., Miyamoto, K., Kawakami, Y., Role of endogenous adenosine in hypoxic ventilatory response in humans: a study with dipyridamole, *J. Appl. Physiol.* **76**(1), 196–203 (1994).
4. Bascom, D.A., Clement, I.D., Cunningham, D.A., Painter, R., Robbins, P.A., Changes in peripheral chemoreflex sensitivity during sustained, isocapnic hypoxia, *Respir. Physiol.* **82**(2), 161–176 (1990).
5. Nishimura, M., Suzuki, A., Nishiura, Y., Yamamoto, H., Miyamoto, K., Kishi, F., Kawakami, Y., Effect of brain blood flow on hypoxic ventilatory response in humans, *J. Appl. Physiol.* **63**(3), 1100–1106 (1987).
6. Iijima, T., Aoyagi, T., Iwao, Y., Masuda, J., Fuse, M., Kobayashi, N., Sankawa, H., Cardiac output and circulating blood volume analysis by pulse dye-densitometry, *J. Clin. Monit.* **13**(2), 81–89 (1997).
7. Slutsky, A.S., Rebuck, A.S., Heart rate response to isocapnic hypoxia in conscious man, *Am. J. Physiol.* **234**(2), H129–132 (1978).
8. Yamamoto, M. Mechanism of ventilatory and heart-rate responses during sustained hypoxia in humans–role of endogenous adenosine, *Hokkaido Igaku Zasshi* **72**(6), 583–96 (1997). [Article in Japanese]
9. Marshall, J.M., Metcalfe, J.D., Analysis of the cardiovascular changes induced in the rat by graded levels of systemic hypoxia, *J. Physiol.* **407**, 385–403 (1988).
10. Kato, H., Menon, A.S., Slutsky, A.S., Mechanisms mediating the heart rate response to hypoxemia, *Circulation* **77**(2): 407–414 (1988).
11. Widdicombe, J.G., Respiratory reflexes in man and other mammalian species, *Clin. Sci.* **21**, 163–170 (1961).
12. Cargill, R.I., Kiely, D.G., Lipworth, B.J., Left ventricular systolic performance during acute hypoxemia, *Chest* **108**(4), 889–891 (1995).
13. Marshall, J.M., Metcalfe, J.D., Effects of systemic hypoxia on the distribution of cardiac output in the rat, *J. Physiol.* **426**: 335–53 (1990).

Respiratory, Cerebrovascular and Pressor Responses to Acute Hypoxia: Dependency on $P_{ET_{CO_2}}$

Philip N. Ainslie and Marc J. Poulin

1. Introduction

Acute hypoxia leads to changes not only in ventilation but also in cardiovascular [1] and cerebral blood flow (CBF) dynamics [2]. However, there seems to be no available data concerning the combined ventilatory, cardiovascular and cerebrovascular responses to acute hypoxia in humans. Further, although hypercapnia may enhance the acute hypoxic ventilatory response (AHVR) [3], it has not been clearly shown how hypercapnia may regulate changes in the cardiovascular and cerebrovascular responses to acute hypoxia. The lack of investigations surrounding the regulation and integration of the ventilatory, cerebrovascular and cardiovascular response by CO_2 to acute hypoxia is somewhat surprising when one considers the important clinical relevance of such responses in health and disease.

The aim of the present study, therefore, was to clarify (1) the ventilatory, cardiovascular and cerebrovascular responses to acute hypoxia and the role PCO_2 plays on these responses and (2) to examine the interrelationships between the respiratory, cerebrovascular and pressor responses to acute hypoxia and their dependence on end-tidal (i.e. arterial) PCO_2 ($P_{ET_{CO_2}}$). The present study supplies further elucidation of the integrated respiratory, cerebrovascular and pressor responses to acute hypoxia and provide insight into the potential relationships between ventilation, CBF and cardiovascular regulation in humans. We tested the hypothesis that increases in ventilatory sensitivities to hypoxia, caused by increasing the $P_{ET_{CO_2}}$ level would also be reflected in, and potentially related to, increases in CBF and cardiovascular sensitivities.

Philip N. Ainslie[1] and Marc J. Poulin[1,2,3] • [1]Departments of Physiology & Biophysics and [2]Clinical Neurosciences, Faculty of Medicine and [3]Faculty of Kinesiology, University of Calgary, 3330 Hospital Drive NW, Calgary Alberta, T2N 4N1 Canada.

Post-Genomic Perspectives in Modeling and Control of Breathing, edited by Jean Champagnat, Monique Denavit-Saubié, Gilles Fortin, Arthur S. Foutz, Muriel Thoby-Brisson. Kluwer Academic/Plenum Publishers, 2004.

2. Methods

2.1. Subjects

Nine healthy male subjects (24.2 ± 3.5 (mean ± SD) years) participated in this study. Participants were not taking any medication, all were non-smokers, and none had any history of cardiovascular, cerebrovascular, or respiratory disease. The research study was approved by the Conjoint Health Research Ethics Board at the University of Calgary.

2.2. Protocol

The experiments were conducted in our laboratory located at 1103 m above sea level, and the average barometric pressure for the study days was 663 ± 12 Torr. Each subject was required to make two visits to the laboratory. During the initial visit, measurements of resting end-tidal gases and estimates of hypoxic and hypercapnic sensitivities were conducted, and the subjects became familiarized with the apparatus and experimental testing procedures. For the next visit, subjects reported to the laboratory at the same time of day with each visit separated by at least 3 days. During the final visit, in a randomized design, subjects conducted 3 separate protocols. Each protocol was separated by a 45 min rest period. Prior to the experimental protocol, the subject's normal $P_{ET_{CO_2}}$ and end-tidal O_2 tension ($P_{ET_{O_2}}$) were measured prior to the experiment for approximately 10 min, as described in detail elsewhere [4].

2.3. Incremental Step Hypoxic Protocol

Each experimental protocol began with an eight-minute period during which the subject breathed normally through a mouthpiece with the nose occluded by a nose clip. Accurate control of the end-tidal gases was achieved using the technique of dynamic end-tidal forcing (BreatheM v2.07, University Laboratory of Physiology, Oxford, UK), as described previously [4]. After an eight minute lead-in period of euoxia ($P_{ET_{O_2}} = 88$ Torr) the hypoxic stimulus was varied by holding the $P_{ET_{O_2}}$ at 7 different predetermined levels ($P_{ET_{O_2}} = 88.0$ 75.2, 64.0, 57.0, 52.0, 48.2, and 45.0 Torr), with each step lasting 90 sec. These levels of hypoxia were calculated to provide equal steps in oxygen saturation of the arterial blood (Sa_{O_2}), by using the relationship described by Severinghaus [5]. Using the same hypoxic ramp protocol, each subject conducted 3 protocols: (1) poikilocapnic hypoxia ($P_{ET_{CO_2}}$ uncontrolled), (2) isocapnic hypoxia ($P_{ET_{CO_2}}$ held 1.0 Torr above resting value) and (3) hypercapnic hypoxia ($P_{ET_{CO_2}}$ held 9.0 Torr above resting value).

2.4. Measurement of Cerebral Blood Flow Velocity, Heart Rate and Blood Pressure:

Backscattered Doppler signals from the right middle cerebral artery (MCA) were measured continuously during the protocol using a 2 MHz pulsed Doppler Ultrasound system (TC22, SciMed, Bristol, England). The MCA was identified by an insonation pathway through the right temporal window just above the zygomatic arch by using search techniques described previously [6]. In this study, the peak blood velocity was acquired every

10 ms and averaged over each heart beat, and this was used as the primary index of CBF[6]. During each test, mean arterial blood pressure (MAP) and heart rate (HR) were measured continuously using finger photoplythesmography and an ECG monitor, respectively.

2.5. Data Analysis

Sensitivities to hypoxia during each of 3 hypoxic conditions were determined by linear regression between the mean ventilation, CBF, heart rate, MAP and arterial oxygen saturation (100-SaO_2) during the final 15 sec of each incremental step of hypoxia. To calculate the percentage (%) change from baseline, data were averaged over the 7 min period of euoxia immediately preceding any changes in $P_{ET_{O_2}}$. Variables are presented as means ± standard deviation (SD). Data were initially tested for normality, before being analyzed by a two-way repeated-measures analysis of variance. Post hoc tests (Tukeys) were performed to isolate significant differences. Relationships between variables were examined using linear regression. Statistical significance was set at $P \leq 0.05$ for all statistical tests.

3. Results

3.1. General Observations

During the poikilocapnic condition, both CBF and MAP remained relatively unchanged (Figure 1). Although the responses were variable within subjects, ventilation tended to increase during the poikilocapnic hypoxic conditions, which was reflected in the subsequent decreases in $P_{ET_{CO_2}}$ (Figure 1).

3.2. Hypoxic Sensitivities (AHVR)

When compared to the poikilocapnic conditions, the AHVR increased significantly by 1.76 and 2.83 l · min^{1} %^{1} in the isocapnic and hypercapnic conditions, respectively.

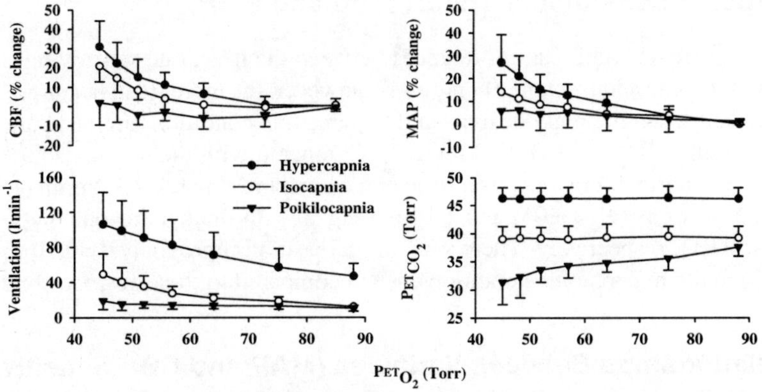

Figure 1. Influence of $P_{ET_{CO_2}}$ on the respiratory, cerebrovascular and pressor responses to acute hypoxia during poikilocapnic, isocapnic and hypercapnic conditions. Values are mean ± SD; n = 9.

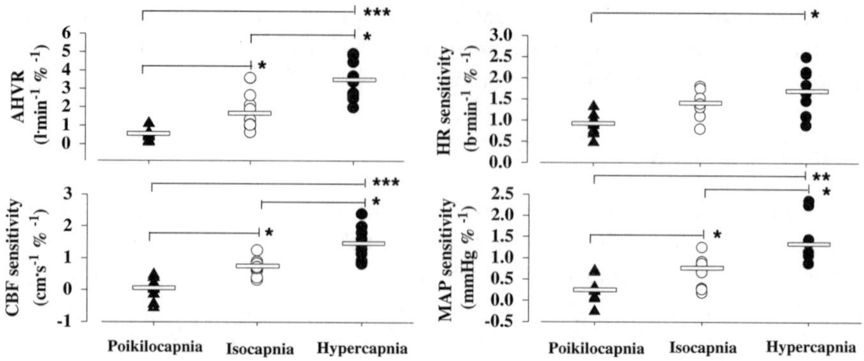

Figure 2. Individual ventilatory (AHVR), cerebral blood flow velocity (CBF), heart rate (HR) and mean arterial blood pressure (MAP) sensitivities to acute hypoxia during poikilocapnic, isocapnic and hypercapnic conditions. Horizontal bars represent mean data. * denotes significant differences (*, $P < 0.05$; **$P < 0.01$; ***$P < 0.001$).

Further, the AHVR was $1.07\,\mathrm{l} \cdot \mathrm{min}^{-1}\ \%^{-1}$ higher in hypercapnic condition when compared with the isocapnic conditions ($P < 0.05$).

3.3. Hypoxic Sensitivities (CBF)

As shown in Figure 2, there were unremarkable changes in the CBF sensitivity during conditions of poikilocapnic hypoxia. Maintaining the $\mathrm{PET}_{\mathrm{CO_2}}$ during the isocapnic conditions increased the CBF sensitivity by approximately $0.59\ \mathrm{cm} \cdot \mathrm{s}^{-1}\ \%^{-1}$ ($P < 0.05$), when compared to the poikilocapnic conditions. Likewise, there was a comparable increase of approximately $0.67\ \mathrm{cm} \cdot \mathrm{s}^{-1}\ \%^{-1}$ ($P < 0.05$) in the hypercapnic conditions, when compared to the isocapnic conditions; this represents an increase of $1.26\ \mathrm{cm} \cdot \mathrm{s}^{-1}\ \%^{-1}$ ($P < 0.001$) in the hypercapnic condition, when compared to the poikilocapnic condition (Figure 2).

3.4. Hypoxic Sensitivities (Heart Rate and MAP)

There were no significant differences between the heart rate sensitivities to hypoxia in the poikilocapnic and isocapnic conditions; however, the increase in heart rate sensitivity was significant when the poikilocapnic and hypercapnic conditions are compared (0.88 vs. 1.48 beats $\cdot\ \mathrm{min}^{-1}\ \%^{-1}$; $P < 0.05$; Figure 2). Compared with the poikilocapnic condition, the MAP sensitivities were increased on average by approximately 0.44 mmHg $\%^{-1}$ in the isocapnic conditions ($P < 0.05$) and 1.14 mmHg $\%^{-1}$ in the hypercapnic hypoxic conditions ($P < 0.01$), respectively. There was an increase of approximately 0.70 mmHg $\%^{-1}$ ($P < 0.05$) in the hypercapnic conditions, when compared to the isocapnic conditions.

3.5. Relationships Between Variables (MAP and CBF Sensitivities)

The similar changes in MAP and CBF sensitivities during each condition were reflected in significant correlations between the MAP and CBF sensitivities in both hypercapnic (r = 0.84; $P < 0.05$) and isocapnic (r = 0.80; $P < 0.05$) conditions but not in the poikilocapnic (r = 0.24; *NS*) condition.

3.6. Relationships Between Variables (AHVR and CBF Sensitivities)

Strong correlations were evident between the AHVR and CBF sensitivities in both hypercapnic (r = 0.83; $P < 0.05$) and poikilocapnic (r = −0.71; $P < 0.01$) conditions but not in the isocapnic (r = 0.1; NS) condition.

4. Discussion

The present study has yielded a number of important and novel findings. Firstly, the ventilatory, CBF and cardiovascular responses to hypoxia are strongly influenced and regulated by $P_{ET_{CO_2}}$. Secondly, the strong correlations between the AHVR and CBF sensitivities suggest (1) a high AHVR decreases the CBF sensitivity during poikilocapnic hypoxia, most probably due to hypocapnic-induced cerebral vasoconstriction and (2) a high AHVR increases CBF sensitivity during periods of hypercapnic hypoxia, possibly mediated by hypercapnic and hypoxic-induced sympathetic activation and/or the combined effects of both hypercapnic and hypoxic-induced cerebral vasodilation.

4.1. Ventilatory and CBF Responses to Hypoxia

CBF rises in proportion to the severity of isocapnic hypoxia in all mammals. However, during poikilocapnic hypoxia, there is considerable variability between individuals and species because of the wide variability in AHVR, which determines how far the P_{CO_2} falls [7]. A strong AHVR and a large fall in P_{CO_2} in experimental subjects led to early reports that hypoxic-induced cerebral vasodilation was a threshold phenomenon, scarcely beginning until Pa_{O_2} had fallen to about 30 Torr [8]. Our results from the poikilocapnic protocol are not inconsistent with this 'threshold phenomenon' i.e. the vasodilatory effects of the hypoxia were balanced out by the vasoconstrictory effects of the hypocapnic, resulting in little change in the CBF.

4.2. Relationship Between Variables (MAP & CBF Sensitivities)

Although hypoxia and hypercapnia are known to attenuate normal effective cerebral autoregulation [9], this was not apparent in the poikilocapnic hypoxia protocol. Blood pressure sensitivity was elevated during both the isocapnic and hypercapnic hypoxia protocols, indicating a greater degree of sympathetic-mediated cardiovascular stress. Cerebral perfusion pressure, the difference between MAP and intracranial pressure when the latter exceeds central venous pressure [10] is a major determinant of CBF. Although we did not measure intracranial pressure or central venous pressure in our study, it is likely that both increased with increasing levels of $P_{ET_{CO_2}}$ [6]. If so, cerebral perfusion pressure would be higher during the isocapnic and hypercapnic hypoxia protocols, when compared with the poikilocapnic hypoxia protocol. This likely increase in cerebral perfusion pressure is potentially the major mediator of the increase in the CBF during the isocapnic and hypercapnic hypoxia protocols. Although not cause and effect, the relationship between the MAP and CBF sensitivities during isocapnia and hypercapnia, when cerebral autoregulation appears to be attenuated, are consistent with the view that increases in MAP would lead to parallel increases in cerebral perfusion pressure and therefore CBF.

4.3. Relationship Between Variables (AHVR and CBF Sensitivities)

During the poikilocapnic condition, a strong negative relationship was found between the AHVR and the CBF sensitivities. This relationship is perhaps not surprising since a higher ventilatory drive to hypoxia per se will elicit greater decreases in $P_{ET_{CO_2}}$. Decreases in $P_{ET_{CO_2}}$ cause a rapid decrease in CBF due to hypocapnic-induced cerebral vasoconstriction [9].

In the present study, it is of interest that there were significant correlations between the AHVR and CBF sensitivities during the hypercapnic condition. We speculate on two possible explanations for this relationship. Firstly, the effects of both hypoxia and hypercapnia will cause increases in the sympathetic-mediated cardiovascular responses [13], resulting in peripheral vasoconstriction and subsequent increases in cardiac output and hence MAP [5]. Since cerebral autoregulation is attenuated during these hypoxic and hypercapnic conditions [1], such increases in MAP are likely to cause increases in cerebral perfusion pressure and therefore CBF. A second possibly for the correlations between the AHVR and CBF sensitivities during the hypercapnic condition may be related to the combined effects of the hypercapnic and hypoxic-induced cerebral vasodilation.

5. Acknowledgements

This project was supported by the Alberta Heritage Foundation for Medical Research (AHFMR), Heart and Stroke Foundation of Alberta, NWT, & Nunavut and the Canadian Institutes of Health Research. PNA is supported by a AHFMR post-doctoral fellowship. We extend our gratitude to Professor PA Robbins for assistance in setting up the dynamic end-tidal forcing technique in Calgary.

References

1. J.E. Brian, Carbon dioxide and the cerebral circulation, *Anesthesiology* **88**, 1365–1386 (1998).
2. R.S. Fitzgerald and S. Lahiri, Reflex responses to chemoreceptor stimulation, in: *Handbook of Physiology*, edited by A.P. Fishman. (American Physiological Society, Bethesda, Maryland, 1986), pp. 313–362.
3. A. Hanada, M. Sander and J. Gonzalez-Alonso, Human skeletal muscle sympathetic nerve activity, heart rate and limb haemodynamics with reduced blood oxygenation and exercise, *J. Physiol.* **551**, 635–647 (2003).
4. K. Ide, M. Eliasziw, and M.J. Poulin, The relationship between middle cerebral artery blood velocity and end-tidal P_{CO_2} in the hypocapnic-hypercapnic range in humans, *J. Appl. Physiol.* **95**, 129–137 (2003).
5. J.M. Marshall, Peripheral chemoreceptors and cardiovascular regulation, *Physiol. Rev.* **74**, 543–594 (1994).
6. J.R. Munis and L.J. Lozada, Giraffes, siphons, and starling resistors. Cerebral perfusion pressure revisited, *J. Neurosurg. Anesthesiol.* **12**, 290–296 (2000).
7. M.J. Poulin and P.A. Robbins, Indexes of flow and cross-sectional area of the middle cerebral artery using doppler ultrasound during hypoxia and hypercapnia in humans, *Stroke* **27**, 2244–2250 (1996).
8. M.J. Poulin and P.A. Robbins, Influence of cerebral blood flow on the ventilatory response to hypoxia in humans, *Exp. Physiol.* **83**, 95–106 (1998).

9. J.W. Severinghaus, Cerebral circulation at high altitude, in: *High Altitude: An Exploration of Human Adaptation*, edited by T.F. Hornbein and R.B. Schoene. (Marcel Dekker, New York, 2001), pp. 343–375.
10. J.W. Severinghaus, Simple, accurate equations for human blood O_2 dissociation computations, *J. Appl. Physiol.* **46**, 599–602 (1979).
11. S. Shimojyo, P. Scheinberg, K. Kogure, and O.M. Reinmuth, The effects of graded hypoxia upon transient cerebral blood flow and oxygen consumption, *Neurology* **18**, 127–133 (1968).
12. A. Weyland, W. Buhre, S. Grund, H. Ludwig, S. Kazmaier, W. Weyland, and H. Sonntag, Cerebrovascular tone rather than intracranial pressure determines the effective downstream pressure of the cerebral circulation in the absence of intracranial hypertension. *J. Neurosurg. Anesthesiol.* **12**, 210–216 (2000).
13. A. Xie, J.B. Skatrud, D.C. Crabtree, D.S. Puleo, B.M. Goodman, and B.J. Morgan, Neurocirculatory consequences of intermittent asphyxia in humans, *J. Appl. Physiol.* **89**, 1333–1339 (2000).

Can Cardiogenic Oscillations Provide an Estimate of Chest Wall Mechanics?

Eve Bijaoui, Daniel Anglade, Pascale Calabrese, André Eberhard, Pierre Baconnier and Gila Benchetrit

1. Introduction

Every time the heart beats, it produces a mechanical deformation of the lungs causing small fluctuations of airway pressure and flow called cardiogenic oscillations (CO). CO have been observed on respiratory signals during pulmonary function tests, during relaxed expiration as well as during apnea, as a mean for differentiating central and obstructive apneas[1]. Finally, we have recently shown that the processing of CO in mouth pressure and airflow can be used as a non-invasive measurement of airway resistance[2].

Ventilation can be assessed non invasively by respiratory inductive plethysmography (RIP) which follows the cross-sectional areas of the ribcage and abdomen as they change during respiration. Several studies have shown that a good approximation of tidal volume[3] and derived airflow[4] can be obtained with RIP using a specific calibration procedure. RIP can be used as a non-invasive technique for computation of respiratory mechanics[5]. CO can also be observed on thoracic and abdominal signals obtained by RIP. The aim of the study was to investigate the use of CO recorded on a RIP signal and in airway pressure as means for determining a parameter related to the mechanical properties of the chest wall.

2. Methods

2.1. Experimental Protocol

We studied 4 human volunteers (2M/2W) with no history of lung disease. Ribcage (*RC*) and abdomen (*AB*) movements were recorded with computer-assisted respiratory inductance plethysmography (*RIP*) (Visuresp, RBI, France). Airflow ($\dot{V}$) was measured

---

Eve Bijaoui, Daniel Anglade, Pascale Calabrese, André Eberhard, Pierre Baconnier and Gila Benchetrit • Laboratoire PRETA-TIMC, Université Joseph Fourier, 38700 La Tronche, France.

Post-Genomic Perspectives in Modeling and Control of Breathing, edited by Jean Champagnat, Monique Denavit-Saubié, Gilles Fortin, Arthur S. Foutz, Muriel Thoby-Brisson. Kluwer Academic/Plenum Publishers, 2004.

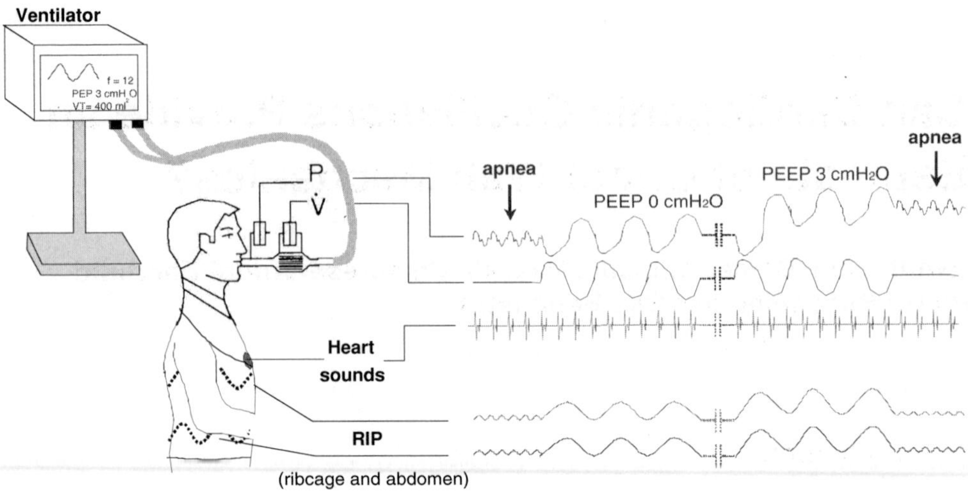

Figure 1. Experimental setup for cardiogenic oscillations recordings. P: mouth pressure; $\dot{V}$: pneumotach flow.

with a pneumotachograph (Fleisch #1, Lausanne, Switzerland) connected to a differential pressure transducer (163PC01D36, Micro Switch). Mouth pressure (P) was measured with a differential pressure transducer (142PC01D, Micro Switch) connected to a side port proximal to the pneumotachograph. Simultaneous recording of heart sounds, by placing a microphone over the chest (Figure 1), allowed us to decompose the signals into individual cardiogenic cycles.

Subjects wore a nose clip and laid down in the supine position while breathing through a mouth-piece. All subjects were mechanically ventilated with the Dräger Evita 4 ventilator in the spontaneous mode. Measurements were made at the end of expiration over a period of up to 12 s during which subjects attempted to relax all respiratory muscles while supporting the cheeks with the hands and keeping the glottis open. The ventilator settings, established by the responsible physician, were kept constant for each subject throughout the experiment, with the exception of PEEP. Six different levels of PEEP in 1-cmH_2O stages and from 0 to 5 cmH_2O, were applied in random order and maintained at least two minutes before taking measurements. Figure 1 gives a representation of the experimental setup for cardiogenic oscillations recordings on P, RC and AB signals.

In order to mimic changes in chest wall compliance, the subjects were asked to perform the same experiment having their arms raised.

All signals were sampled at 100 Hz with a 16-bit analog-digital converter (PowerLab, ADInstruments, Australia). Data acquisition was performed using the Chart^{TM} data acquisition software (ADInstruments, Australia) and signals were stored on a PC computer using for further analysis.

2.2. Data Processing

Data analysis was carried out using the Matlab^{TM} 6.5 mathematical software (The Mathworks, Nattick, MA).

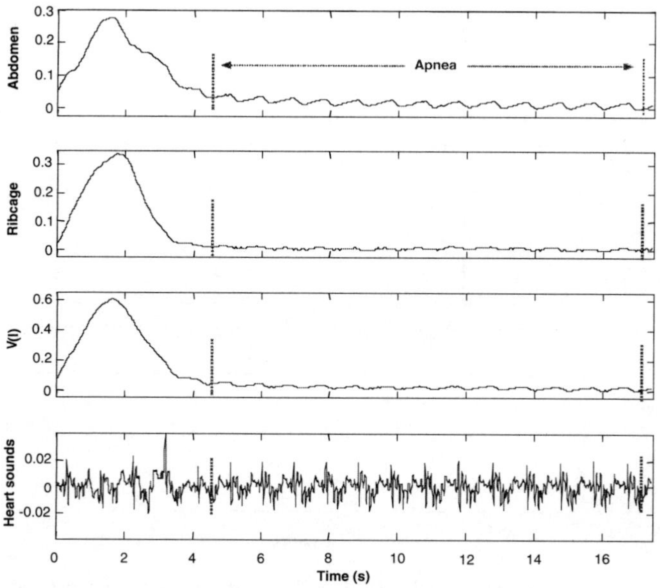

Figure 2. Representative example of abdomen, ribcage signals and computed RIP volume V measured during a voluntary breath-hold episode, together with the heart sounds signals.

The method used for calibration of the *RIP* signal in order to get an estimate of the absolute volume has been described in details elsewhere[4]. Briefly, a least-squares method was applied over one breath to obtain a *RIP* volume (V) by a specific combination of *RC* and *AB* signals compared to the integrated flow signal. Figure 2 shows a representative example of *RC*, *AB* and V signals obtained in one subject during a breath-hold episode and demonstrates that the presence of cardiogenic oscillations on both *RC* and *AB* as well as on the computed V.

We first used the heart sound signal to divide the computed V and P into individual cardiogenic cycles[2]. The first heart sound, which had the largest amplitude and identifies the onset of ventricular systole, was used as the reference for each cardiogenic oscillations in V or P. The volume changes (ΔV_{CO}) and pressure changes (ΔP_{CO}) caused by cardiogenic oscillations were then calculated for each cycle (Figure 3). For each subject and each level of PEEP, ten cardiogenic cycles were analyzed and the ratio C_{CO} (l/cmH_2O) of ΔV_{CO} to ΔP_{CO} was computed for each cycle. Finally, a mean value of C_{CO} was calculated for each subject at each PEEP level.

3. Results

Figure 4 shows the effects of PEEP on C_{CO} for all subjects. We found that there was no marked change in the value of C_{CO} over the PEEP range investigated. Subject #4 was not able to relax and keep the glottis open at 0 cmH_2O PEEP therefore we did not get any good C_{CO} value for this PEEP level.

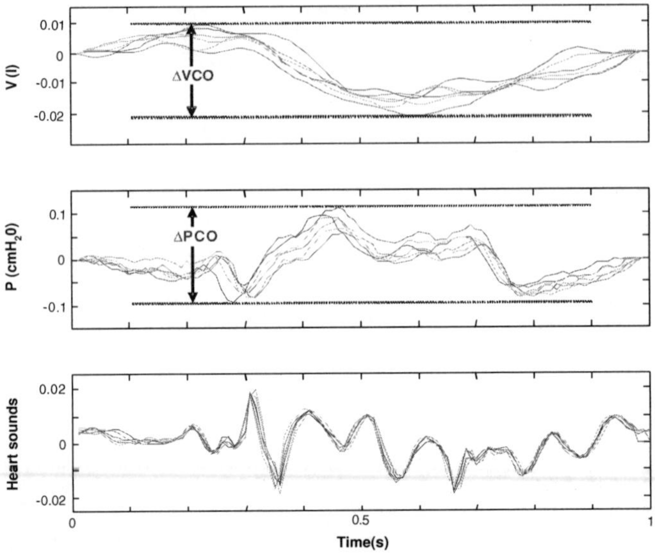

Figure 3. Cycle-by-cycle cardiogenic oscillations on RIP volume (top) and mouth pressure (middle), together with cycle-by-cycle heart sounds (bottom panel). ΔV_{CO}: variations in RIP volume caused by CO; ΔP_{CO}: variations in mouth pressure caused by CO

Figure 5 shows that repeating the experiment with the subject having the arms raised resulted in a marked decrease in C_{CO}. Values of C_{CO} are taken at 0 cmH_2O PEEP, except for subject #4, in whom values shown were obtained at 1 cmH_2O PEEP.

4. Discussion

Our goal was to investigate the use of CO in volume measured by RIP technique and pressure as means for determining a parameter related to chest wall mechanics. This involved first obtaining and processing *P* and *V* in an appropriate way, and then interpreting the resulting C_{CO} in physiological terms. The data acquisition and processing steps involved a number of assumptions.

We have previously discussed some of the issues about using cardiogenic oscillations in pressure and flow[2]. An important point is the effect of an apnea on cardiac activity. The changes in heart rate throughout the breath-hold maneuver were small enough to be considered negligible and the individual cycles of *P* and *V* were highly reproducible (Figure 3). Furthermore, the computation of individual C_{CO} values at each PEEP level showed good reproducibility (Figure 5), thus ensuring the robustness of the method.

The use of calibrated RIP tries to quantitatively relate changes in RC and AB signals to tidal volume contributions from the ribcage and abdominal compartments, respectively. However, the calibration factors involved are sensitive to posture[5]. Our subjects were in supine position for the entire experiment duration therefore we did not expect our calibration factors to change throughout the recording session.

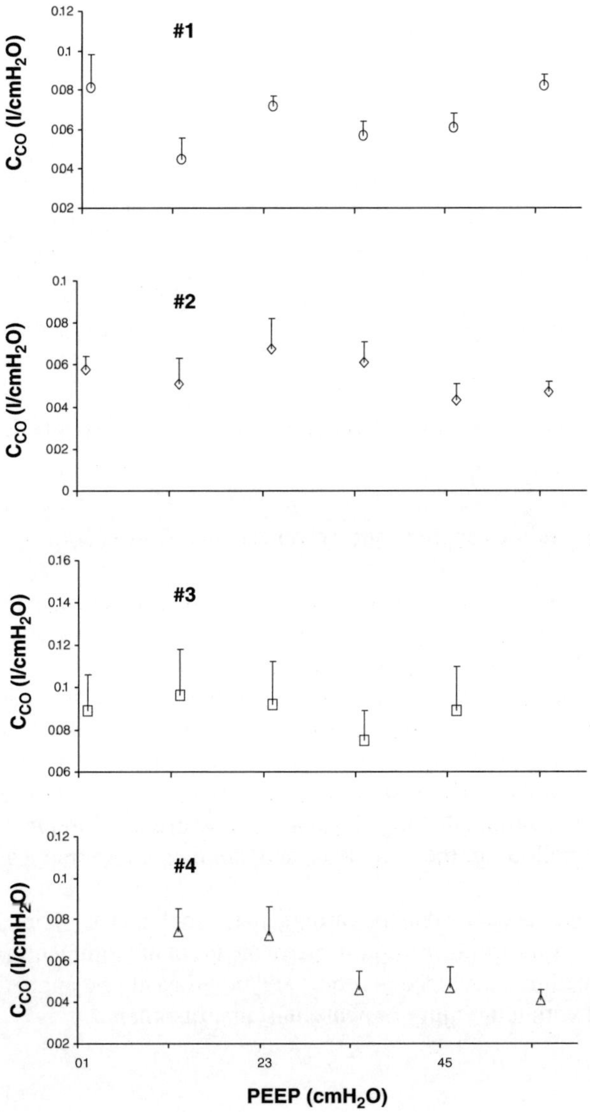

Figure 4. C_{CO} versus PEEP levels for all subjects (#1–#4). Each symbol represents the mean values for each PEEP level. Error bars are SDs.

We found that there was no marked change in the value of C_{CO} over the PEEP range investigated (Figure 5). In addition, the values are similar to chest wall compliance values reported by Pelosi et al.[6]. Indeed, they found that in normal subjects, chest wall compliance remained unaffected by changes in PEEP varying from 0 to 5 cmH$_2$0.

We also found that changing arm position lead to a decrease in C_{CO} values. Arm elevation is likely to "stiffen" the chest wall along with modifications in ribcage and abdominal movements[7]. An increase in chest wall stiffness is characterized by a decrease in chest wall

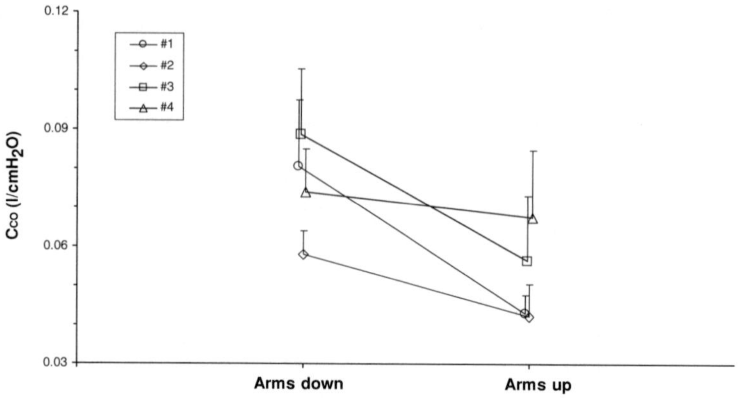

Figure 5. Effect of raising the arms on C_{CO}. Each symbol represents the mean of 10 cardiac cycles; Errors bars are SDs of the mean.

compliance. These findings support our hypothesis that C_{CO} is somehow related to chest wall mechanics.

Although the lung contribution to respiratory system mechanics is primarily altered in numerous respiratory diseases, its chest wall contribution is not to be neglected[6]. For instance, chest wall compliance is decreased in patients with acute respiratory distress syndrome and acute lung injury. Chest wall compliance is also decreased in restrictive diseases, such as neuromuscular and thoracic wall disorders[8]. Chest wall mechanics can be obtained in spontaneously breathing subjects using whole body plethysmography, which requires the subject to be alert. In mechanically ventilated patients, chest wall mechanics are determined using esophageal pressure measurements, which requires the placement of a balloon in the esophagus and can thus be impractical in some clinical situations.

In summary, we speculate that the cardiogenic compliance C_{CO} obtained from cardiogenic oscillations may be useful as a non-invasive means of obtaining information chest wall compliance. The method developed is robust and could easily be applied in mechanically ventilated subjects with lung injury or neuromuscular disorders.

References

1. Ayappa I., Norman R.G. and Rapoport D.M. Cardiogenic oscillations on the airflow signal during continuous positive airway pressure as a marker of central apnea. *Chest.* 116(3): 660–666 (1999)
2. Bijaoui E., Baconnier P.F. and Bates J.H.T. Mechanical output of the lung using cardiogenic oscillations. *J. Appl. Physiol.* 91(2): 859–865 (2001)
3. Banzett R.B., Mahan S.T., Garner D.M., Brughera A. and Loring S.H. A simple and reliable method to calibrate respiratory magnometers and respitrace. *J. Appl. Physiol.* 79(6): 2169–2176 (1995)
4. Eberhard A., Calabrese P., Baconnier P.F. and Benchetrit G. Comparison between the respiratory inductance plethysmography signal derivative and the airflow. *Adv. Exp. Med. Biol.* 499: 489–494 (2001)
5. Habib R.H., Pyon K.H. and Courtney S.E. Optimal high-frequency oscillatory ventilation settings by nonlinear lung mechanics analysis. *Am. J. Respir. Crit. Care Med.* 166: 950–953 (2002)

6. Pelosi P., Cereda M., Foti G., Giacomini M. and Pesenti A. Alterations of lung and chest wall mechanics in patients with acute lung injury: effects of positive end-expiratory pressure. *Am. J. Resp. Crit. Care Med.* 152: 531–537 (1995)
7. Alison J.A., Reguis J.A., Donnelly P.M., Adams R.D, Sullivan C.E. and Bye P.T. End-expiratory lung volume during arm and leg exercise in normal subjects and patients with cystic fibrosis. *Am. J. Resp. Crit. Care Med.* 158(5):1450–1458 (1998)
8. Papastamelos C., Panitch H.B. and Allen J.L. Chest wall compliance in infants and children with neuromuscular disease. *Am. J. Resp. Crit. Care Med.* 154(4):1045–1048 (1996)

Nonlinear Modeling of the Dynamic Effects of Arterial Pressure and Blood Gas Variations on Cerebral Blood Flow in Healthy Humans

Georgios D. Mitsis, Philip N. Ainslie, Marc J. Poulin, Peter A. Robbins and Vasilis Z. Marmarelis

1. Introduction

Cerebrovascular resistance is controlled by multiple homeostatic mechanisms, which regulate cerebral blood flow (CBF), maintaining it relatively constant despite changes in cerebral perfusion pressure[1–2]. The regulation of CBF was long viewed as a static phenomenon, whereby the "steady-state" pressure-flow relationship is described by a sigmoidal curve with a wide plateau, suggesting that CBF remains constant despite changes in pressure within certain bounds. However, with the development of Transcranial Doppler (TCD) ultrasonography for the noninvasive, high-temporal resolution measurement of CBF velocity (CBFV), it has been shown that CBFV can vary rapidly in response to variations of systemic arterial blood pressure (ABP) over various time scales[3–4].

The availability of such data has revealed information about the dynamic properties of cerebral autoregulation and the quantitative manner in which rapid changes in pressure induce rapid changes in flow. It has also cast doubt on the validity of the notion of "steady-state", since no such "steady-state" is ever observed in the natural operation of cerebral circulation. In addition to controlled experiments, spontaneous fluctuations of beat-to-beat mean ABP (MABP) and mean CBFV (MCBFV), which possess broadband characteristics, have been recently employed for the study of dynamic cerebral autoregulation, using linear[5–7] and nonlinear[8–9] modeling methods. Impulse response or transfer function estimates in linear analysis and Volterra models in nonlinear analysis have

Georgios D. Mitsis and Vasilis Z. Marmarelis • Department of Biomedical Engineering, University of Southern California, OHE 500, University Park, Los Angeles CA 90089-1451, USA. **Philip N. Ainslie and Marc J. Poulin** • Department of Physiology & Biophysics, Faculty of Medicine, University of Calgary, HMRB-212, 3330 Hospital Drive NW, Calgary Alberta T2N 4N1, Canada. **Peter A. Robbins** • The University Laboratory of Physiology, University of Oxford, Oxford OX1 3PT, UK.

Post-Genomic Perspectives in Modeling and Control of Breathing, edited by Jean Champagnat, Monique Denavit-Saubié, Gilles Fortin, Arthur S. Foutz, Muriel Thoby-Brisson. Kluwer Academic/Plenum Publishers, 2004.

demonstrated that cerebral autoregulation is more effective in the low-frequency range (below 0.1 Hz), where most of the MABP spectral power resides (i.e., most spontaneous MABP changes do not cause large MCBFV variations). The presence of significant nonlinearities, which are more prominent in this low frequency range, was indicated in [9].

It is also well established that arterial CO_2 and O_2 produce vascular responses in cerebral vessels[1–2], with arterial CO_2 tension being one of the strongest physiologic modulators of CBF[1]. A number of studies have examined CBFV responses to step changes in CO_2 and O_2 tension[10–13], and it was shown that this response is not instantaneous but lags the gas tension changes by several seconds[12–13].

The reactivity of blood vessels to CO_2 and O_2 can be assessed by breath-to-breath measurements of end-tidal CO_2 and O_2 (P_{ETCO2} and P_{ETO2}); hence spontaneous variations of the latter may be employed in the study of CBF regulation, in a similar manner to MABP variations. It is important to note that in this way the dynamics of autoregulation are studied under normal operating conditions, in contrast to controlled experimental protocols, where the system dynamics are probed in a specific manner. To our knowledge, this has been done only in one study so far[14], whereby spontaneous breath-to-breath P_{ETCO2} variations were employed to assess the effect of arterial CO_2 on MCBFV. However, the effect of MABP and P_{ETCO2} on MCBFV was assumed to be linear and nonlinearities were included solely as second-order interaction terms, neglecting the strongly nonlinearities present in autoregulation[9]. Hence in this paper, a novel methodology for modeling multiple-input nonlinear systems, termed the Laguerre-Volterra network (LVN), is used to assess the nonlinear dynamic effects of MABP, P_{ETCO2} and P_{ETO2} (and their interactions) on MCBFV in an appropriate way.

2. Methods

2.1. Experimental Methods

The experimental data were obtained from ten healthy subjects (age: 29.1 ± 4.8 years [mean $\pm$ SD], weight 78.5 ± 4.8 kg, height: 175.4 ± 5.5 cm) under normal, free-breathing conditions. None of the participants was on any medication and all of them were normotensive and did not have a history of any cardiovascular, pulmonary or cerebrovascular diseases. ABP and CBFV were measured by finger photoplethysmography and by transcranial Doppler ultrasonography in the right middle cerebral artery respectively, while P_{ETCO2} and P_{ETO2} were measured by mass spectrometry. The surrogate Doppler signal for CBF is the mean value of the velocity corresponding to the maximum Doppler shift $\bar{V}_P$, averaged over the entire cardiac cycle. Although it does not account for changes in vessel diameter, it has been found to represent CBF well in almost all practical cases[15], especially for the conditions of this study. All experimental variables were sampled every 10 ms. Real time beat-to-beat values of MABP and MCBFV were calculated by integrating the waveform of the sampled signals within each cardiac cycle (R-R interval). The beat-to-beat values were then interpolated and resampled at 1 Hz to obtain equally spaced time series of MABP and MCBFV data. The breath-to-breath data were interpolated in order to obtain values every 1 sec.

2.2. Mathematical Methods

The multiple-input LVN (MI-LVN) is a novel variant of the Volterra-Wiener approach for modeling multiple-input nonlinear systems[16]. Its efficiency in modeling dynamic pressure autoregulation has been demonstrated previously[9]. The MI-LVN combines Laguerre expansions with Volterra-type networks, whereby each of the system inputs is preprocessed by a different Laguerre filter-bank, the outputs of which are fully connected to a layer of hidden units with polynomial activation functions. The MI-LVN representation of a system is equivalent to the general Volterra model[16], given below for a two-input, Q-th order system:

$$y(n) = k_0 + \sum_{n=1}^{Q} \sum_{i_1=1}^{2} \cdots \sum_{i_2=1}^{2} \left\{ \sum_{m_1} \cdots \sum_{m_n} k_{i_1 \ldots i_n}(m_1, \ldots, m_n) x_{i_1}(n - m_1) \cdots x_{i_n}(n - m_n) \right\} \tag{1}$$

where $x_1(n)$, $x_2(n)$ are the system inputs, $y(n)$ *is* the system output and $k_{i1\ldots i}$ denotes the n-th order Volterra kernel of the system. If $i_1 = \cdots = i_n$, $k_{i\ldots i}$ denotes the i-th input, n-th order self-kernel, which describes the linear ($n = 1$) and nonlinear ($n > 1$) effects of the i-th input on the output, whereas if some of $i_1, \ldots, i_n$ are different, it denotes the n-th order cross-kernel of the system, which describes the nonlinear interactions between the two inputs.

The Volterra kernels in (1) are obtained in terms of the parameters of the MI-LVN, which are in turn determined by training the network with the input-output data via an iterative gradient descent scheme. For more details, the reader is referred to[16]. In this case, the input signals of the MI-LVN are the spontaneous MABP, P_{ETCO2} and P_{ETO2} fluctuations, while the output signal is the MCBFV variations. Six-minute data segments (i.e., with a length of 360 points) are used to train the network.

3. Results

The mean values for the experimental data, averaged over the 45 min recordings, are equal to 78.9 ± 8.2 mm Hg, 38.6 ± 2.6 mm Hg, 86.1 ± 4.6 mm Hg and 55.5 ± 4.4 cm/sec for MABP, P_{ETCO2}, P_{ETO2} and MCBFV respectively. Most of the signals power lies below 0.1 Hz, while MCFV exhibits some power up to 0.3 Hz. Note that the mean values of P_{ETCO2}, P_{ETO2} are lower than their normal values (40 mm Hg and 100 mm Hg respectively), since the study was conducted at a high altitude ($\sim$1103 m above sea level, barometric pressure $\sim$660 mmHg).

The performance of the MI-LVN model is assessed in terms of the achieved output prediction Normalized Mean Square Error (NMSE), which is defined as the sum of squares of the residuals of the model prediction (i.e., difference from the true output) over the sum of squares of the true output. The NMSEs for one-input (whereby the model input is MABP, P_{ETCO2} or P_{ETO2}) and two-input (whereby the first input is MABP and the second is P_{ETCO2} or P_{ETO2}) models are given in Table 1. It should be noted that all the NMSEs correspond to models with the same number of free parameters. While MABP variations are the main determinant of MCBFV variations, the incorporation of P_{ETCO2} or P_{ETO2} as additional model inputs results in a considerable reduction of the achieved prediction NMSE

Table 1. Mean prediction NMSEs (±standard deviations) for one- and two-input models.

	Model input(s)/NMSE [%]				
Model order Q	MABP	P_{ETCO2}	P_{ETO2}	MABP & P_{ETCO2}	MABP & P_{ETO2}
1 (Linear)	32.8 ± 13.2	71.6 ± 12.1	73.3 ± 15.3	24.8 ± 11.1	24.2 ± 10.6
3 (Nonlinear)	20.0 ± 9.2	51.5 ± 10.4	51.8 ± 13.6	14.5 ± 6.9	14.9 ± 6.7

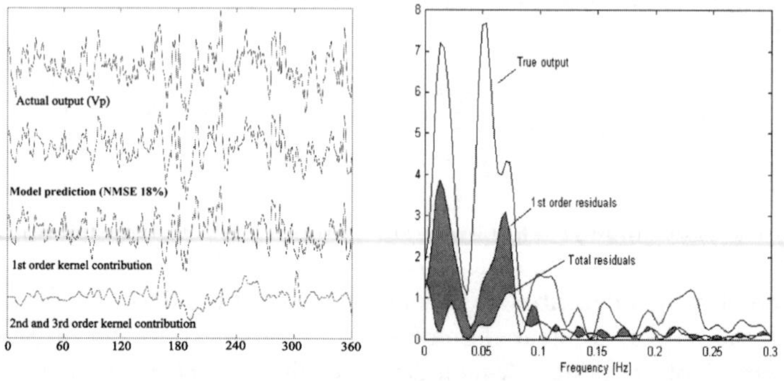

Figure 1. Left panel: Actual MCBFV output and model predictions (total, linear and nonlinear terms) for a typical data segment. Right panel: Spectra of actual output and model residuals (linear and total).

(over 5%). The obtained results for P_{ETCO2} and P_{ETO2} are similar, reflecting the fact that the two signals are correlated; hence their dynamic effects on MCBFV are analogous. The NMSE reduction achieved when nonlinear (third-order) models are employed is significant (over 10% compared to linear models).

The performance of the MI-LVN model for a typical data segment, when MABP and P_{ETCO2} are used as model inputs is illustrated further in Fig. 1, where the actual output (MCBFV) is shown along with the model prediction, and the decomposition of the latter into its linear and nonlinear components (left panel). In the right panel, the spectrum of the MCBFV output is shown along with the spectra of the linear and the total model residuals. The model prediction is close to the true output and the contribution of the nonlinear terms, as denoted by the shaded area, is prominent below 0.08 Hz.

The decomposition of the model output into its corresponding MABP, P_{ETCO2} and cross-term components is shown in the left panel of Fig. 2, while the spectra of the output, MABP and total residuals are shown in the right panel. The contribution of the P_{ETCO2} terms and the cross-terms is significant and lies in the low frequencies (below 0.08 Hz and especially below 0.04 Hz—denoted by the shaded area). By comparing Figs. 1 and 2 it can be inferred that the relative nonlinear-to-linear effect is greater for P_{ETCO2} than MABP. When P_{ETO2} is used as a second model input, the results are very similar (i.e., its contribution lies mainly in the low frequency range) and are not shown separately in the interest of space.

The first-order MABP and P_{ETCO2} kernels (k_1 and k_2 in Eq. 1 respectively) for one subject, averaged over the 45 min recording, are shown in Fig. 3 in the time and frequency domains (log-linear plots). The MABP kernel is very consistent among different segments,

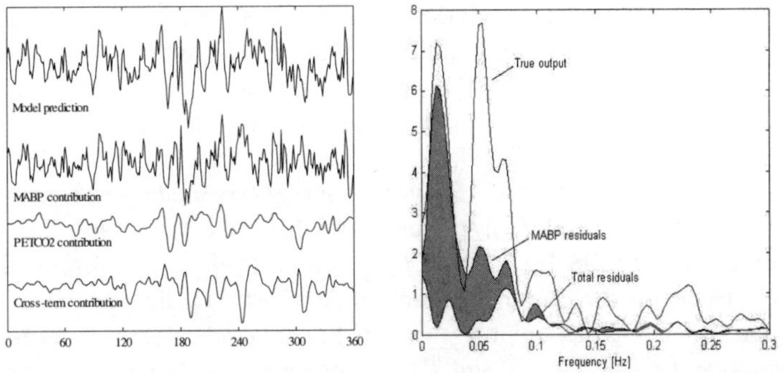

Figure 2. Left panel: Actual output, model prediction and contributions of MABP, P_{ETCO2} and nonlinear interaction terms for the data segment of Fig. 3. Right panel: Spectra of output, MABP and total residuals.

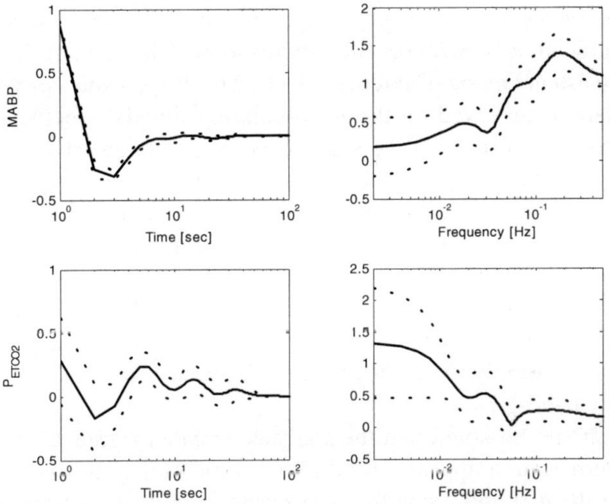

Figure 3. Average first-order MABP and P_{ETCO2} kernels (solid line) and standard deviations (dotted lines) for one subject. Left panel: time domain, right panel: FFT magnitude.

as denoted by the tight standard deviation bounds, and exhibits a high-pass characteristic, while the P_{ETCO2} kernel is more variable and exhibits a low-pass characteristic. The high variance in the initial time lags, which is consistently observed (for both the linear and nonlinear kernels), is viewed as a pure delay on the order of 3–4 sec. The latter is also observed for the linear P_{ETO2} kernel (not shown), which resembles its P_{ETCO2} counterpart, reflecting the significant correlations between P_{ETCO2} and P_{ETO2}.

4. Discussion

Spontaneous fluctuations of MABP and MCBFV have been proven to be useful in studying the dynamic characteristics of cerebral autoregulation[5–9]. Hereby, the same is shown for spontaneous fluctuations of P_{ETCO2} and P_{ETO2}, which reflect the dynamic effects of blood gas variations on CBF regulation. The obtained results describe the function of the system under normal operating conditions without any intervention, unlike previous studies[10–13]. Since the presence of nonlinearities in autoregulation is well established,[8–9] the models are obtained in an appropriate nonlinear, multiple-input context, employing a recently developed methodology[16].

The results demonstrate that MABP variations explain the largest fraction of MCBFV variations and that P_{ETCO2} and P_{ETO2} fluctuations have a considerable effect in the low frequency range (below 0.08 Hz). Since P_{ETCO2} and P_{ETO2} are significantly correlated, their dynamic effects on the output are similar. As shown by the spectra of the residuals (Figs. 1–2), MABP terms act mainly through linear mechanisms, whereas the relative nonlinear component of the P_{ETCO2} and P_{ETO2} contribution is larger. This is shown also by the achieved NMSEs when one-input models are considered (columns 1–3 in Table 1).

The linear MABP dynamics exhibit a high-pass characteristic, which implies that slow MABP changes are attenuated more effectively, i.e., autoregulation of pressure variations is more effective in the low-frequency range. The negative undershoot observed in the time domain is indicative of an autoregulatory response. The form of the MABP first-order kernel is very consistent among different subjects. The P_{ETCO2} and P_{ETO2} first-order kernels reflect the correlations observed for the corresponding signals; specifically the P_{ETO2} first-order kernel is roughly a reversed version of its P_{ETCO2} counterpart in the time domain and exhibits very similar low-pass characteristics in the frequency domain. They are also more variable with respect to different subjects. A pure delay of 3–4 sec, roughly equal for the P_{ETCO2} and P_{ETO2} responses, is associated with the high variance observed for the initial time lags of the corresponding first and second-order kernels. Most of the power of the second-order self-kernels and cross-kernels lies in the low frequency range, with the peaks being related to the corresponding linear frequency response peaks (in the case of cross-kernels this may imply the presence of nonlinear interactions which are active at the specific frequency bands).

Finally, it should be noted that the analysis revealed significant nonstationarities of no apparent pattern, which appear over different time scales, for the frequency response characteristics of the first and second-order kernels. This is an important issue that should be addressed in the future with appropriate time-frequency analysis methods.

References

1. D.D. Heistad and H.A. Kontos, Cerebral circulation, in: *Handbook of Physiology. The Cardiovascular System. Peripheral Circulation and Organ Blood Flow*, (Am. Physiol. Soc., Bethesda, MD, 1983), pp. 137–182.
2. L. Edvinsson and D.N. Krause, *Cerebral Blood Flow and Metabolism*, (Lippincott Williams and Wilkins, Philadelphia, PA, 2002).
3. D.J. Marsh, J.L. Osborn and W.J. Cowley, 1/f fluctuations in arterial pressure and regulation of renal blood flow in dogs, *Am. J. Physiol.* **258**, F1394–1400, (1990).
4. R. Zhang, J.H. Zuckerman and B.D. Levine, Spontaneous fluctuations in cerebral blood flow velocity: insights from extended duration recordings in humans, *Am. J. Physiol.* **278**, H1848–1855, (2000).
5. C.A. Giller, The frequency-dependent behavior of cerebral autoregulation, *Neurosurgery* **27**, 362–368, (1990).
6. R.B. Panerai, A.W.R. Kelsall, J.M. Rennie and D.H. Evans, Frequency-domain analysis of cerebral autoregulation from spontaneous fluctuations in arterial blood pressure, *Med. Biol. Eng. Comput.* **36**, 315–322, (1998).

7. R. Zhang, J.H. Zuckerman, C.A. Giller and B.D. Levine, Transfer function analysis of dynamic cerebral autoregulation in humans, *Am. J. Physiol.* **274**, H233–H241, (1998).
8. R.B. Panerai, S.L. Dawson and J.F. Potter, Linear and nonlinear analysis of human dynamic cerebral autoregulation, *Am. J. Physiol.* **277**, H1089–H1099, (1999).
9. G.D. Mitsis, R. Zhang, B.D. Levine and V.Z. Marmarelis, Modeling of nonlinear systems with fast and slow dynamics. II. Application to cerebral autoregulation in humans, *Ann. Biomed. Eng.* **30**, 555–565, (2002).
10. D.A. Wilson, R.J. Traystman and C.E. Rapela, Transient analysis of the canine cerebrovascular response to carbon dioxide, *Circ. Res.* **56**, 596–605, (1985).
11. I. Ellingsen, A. Hauge, G. Nicolaysen, M. Thoresen and L. Walloe, Changes in human cerebral blood flow due to step changes in P_{AO2} and P_{ACO2}, *Acta Physiol. Scand.* **129**, 157–163, (1987).
12. M.J. Poulin, P.-J. Liang and P.A. Robbins, Dynamics of the cerebral blood flow response to step changes in end-tidal P_{CO2} and P_{O2} in humans, *J. Appl. Physiol.* **81**, 1084–1095, (1996).
13. M.J. Poulin, P.-J. Liang and P.A. Robbins, Fast and slow components of cerebral blood flow response to step decreases in end-tidal P_{CO2} in humans, *J. Appl. Physiol.* **85**, 388–397, (1998).
14. R.B. Panerai, D.M. Simpson, S.T. Deverson, P Mahony, P. Hayes and D.H. Evans, Multivariate dynamic analysis of cerebral blood flow regulation in humans, *IEEE Trans. Biomed. Eng.* **47**, 419–421, (2000).
15. M.J. Poulin and P.A. Robbins, Indexes of flow and cross-sectional area in the middle cerebral artery using Doppler ultrasound during hypoxia and hypercapnia in humans, *Stroke* **27**, 2244–2250, (1996).
16. G.D. Mitsis and V.Z. Marmarelis, Modeling of nonlinear systems with fast and slow dynamics. I. Methodology, *Ann. Biomed. Eng.* **30**, 272–281, (2002).

7

VENTILATORY RESPONSE TO EXERCISE

Mixed Venous CO_2 and Ventilation During Exercise and CO_2-Rebreathing in Humans

Toru Satoh, Yasumasa Okada, Yasushi Hara, Fumio Sakamaki, Shingo Kyotani, and Takeshi Tomita

1. Introduction

The physiological mechanism of exercise-induced hyperpnea is an important and long-standing issue in respiratory physiology research.[1] However, the precise mechanism of exercise-induced hyperpnea still is not understood. It has been confirmed that neural central command from the hypothalamus plays an important role in respiratory control during exercise.[2] Activation of receptors in working muscle[3,4] and stimulation of the carotid body by elevated plasma potassium[5] have both been shown to contribute to exercise-induced hyperpnea. However, most of the other hypotheses to explain exercise-induced hyperpnea have been based on the observed close relationship between ventilation and the level of metabolic work; these hypotheses include (1) sensing of CO_2 by receptors in the pulmonary circulation,[6–9] (2) sensing of arterial CO_2 partial pressure ($PaCO_2$) as well as $PaCO_2$ oscillation by the carotid body[10,11] and (3) modulation of stretch-sensitive afferent activity by CO_2.[12,13] Thus, it has been widely assumed that metabolically produced CO_2 is the most important element in the mechanism of exercise-induced hyperpnea. Therefore, the relationship between ventilation (VE) and mixed venous CO_2 partial pressure ($PvCO_2$) as well as the relationship between VE and $PaCO_2$ needs to be fully analyzed in order to understand the contribution of metabolically produced CO_2 to exercise-induced hyperpnea. However, most of the previous reports on $PvCO_2$ dynamics during exercise in humans were not based on direct measurement, but rather on the estimation of $PvCO_2$ by CO_2 rebreathing,[14,15] and $PvCO_2$ has been directly measured during exercise in humans only in

---

Toru Satoh, Yasushi Hara, Fumio Sakamaki, Shingo Kyotani and Takeshi Tomita • Division of Cardiology and Pulmonary Circulation, Department of Medicine, National Cardiovascular Center, Suita, Osaka 565-8565 Japan. **Yasumasa Okada** • Department of Medicine, Keio University Tsukigase Rehabilitation Center, Shizuoka 410-3215 Japan.
Correspondence (Present Address): T. Satoh, Department of Medicine, Keio University Shool of Medicine, Shinanomachi 35, Shinjuku-ku, Tokyo 160-8582 Japan.

Post-Genomic Perspectives in Modeling and Control of Breathing, edited by Jean Champagnat, Monique Denavit-Saubié, Gilles Fortin, Arthur S. Foutz, Muriel Thoby-Brisson. Kluwer Academic/Plenum Publishers, 2004.

a few studies.[16] This is mainly because sampling of mixed venous blood during exercise in humans is technically difficult. In order to further understand the significance of metabolic CO_2 production in the physiological mechanism of exercise-induced hyperpnea, we collected data on the relationship between VE and $PvCO_2$, as well as on the relationship between VE and $PaCO_2$. Furthermore, in order to evaluate these data more thoroughly, we compared this VE-$PvCO_2$ relationship during exercise with that during a different CO_2-loading condition, i.e. during CO_2 rebreathing.

2. Methods

2.1. Subjects

The study subjects were 15 patients with cardiac diseases, who underwent a pulmonary hemodynamic examination for the evaluation of their pulmonary hemodynamic function. Eleven had mitral valvular heart diseases (4 dominant mitral stenosis, 4 dominant mitral regurgitation and 3 combined mitral stenosis and regurgitation), 2 had dilated cardiomyopathy, and 2 had chronic pulmonary thromboembolism. No patient had a ventilatory disorder. Their ages were 51 ± 15 years old (mean ± S.D.). Eight patients were male and 7 were female. The purpose, protocol and risks of the study were fully explained to the subjects, and written informed consent was obtained from each.

2.2. Monitoring of Hemodynamic and Ventilatory Parameters

A Swan-Ganz catheter was inserted via the right internal jugular vein into the pulmonary artery, and a fine arterial catheter was inserted into the left radial artery. Continuous monitoring of hemodynamic parameters including arterial and pulmonary arterial pressures was conducted. CO_2 concentration in the expired gas was monitored with a gas analyzer (AE280, Minato Medical Instruments, Osaka). VE was measured with a hot-wire flowmeter. Arterial and pulmonary arterial blood samples were collected based on the protocol, and blood gas values were measured using a conventional blood gas analyzer. Cardiac output was determined using Fick's equation.

2.3. Protocol of Exercise and CO_2 Rebreathing

For the exercise test, the subject was asked to get on an upright cycle ergometer and pedal at a speed of 55 rpm (without any added load) for one minute. Then the load was increased by 1 watt/4 sec (15 watts/minute) to the symptom-limited maximum. Expired gas was analyzed every 6 seconds throughout the period of exercise. Arterial and pulmonary arterial blood was sampled before exercise and every minute during exercise for blood gas analysis. On the same day, the subject was tested for CO_2 rebreathing using a bag containing 6 liters of air, with hemodynamic, expired gas and blood gas analyses performed as during exercise. O_2 consumption was determined in advance, and an equal amount of O_2 was supplemented into the rebreathing bag to maintain a constant inspired O_2 concentration throughout the rebreathing test. After a 30 minutes rest, a second rebreathing test was conducted using a bag containing 6 liters of 100% O_2 to suppress the carotid body function[17] after breathing 100% O_2 for 2 minutes.

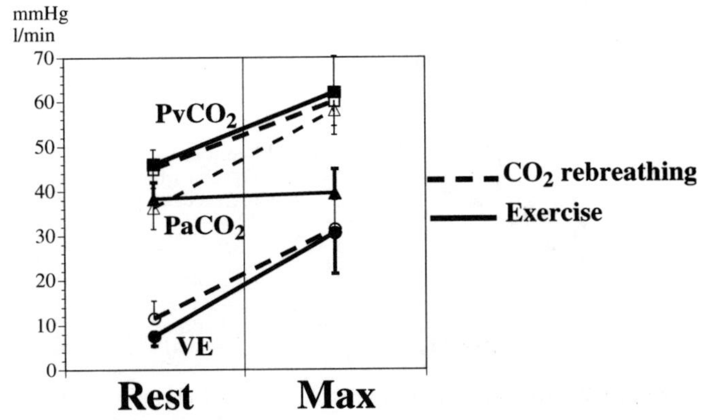

Figure 1. Blood and ventilatory gas changes during exercise and CO_2 rebreathing.

2.4. Statistical Analysis

The relationship between VE and $PvCO_2$ was evaluated by linear regression analysis. To compare the regression line between VE and $PvCO_2$ during exercise with that during CO_2 rebreathing in every subject, slopes and y-intercepts of the regression lines for 15 patients were evaluated by linear regression analysis. Comparison of the two regression lines was performed using an analysis of covariance. $P < 0.05$ was considered to be significant.

3. Results

Figure 1 depicts changes in VE, $PaCO_2$ and $PvCO_2$ at rest, and at the respiratory compensation point during exercise, or at the maximal response during CO_2 rebreathing. $PaCO_2$ did not change during exercise (38.3 ± 3.7 to 39.5 ± 5.4 mmHg), but was notably raised during CO_2 rebreathing (36.3 ± 4.9 to 58.0 ± 5.5 mmHg). $PvCO_2$ was greatly elevated during both exercise (45.6 ± 3.3 to 61.9 ± 8.2 mmHg) and CO_2 rebreathing (45.3 ± 4.2 to 60.2 ± 5.5 mmHg). VE increased from 7.5 ± 2.2 to 30.5 ± 9.1 l/min during exercise and from 11.5 ± 4.0 to 31.4 ± 6.7 l/min during CO_2 rebreathing.

An example of the relationship between VE and $PvCO_2$ during exercise and two different CO_2-rebreathing tests is depicted in Fig. 2. VE during CO_2 rebreathing using air is slightly greater than that observed during exercise. When 100% O_2 was inhaled instead of air, VE was reduced and the regression line shifted toward and lay near the regression line for exercise, resulting in equal slopes for the two regression lines.

This analysis was applied to all 15 patients. Statistical equality of the three regression lines for the relation between VE and $PvCO_2$ was ascertained (Fig. 3).

Next, to test whether these regression lines were substantially equal or not, an analysis of covariance was performed between the regression lines for the three tests. The three regression lines were similar, and the regression lines for exercise and for CO2 rebreathing using O_2 were effectively equivalent. A separate statistical analysis was performed on the data from each of the 15 subjects. The regression lines for exercise and CO_2 rebreathing using O_2 were substantially equal for all 15 subjects (Fig. 4). The P-value for the difference

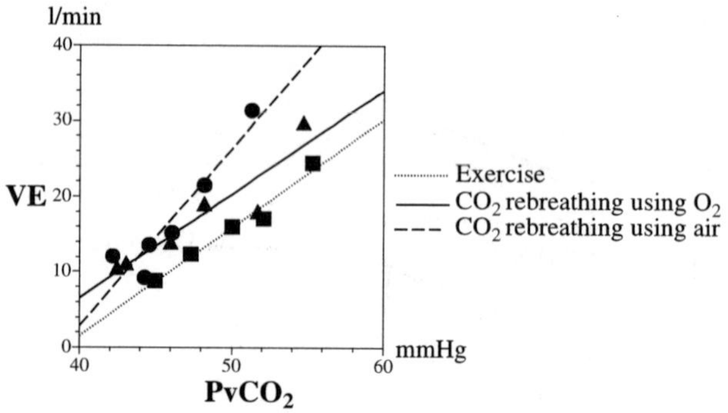

Figure 2. Representative example of the VE-$PvCO_2$ relationship.

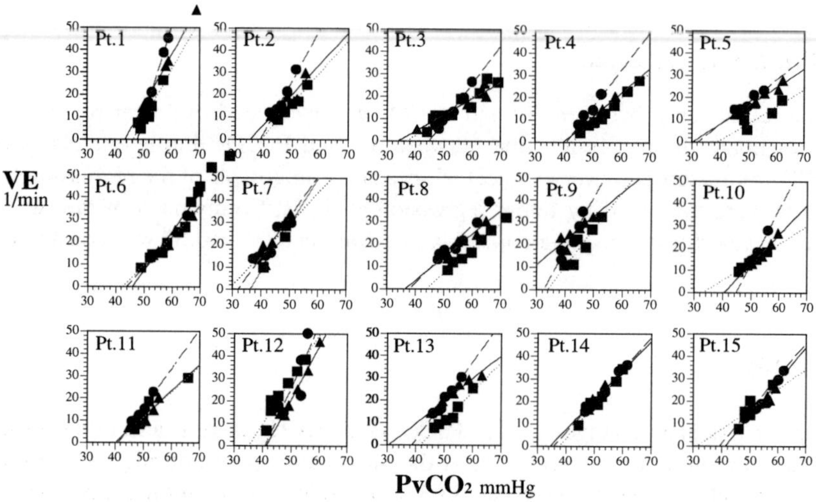

Figure 3. VE-$PvCO_2$ relationship in 15 patients.

in regression lines between exercise and CO_2 rebreathing using air among the 15 subjects was 0.08 for the slope and 0.16 for the y-intercept, and those for the difference between CO_2 rebreathing using O_2 and exercise were 0.87 and 0.69, respectively.

4. Discussion

We have analyzed the human ventilatory response to exercise as a function of $PvCO_2$, and compared it with the ventilatory response for CO_2 rebreathing. This is the first study to directly measure $PvCO_2$ during both exercise and CO_2 rebreathing in humans. We have found that the VE-$PvCO_2$ relationships during exercise and CO_2 rebreathing coincided almost exactly.

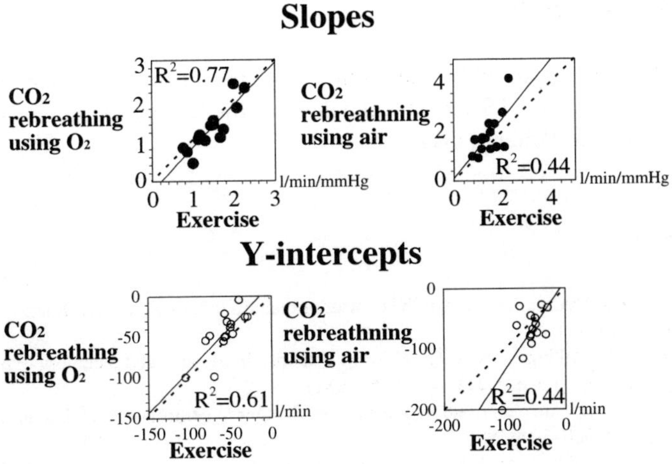

Figure 4. Slopes and y-intercepts of the regression lines between $PvCO_2$ and VE for 15 patients.

4.1. Physiological Significance of $PvCO_2$ in Respiratory Control

If ventilatory control during exercise and CO_2 rebreathing is mainly dependent on CO_2 chemosensitivity and if CO_2 chemoreceptors are located only in the arterial system, there must be two different scales for receptor sensitivity in hypercapnic and eucapnic states, because $PaCO_2$ changes little during exercise but markedly during CO_2 rebreathing. The close coupling of VE and $PvCO_2$ in two different conditions in our study could be derived from arterial chemoreception.[10,11,18] However, if CO_2 chemoreceptors exist in the venous system, this close coupling can be explained as the result of a cause-effect relationship. So far, the existence of venous CO_2 chemoreceptors in mammals has not been widely accepted. Cropp and Comroe[19] and Sylvester et al.[20] have denied the existence of venous CO_2 chemoreceptors in the dog, because infusion of CO_2-equilibrated blood did not initiate ventilatory responses until the infused stimulus reached the systemic arterial circulation. Orr et al.[21] also have claimed that venous CO_2 chemoreceptors do not exist in the cat, on the basis of their observation that venous CO_2 loading did not induce respiratory increase unless $PaCO_2$ was raised. However, Kollmeyer and Kleinman[6] changed $PvCO_2$ independently of arterial blood composition using an extracorporeal shunting system in the young puppy, and showed that CO_2 in venous blood stimulates ventilation. Sheldon and Green[8] separated the systemic and pulmonary circulation in the dog, and demonstrated respiratory increase by selectively elevating $PvCO_2$. We conducted electrical stimulation of the rabbit pulmonary artery, and found that neural respiratory output increased when we stimulated the proximal dorsal surface of the pulmonary trunk.[22] These reports suggest the existence of CO_2 chemoreceptors in the venous system.

Nevertheless, in the present study we identified substantial equality between the VE-$PvCO_2$ relationships during exercise and CO_2 rebreathing in each of our subjects. This observation may indicate that CO_2 in mixed venous blood is involved in human respiratory control under conditions in which $PvCO_2$ is elevated to a certain high level, such as during exercise and CO_2 rebreathing. Further study is needed to more directly examine the possibility of venous CO_2 chemoreception in humans.

5. Acknowledgements

This study was supported by the Grants from the Japanese Ministry of Education, Science, Sports and Culture (TS and YO) and the Specific Diseases Grant from the Japanese Ministry of Health and Welfare (YO).

References

1. J. H. Mateika, and J. Duffin, A review of the control of breathing during exercise, *Eur. J. Appl. Physiol.* **71**, 1–27 (1995).
2. F. L. Eldridge, D. E. Millhorn, and T. G. Waldrop, Exercise hyperpnea and locomotion: parallel activation from the hypothalamus, *Science*, **211**, 844–846 (1981).
3. H. V. Forster, and L. G. Pan, Contribution of acid-base changes to control of breathing during exercise, *Can. J. Appl. Physiol.* **20**, 380–394 (1995).
4. D. A. Oelberg, A. B. Evans, M. I. Hrovat, P. P. Pappagianopoulos, S. Patz, and D. M. Systrom, Skeletal muscle chemoreflex and pHi in exercise ventilatory control, *J. Appl. Physiol.* **84**, 676–682 (1998).
5. D. J. Paterson, P. A. Robbins, and J. Conway, Changes in arterial plasma potassium and ventilation during exercise in man, *Respir. Physiol.* **78**, 323–330 (1989).
6. K. R. Kollmeyer, and L. I. Kleinman, A respiratory venous chemoreceptor in the young puppy, *J. Appl. Physiol.* **38**, 819–826 (1975).
7. E. A. Phillipson, G. Bowes, E. R. Townsend, J. Duffin, and J. D. Cooper, Role of metabolic CO_2 production in ventilatory response to steady-state exercise, *J. Clin. Invest.* **68**, 768–774 (1981).
8. M. I. Sheldon, and J. F. Green, Evidence for pulmonary CO_2 chemosensitivity: effects on ventilation. *J. Appl. Physiol.: Respir. Environ. Exerc. Physiol.* **52**, 1192–1197 (1982).
9. E. R. Schertel, D. A. Schneider, L. Adams, and J. F. Green, Effect of pulmonary arterial P_{CO_2} on breathing pattern, *J. Appl. Physiol.* **64**, 1844–1850 (1988).
10. B. A. Cross, A. Davey, A. Guz, P. G. Katona, M. MacLean, K. Murphy, S. J. G. Semple, and R. P. Stidwell, The pH oscillations in arterial blood during exercise: a potential signal for the ventilatory response in the dog, *J. Physiol. (Lond.)* **329**, 57–73 (1982).
11. S. A. Ward, Peripheral and central chemoreceptor control of ventilation during exercise in humans, *Can. J. Appl. Physiol.* **19**, 305–333 (1994).
12. J. F. Green, E. R. Schertel, H. M. Coleridge, and J. C. Coleridge, Effect of pulmonary arterial P_{CO_2} on slowly adapting pulmonary stretch receptors, *J. Appl. Physiol.* **60**, 2048–2055 (1986).
13. E. S. Schelegle, and J. F. Green, An overview of the anatomy and physiology of slowly adapting pulmonary stretch receptors, *Respir. Physiol.* **125**, 17–31 (2001).
14. J. H. Auchincloss, R. Gilbert, M. Kuppinger, and D. Peppi, Mixed venous CO_2 tension during exercise, *J. Appl. Physiol.: Respir. Environ. Exerc. Physiol.* **48**, 933–938 (1980).
15. G. Alves da Silva, A. el-Manshawi, G. J. Heigenhauser, and N. L. Jones, Measurement of mixed venous carbon dioxide pressure by rebreathing during exercise, *Respir. Physiol.* **59**, 379–392 (1985).
16. R. Casaburi, J. Daly, J. E. Hansen, and R. M. Effros, Abrupt changes in mixed venous blood gas composition after the onset of exercise, *J. Appl. Physiol.* **67**, 1106–1112 (1989).
17. B. F Whipp, and K. Wasserman, Carotid bodies and ventilatory control dynamics in man. *Fed. Proc.* **39**, 2668–2673 (1980).
18. Y. Okada, Z. Chen, and S. Kuwana S, Cytoarchitecture of central chemoreceptors in the mammalian ventral medulla. *Respir. Physiol.* **291**, 13–23 (2001).
19. G. J. A. Cropp, and J. H. J. Comroe, Role of mixed venous blood P_{CO_2} in respiratory control. *J. Appl. Physiol.* **16**, 1029–1033 (1961).
20. J. T. Sylvester, B. J. Whipp, and K. Wasserman, Ventilatory control during brief infusions of CO_2-laden blood in the awake dog. *J. Appl. Physiol.* **35**, 178–186 (1973).
21. J. A. Orr, M. R. Fedde, H. Shams, H. Roskenbleck, and P. Scheid, Absence of CO_2-sensitive venous chemoreceptors in the cat. *Respir. Physiol.* **73**, 211–224 (1988).
22. Y. Okada, T. Satoh, S. Kuwana, M. Kashiwagi, and T. Kusakabe, Electrical stimulation of the rabbit pulmonary artery increases respiratory output, *Respir. Physiol. Neurobiol.* **140**, 209–217 (2004).

Effects of Pain and Audiovisual Stimulation on the Hypoxic Ventilatory Response

Suzanne B. Karan, J. Russell Norton, William Voter, Linda Palmer, and Denham S. Ward

1. Introduction

The non-chemoreflex drive to breathe is complex and influenced by the cortical, limbic and reticular activating systems. This reflex can be facilitatory or inhibitory and may have different modulating effects on normoxic ventilation and the acute hypoxic response (AHR). This interaction can be particularly important when considering the effects of the non-chemoreflex drive on anesthetic induced ventilatory depression.[1–5] This study clarifies the effects of pain and audiovisual stimulation (AVS) on ventilation and on the AHR.

2. Methods

We enrolled 25 (11 male; age 23.4 ⊥ 5.8 yr., mean ⊥ SD) healthy volunteers to complete a stimulation-randomized (tests: AVS, Pain, Rest) protocol during the measurement of their AHR. Subjects visited the lab prior to the experiment in order to acclimate to the apparatus. All breathing experiments were performed with subjects breathing into an oronasal facemask in a semi-reclining position.

AVS was a computer game entitled "You Don't Know Jack® The Ride" (Berkeley Systems©, Jellyvision Inc., 1998) that entailed the subjects' answering trivia questions by typing into a game controller. Pain (Visual Analog Scale of 3–6) was caused by a thermode (Precision Pain Source (PPS-3), Cygnus, Paterson, N.J.) applied to the ventral forearm skin following pre-sensitization with capsaicin cream[6] ($T_{max} = 47°C$). Rest implies that the subject had closed eyes with headphones (no noise) on the ears.

Suzanne B. Karan, J. Russell Norton, William Voter, Linda Palmer, and Denham S. Ward • Department of Anesthesiology, University of Rochester School of Medicine and Dentistry, Rochester, New York, 14642.

Post-Genomic Perspectives in Modeling and Control of Breathing, edited by Jean Champagnat, Monique Denavit-Saubié, Gilles Fortin, Arthur S. Foutz, Muriel Thoby-Brisson. Kluwer Academic/Plenum Publishers, 2004.

The acute hypoxic response was determined by utilizing a step into five minutes of hypoxia (saturation $= 81.4 \pm 1.3$) after five minutes of normoxia. A computer-controlled gas-mixing system, using the dynamic end-tidal forcing technique, controlled the inspired gas concentration on a breath-by-breath basis in order to achieve the desired end-tidal concentrations of CO_2 and O_2. The control AHR was determined at a $P_{ET}CO_2$ level slightly above resting levels to ensure that ventilation had significant chemoreceptor input. The $P_{ET}CO_2$ used in a particular subject was selected by a prior period of breathing 1% inspired CO_2 and measuring the average $P_{ET}CO_2$ over five minutes after steady-state was reached.

Minute ventilation ($\dot{V}_m$) can be expressed either as the product of tidal volume and breathing frequency or mean inspiratory flow and inspiratory time duty cycle, see equation (1). Mean inspiratory flow rate is a good indication of central neural 'output', and duty cycle indicates features such as bulbopontine control and pulmonary and somatic afferent influences.[7] Together, these measures provide a more fundamental indication of the control process than tidal volume and frequency.[8]

Since the acute hypoxic ventilatory response, AHR, is conveniently expressed as the ratio of the change in ventilation to the change in saturation (sat), there results four terms when the AHR is separated into its components, equations (2) and (3).

$$\dot{V}_m = V_T \cdot f = \left(\frac{V_i}{T_i}\right) \cdot \left(\frac{T_i}{T_{tot}}\right) \tag{1}$$

$$\frac{\Delta \dot{V}_m}{\Delta sat} = f \cdot \left(\frac{\Delta V_T}{\Delta sat}\right) + V_T \cdot \left(\frac{\Delta f}{\Delta sat}\right) \tag{2}$$

$$\frac{\Delta \dot{V}_m}{\Delta sat} = \left(\frac{T_i}{T_{tot}}\right) \cdot \left(\frac{\Delta\left(\frac{V_i}{T_i}\right)}{\Delta sat}\right) + \left(\frac{V_i}{T_i}\right) \cdot \left(\frac{\Delta\left(\frac{T_i}{T_{tot}}\right)}{\Delta sat}\right) \tag{3}$$

Data is reported as mean ± standard deviation except as noted. Statistical analysis (STATA statistical package, Stata Corp., College Station, TX) was performed by analysis of variance with these factors: gender, subject nested within gender, test, and test by gender interaction. When a significant test effect was found, *post hoc* Wald test was used to isolate the differences. When the distributions were skewed, a log transformation was used. P values without correction for multiple comparisons are reported. One-sided t-test was used to test the differences of the sensitivities from zero. P values greater than 0.05 are reported as not significant.

3. Results

The table shows the values of the AHR and its components. $P_{ET}CO_2$ was well controlled across tests and across normoxia (39.4 ± 3.8 mmHg) and hypoxia (39.4 ± 3.6 mmHg). The calculated sensitivities for the components of ventilation were significantly different from zero, except for the frequency sensitivity during pain (the frequency sensitivity at rest showed marginal statistical significance). When a significant test effect is indicated, *post hoc* pair-wise analysis always showed a difference between AVS and Pain with the values for Rest in between.

Table 1. Ventilatory measurements during normoxia and the sensitivities. Vent, Freq, V_T, T_i/T_{tot}, and, V_i/T_i calculated during the last two minutes of normoxia. Freq sens, V_T sens, T_i/T_{tot} sens, and T_i/T_i sens are the change in the measurement from normoxia (averaged over minute 3–4) to hypoxia (averaged over minute 7–8) divided by the change in saturation. * = $P < 0.01$.

	AHR ($l \cdot min^{-1} \cdot \%sat^{-1}$)	Vent (l/min)	Freq (min^{-1})	Freq sens ($l \cdot min^{-1} \cdot \%sat^{-1}$)	V_T (mL)	$V_T$sens (ml/%sat)	T_i/T_{tot}	T_i/T_{tot}sens ($\%sat^{-1}$)	V_i/T_i (l/sec)	V_i/T_i sens ($l \cdot sec^{-1} \cdot \%sat^{-1}$)
AVS	0.53 ±0.35	12.9 ±3.6	17.9 ±2.3	0.07 ±0.14	733 ±146	26.8 ±15.2	0.40 ±0.04	−1.52 ±1.91	0.58 ±0.13	0.020 ±0.013
Pain	0.36 ±0.27	10.5 ±2.7	14.8 ±2.7	0.01 ±0.12	735 ±200	24.5 ±19.1	0.38 ±0.04	−2.59 ±1.94	0.48 ±0.10	0.012 ±0.008
Rest	0.45 ±0.33	9.8 ±1.7	14.2 ±2.9	0.07 ±0.18	732 ±189	26.8 ±20.4	0.38 ±0.05	−1.65 ±2.86	0.46 ±0.09	0.018 ±0.018
Test sig.	*	*	*	NS	NS	NS	P = 0.06	NS	*	*
Sex sig.	NS	*	NS	P = 0.04	*	NS	NS	NS	*	NS

The stimulatory effect of AVS on minute ventilation occurred during both normoxia and hypoxia. During normoxia, this was attributed to both increased frequency of ventilation and increased inspiratory drive (V_i/T_i). For AVS, the AHR was increased because of an increase in respiratory frequency without a change in the tidal volume sensitivity (eq. 2) or by an increase in V_i/T_i without a change in the V_i/T_i sensitivity (eq. 3). For Pain, the AHR was not significantly increased. However, analysis of the components of AHR showed a significant decrease in the V_i/T_i sensitivity.

There did not seem to be any important sex differences in the results. The significant difference in the normoxic ventilation and tidal volume could be because of the difference in body mass between men and women.

4. Discussion

The effects of pain on respiration are not simple. Clinical observation confirms that pain is often a respiratory stimulant and this is supported by the literature.[9] The interaction of pain and hypoxia, however, has not been clearly elucidated. Studies examining the respiratory effects of post-operative pain relief have shown the most common agents to be associated with significant hypoxemic events[10–12] that are not always predictably based on patient or drug characteristics. Furthermore, the pathways that modulate these responses may not involve chemoreflexes[13] and may even bypass cortical structures.[14]

In this study, painful thermal stimulation did not significantly increase normoxic ventilation, nor did it augment the acute hypoxic response. This agrees with previous experiments done by Sarton et al.[1] in which a painful electrical stimulus (to a VAS of 4.5–5.5) did not augment the AHR. However, in contrast to our study, Sarton's painful stimulus did cause a significant increase in normoxic minute ventilation (with non-significant increases in normoxic tidal volume and respiratory rate). Electrical stimulation may have a different effect on ventilation than thermal stimulation, although both studies did control for similar VAS scores.

In both studies, the painful stimulus caused a non-significant but downward trend of the AHR. In our study, this inhibition of the AHR may be explained by further dissecting out the components of the AHR. In fact, the inspiratory drive sensitivity (V_i/T_i sens) was significantly reduced during pain compared with rest. According to equation (3) above, the ability to enhance the AHR relies upon the inspiratory drive. It is possible that Sarton observed a similar trend, though this data was not published. The mechanism that underlies this phenomenon may be a function of the subjects' volitional control over ventilation. In a study by Spicuzza and colleagues,[15] the hypoxic ventilatory response was considerably lower in yoga trainees compared with controls, with similar heart rate and blood pressure responses between groups. This inhibition was mainly the result of a decreased frequency of respiration.

In contrast, we found that AVS was a potent stimulant for ventilation during both normoxia and hypoxia over both Pain and Rest. In comparison with the experiment of Chin et al.,[16] the AVS that we chose entailed a degree of mental coordination and caused the increase in normoxic minute ventilation through an increase in respiratory frequency (f) and inspiratory drive (V_i/T_i). Our normoxic observations are also in line with Mador et al.,[17] who found that audiovisual stimulation (watching television) did not increase minute ventilation as much as mental arithmetic. This augmentation of minute ventilation was achieved primarily via an increase in respiratory frequency and inspiratory drive.

During hypoxia, the ventilatory response was significantly augmented during AVS when compared with Pain and Rest. This is in contrast to the observations of van den Elsen et al.,[3] though their AVS (music videos) was different from ours. Furthermore, Pandit et al. (this volume), using a similar AVS to van den Elsen (listening and watching a video of one's choice), also did not observe an increase in the AHR.

Although it has been theorized that the difference between the studies of the drug effects on the AHR lies mostly in the difference in experimental conditions[5] as it relates to the use of volatile anesthetics, one cannot disclaim the importance of the choice of AVS as well. It is possible that the state of arousal while watching a video of one's choice is unlike the "wakefulness" state induced by the mental interaction involved in playing a computer game. The interaction of arousal and hypoxia may have more to do with the volitional control of breathing. Though the results of our study are comparable to those of Chin[16] and Mador[17], other studies challenge these findings. In a study on the effects of emotions on respiratory pattern, Boiten[18] examined specific respiratory cycle responses to varying emotions, mental activity, and pain (cold pressor test). He found the most significant ventilatory changes during excitement. In contrast, respiratory pattern did not deviate significantly from baseline while subjects performed only the mental activity.

Although there was no quantitative measure of dyspnea during our study, anecdotally, subjects consistently reported less dyspnea during the AVS-AHR experiment when compared with Pain and Rest. From animal studies, we know that there are suprapontine stimulatory and inhibitory influences on ventilation.[19] It is possible that during the Pain and Rest experiments our subjects exerted a "subconscious volitional" respiratory pattern control. In contrast, the augmentation of the AHR during AVS could have been the result of suprapontine disinhibition, caused by distraction from our choice of AVS.

What is clear is that experimental states of arousal and audiovisual stimulation need to be more clearly defined and consistently measured, especially when comparing the conflicting results concerning the hypoxic drive and its interaction with pain and arousal.

5. Acknowledgments

This study was supported by a research fellowship grant (S.Karan) from the Foundation for Anesthesia Education and Research.

References

1. Sarton E, Dahan A, Teppema L, van den Elsen M, Olofsen E, Berkenbosch A, van Kleef J, Acute pain and central nervous system arousal do not restore impaired hypoxic ventilatory response during sevoflurane sedation, *Anesthesiology*, 85, 295–303 (1996)
2. Lam AM, Clement JL, Knill RL, Surgical stimulation does not enhance ventilatory chemoreflexes during enflurane anaesthesia in man, *Can. Anaesth. Soc. J.*, 27, 22–8 (1980)
3. van den Elsen MJ, Dahan A, Berkenbosch A, DeGoede J, van Kleef JW, Olievier IC, Does subanesthetic isoflurane affect the ventilatory response to acute isocapnic hypoxia in healthy volunteers? *Anesthesiology*, 81, 860–7 (1994)
4. Robotham JL, Do low-dose inhalational anesthetic agents alter ventilatory control? *Anesthesiology*, 80, 723–6 (1994)
5. Pandit JJ, The variable effect of low-dose volatile anaesthetics on the acute ventilatory response to hypoxia in humans: a quantitative review, *Anaesthesia* 57(7), 632–43 (2002)
6. Petersen KL, Rowbotham MC, A new human experimental pain model: the heat/capsaicin sensitization model, *Neuroreport*, 10, 1511–6 (1999)
7. Dockery MP, Drummond GB, Respiratory response to skin incision during anaesthesia with infusions of propofol and alfentanil, *Br. J. Anaesth.*, 88, 649–52 (2002)
8. Gautier H, Control of the pattern of breathing, *Clin.Sci.* (Lond), 58, 343–8 (1980)
9. Borgbjerg FM, Nielsen K, Franks J, Experimental pain stimulates respiration and attenuates morphine-induced respiratory depression: a controlled study in human volunteers, *Pain*, 64, 123–8 (1996)
10. Madej TH, Wheatley RG, Jackson IJ, Hunter D, Hypoxaemia and pain relief after lower abdominal surgery: comparison of extradural and patient-controlled analgesia, *Br. J. Anaesth.*, 69, 554–7 (1992)
11. Wheatley RG, Shepherd D, Jackson IJ, Madej TH, Hunter D, Hypoxaemia and pain relief after upper abdominal surgery: comparison of i.m. and patient-controlled analgesia, *Br. J. Anaesth.*, 69, 558–61 (1992)
12. Stone JG, Cozine KA, Wald A, Nocturnal oxygenation during patient-controlled analgesia, *Anesth. Analg*, 89, 104–10 (1999)
13. Duranti R, Pantaleo T, Bellini F, Bongianni F, Scano G, Respiratory responses induced by the activation of somatic nociceptive afferents in humans, *J. Appl. Physiol*, 71, 2440–8 (1991)
14. Waldrop TG, Millhorn DE, Eldridge FL, Klingler LE, Respiratory responses to noxious and nonnoxious heating of skin in cats, *J. Appl. Physiol*, 57, 1738–41 (1984)
15. Spicuzza L, Gabutti A, Porta C, Montano N, Bernardi L, Yoga and chemoreflex response to hypoxia and hypercapnia, *Lancet*, 356, 1495–6 (2000)
16. Chin K, Ohi M, Fukui M, Kita H, Tsuboi T, Hirata H, Noguchi T, Mishima M, Kuno K, Intellectual work using a video game inhibits post hyperventilation hyperpnoea following voluntary hyperventilation while it stimulates breathing at rest, in Modeling and Control of Ventilation, S. G. Semple, L. Adams, B. J. Whipp, eds., Plenum Press, New York, 1995, pp. 81–84
17. Mador MJ, Tobin MJ, Effect of alterations in mental activity on the breathing pattern in healthy subjects, *Am. Rev. Respir. Dis.*, 144, 481–7 (1991)
18. Boiten FA, Frijda NH, Wientjes CJ, Emotions and respiratory patterns: review and critical analysis, *Int. J. Psychophysiol.*, 17, 103–28 (1994)
19. Tenney SM, Ou LC, Ventilatory response of decorticate and decerebrate cats to hypoxia and CO_2, *Respir. Physiol.*, 29, 81–92 (1977)

Effect of Progressive Hypoxia with Moderate Hypercapnia on Ventilatory vs. Respiratory Sensation Responses in Humans

Yoshikazu Sakakibara, Atsuko Masuda, Toshio Kobayashi, Shigeru Masuyama, and Yoshiyuki Honda

1. Introduction

Ventilatory response to CO_2 combined with hypoxic stimulation has been well documented as exhibiting a positive interaction between the two stimuli in humans[1], and in cats[2,3]. Furthermore, Mohan and Duffin[4] examined the effect of hypoxia on ventilatory response to CO_2 using a modified Read's rebreathing method, covering wide range of CO_2 including hypocapnic region following prior hyperventilation.

Although respiratory sensation such as dyspnea and breathlessness are known to be elicited by CO_2 and hypoxia[5,6], it is not well known if there is a positive interaction between both chemical stimuli in the same manner as in ventilation. We previously[7] examined both ventilation and respiratory sensation during hypercapnia conducted by modified CO2 rebreathing with 4 different levels of oxygenation. In that study, we found that, irrespectively, ventilatory responses showed a positive interaction during these multivariable chemical stimuli, and respiratory sensation responses to hypercapnia showed only parallel upward shift with increasing hypoxic stimulation. The purpose of the present study was therefore to confirm this different profile between ventilation and respiratory sensation by applying progressive hypoxia with varying hypercapnia levels, method opposite to our previous one.

2. Methods

The study was carried out on 15 young healthy adults (4 males and 11 females). Their age, height, weight and body surface area were (22.3 ± 4.0) yrs, (161.8 ± 6.5) cm,

---

Yoshikazu Sakakibara • Kanazawa Inst Technol, Kanazawa. **Atsuko Masuda** • Tokyo Med & Dent Univ, Tokyo, 113-8519, Japan. **Toshio Kobayashi** • Hiroshima Univ, Hiroshima. **Shigeru Masuyama and Yoshiyuki Honda** • Chiba Univ, Chiba.

Post-Genomic Perspectives in Modeling and Control of Breathing, edited by Jean Champagnat, Monique Denavit-Saubié, Gilles Fortin, Arthur S. Foutz, Muriel Thoby-Brisson. Kluwer Academic/Plenum Publishers, 2004.

(54.0 ± 5.0) kg and (1.6 ± 0.1) m^2(mean ± S.D.), respectively. They did not smoke and had no cardiopulmonary disorders. Although all subjects were familiar with the laboratory equipment and ventilatory response tests, they did not know the purpose or results until all tests were completed. The study protocol was approved by the local ethics committee, and all subjects gave their informed consent before participating. The subjects were kept free of food and caffeine intake for at least 2 hours prior to the experiment.

The subjects were exposed to progressive hypoxia test with three different CO_2 levels. We maintained the $P_{ET}CO_2$ level at normocapnia, 2 and 4 Torr higher than normocapnic level during progressive hypoxia test, they were designated as HC0, HC2 and HC4, respectively. Before conducting progressive hypoxia test, the rebreathing bag was filled with 6–8 L of room air. The subject initially breathed room air in a resting condition through an open respiratory circuit for 2–3 min. After this, they started rebreathing room air from the bag. During rebreating, we maintained the $P_{ET}CO_2$ level at normocapnia by adjusting the by-pass flow to the CO_2 absorber and defined this test as HC0 run. Under hypercapnic condition, i.e., the HC2 and HC4 runs, we supplied an adequate volume of oxygen via a small tube inserted into the inspiratory portion of the respiratory valve in order to prevent a decline in oxygen level for about 5 minutes after $P_{ET}CO_2$ level reached 2 or 4 Torr higher than normocapnic condition. Then we discontinued oxygen administration and kept these $P_{ET}CO_2$ levels by adjusting the by-pass flow to the CO_2 absorber. During all the runs, we monitored $P_{ET}O2$ and $P_{ET}CO_2$ by rapid response O_2-CO_2 analyzer (San-ei, H-21, Tokyo, JAPAN) and arterial oxygen saturation (SaO_2) and heart rate (HR) by pulse oxymeter (Ohmeda, Biox III). Respiratory sensation was continuously rated on a visual analog scale (VAS) 15 cm in length, that was marked with a regular intervals and word clarification were put aside some of these marks. Ventilatory parameters such as minute ventilation (VE), tidal volume (VT) and respiratory frequency (f) as well as gas fractions were measured with an electric spirometer (Minato Aeromonitor AE-280, Osaka, JAPAN). VAS, electrocardiogram and VE, one output of AE-280, were sampled at a rate of 1 kHz simultaneously and stored into a PC. The experiment was terminated 3–4 min after the start of rebreathing or when SaO_2 approached to 80%. VE and VT were normalized by body surface area. Every parameter was averaged within 6 different sections of data such as during air breathing, just before conduction of hypoxia, and 4 stages of progressive hypoxia, 96, 91, 86, 81% in SaO2. VE and VAS scores were plotted against SaO_2, and the slopes of VE/SaO_2 (l/min/m^2/%) and VAS/SaO_2(/%) were taken as hypoxic sensitivity of ventilation (HVR; hypoxic ventilatory response) and respiratory sensation to hypoxia, respectively. For each subject three runs were performed in the order of HC0, HC2, and HC4 with intervals longer than 60 min. The same set of experiments was conducted twice in the morning and afternoon of the same day.

Statistical difference of mean among the different groups were examined by using Dunnett's multiple comparison as well as the paired t test, both tailed. A p-value less than 0.05 was considered significant.

3. Results

$P_{ET}CO_2$ and ventilatory parameters before and during progressive hypoxia were presented in Figure 1. $P_{ET}CO_2$ was controlled to reach almost exactly the desired level. It is to be

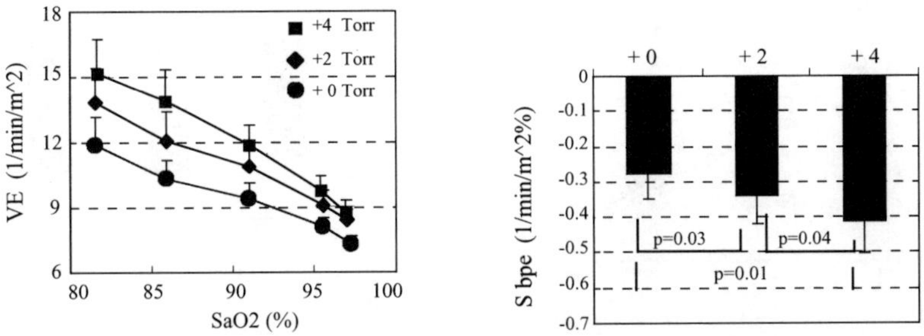

Figure 1. Hypoxic ventilatory response curves (left) under 3 different CO_2 levels (+0, +2, and +4 Torr above normocapnia), and their slopes (right). Values are mean ± SE (n = 15).

noted that f showed little response to hypercapnia, whereas VT increased with hypercapnia (Table 1). As expected, the slope of the SaO_2-ventilation response curve became steeper in accordance with increasing CO_2 level as expected (Figure 1). There was a significant increase in the steepness of HVR in the HC4 (+4 in Figure 1) compared to the HC0 run (+0 in Figure 2) ($p = 0.01$). On the other hand, the slope of the SaO_2-VAS response curve remained almost unchanged (Figure 3) and the curve appeared to shift parallel and upward with intensifying hypercapnia (Figure 1); its magnitude assessed by the intercept of regression equation of VAS to SaO_2 did not reach statistical significance ($p = 0.15$) between HC0 and HC4 runs.

Mean VAS during progressive hypoxia with different CO_2 levels was plotted against mean ventilation (VE in Figure 3). The VAS-VE curves appeared to shift rightward with increasing CO_2 level from normocapnia (+0 line in Figure 3) through 2 and 4 Torr higher than normocapnic $P_{ET}CO_2$ (+2, and +4 line in Figure 3, respectively). The mean values plotted in Figure 3 represented all the results including the two VAS data with no-response. However, we excluded these two data from the calculation of the linear regression lines of

Table 1. Ventilatory responses to progressive hypoxia with 3 different hypercapnia

SaO2 (%)	97	96	91	86	81
HC0					
CO2 (Torr)	40.9 ± 4.19	41.0 ± 3.83	41.0 ± 3.83	40.6 ± 3.81	40.7 ± 3.70
f (cpm)	14.2 ± 4.22	15.2 ± 5.38	15.8 ± 6.82	15.9 ± 7.09	17.0 ± 7.54
VT (ml/m^2)	597.3 ± 248.91	636.9 ± 249.36	649.5 ± 167.65	721.7 ± 220.46	745.0 ± 202.45
HC2					
CO2 (Torr)	42.9 ± 4.50	43.0 ± 3.92	42.3 ± 3.78	42.2 ± 3.98	42.6 ± 3.98
f (cpm)	14.5 ± 5.40	15.3 ± 5.07	17.1 ± 7.92	17.3 ± 8.18	18.3 ± 8.17
VT (ml/m^2)	654.1 ± 231.61	661.7 ± 226.96	698.9 ± 203.20	765.7 ± 198.86	820.9 ± 224.44
HC4					
CO2 (Torr)	44.9 ± 4.08	44.6 ± 3.84	44.3 ± 3.75	43.5 ± 3.65	44.7 ± 3.64
f (cpm)	14.7 ± 5.63	15.6 ± 5.84	16.4 ± 6.16	17.3 ± 7.85	18.8 ± 7.98
VT (ml/m^2)	670.9 ± 218.43	703.0 ± 206.77	808.5 ± 265.70	864.7 ± 200.04	866.0 ± 196.10

HC0, HC2, and HC4 are normocapnic run, and hypercapnic runs with 2, or 4 Torr higher than normocapnic $P_{ET}CO_2$, respectively. CO2, f and VT are end tidal PCO_2, respiratory frequency, and tidal volume, respectively. Data are presented as mean ± SD (n = 15).

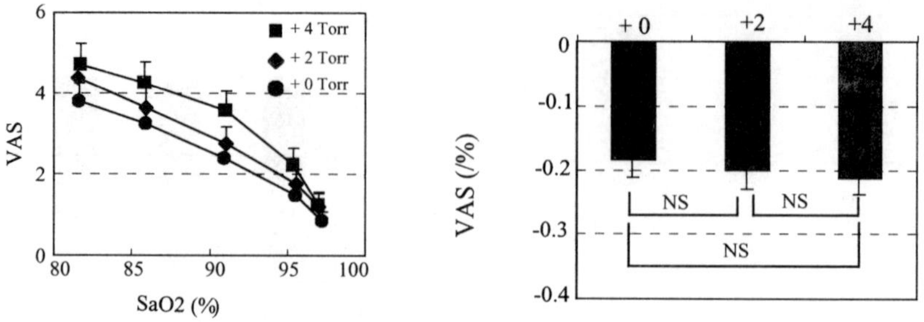

Figure 2. Hypoxic VAS response curves (left) under 3 different CO_2 loads (+0, +2, and +4 Torr above the normocapnia), and their slopes (right). Values are mean ± SE (n = 15).

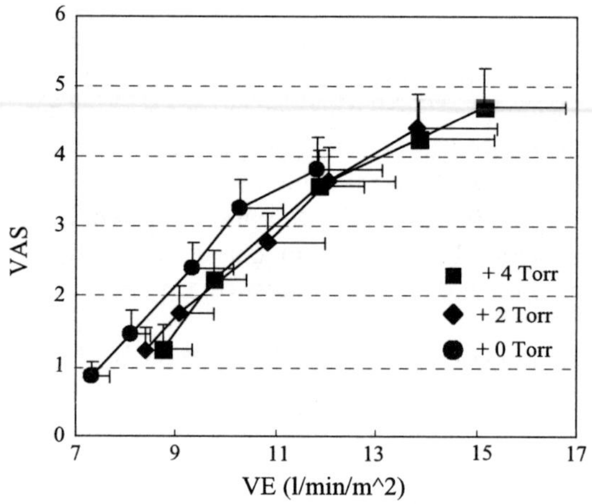

Figure 3. Relationships between VAS and ventilation during progressive hypoxia with 3 different CO_2 loads. Values are mean ± SE (n = 15).

VAS to ventilation. The slopes of these regression lines were 0.81 ± 0.20, 0.47 ± 0.20 and 0.65 ± 0.10, during HC0, HC2, and HC4, respectively (mean ± SE, n = 13). As the paired t test showed borderline significance between HC0 and HC4 ($p = 0.06$), we further predicted VAS at ventilation 10 ($l/min/m^2$) by using these regression lines. Their mean ± S.E. were 3.6 ± 0.57, 2.2 ± 0.80 and 2.8 ± 0.46, during HC0, HC2, and HC4, respectively.

There was statistical significance between HC0 and HC4 ($p = 0.01$) when used paired t tests, but no statistical difference was seen when used Dunnett's way.

4. Discussion

In the present study, we exposed subjects to progressive hypoxia with moderate and almost steady hypercapnia and found a positive interaction in the ventilatory responses. On the other hand, respiratory sensation during these chemical challenges showed a parallel shift. Thus we could confirm our previous results, where a modified Read's rebreathing method was used to give a hypercapnic stimulation under 4 different steady hypoxia levels[7].

The present results clearly indicate that the mechanism underlying respiratory sensation response to hypercapnic hypoxia, or *vice versa* differs from that of ventilation, although either of these chemical stimuli is able to arouse respiratory sensation[5,6]. This also leads to assume that our results are not fully explainable by the central command hypothesis [8,9].

Pulmonary stretch receptors excited by inhaled furosemide alleviate the dyspneic sensation induced by loaded breathing[10]. Our present results demonstrate that ventilatory responses to chemical stimuli were largely depended on VT (Table 1). As it is reasonable to assume that such larger VT during hypercapnia activate more pulmonary stretch receptors, most of the decreased amount of respiratory sensation can be ascribed to CO_2 loads in the present experiments. The mechanism involving pulmonary stretch receptors, however, is not considered to exclude the central command hypothesis[10].

In conclusion, we confirmed the contrasting profile between ventilatory vs. respiratory sensation that was, in the present study, induced by progressive hypoxia with varying steady hypercapnia. The underlying mechanism of the respiratory sensation to chemical stimuli is not fully explained neither by the central command theory, or by pulmonary stretch receptors.

References

1. M. Nielsen and H. Smith, Studies on the regulation of respiration in acute hypoxia. Act Physiol. 24, 293–313 (1952).
2. T. F. Hornbein, Z. Griffo, A. Roos, Quantification of chemoreceptor activity: intercation of hypoxia and hypercapnia. J. Neurophysiol. 24, 561–568 (1961).
3. R.S. Fitzgerald, D.C. Parks, Effect of hypoxia on carotid chemoreceptor response to carbon dioxide in cats, Respir. Physiol. 12, 218–229 (1971).
4. R. Mohan and J. Duffin, The effect of hypoxia on the ventilatory response to carbon dioxide in man. Respir. physiol. 108, 101–115 (1997).
5. R. B. Banzett, R. W. Lansing, K. C. Evans, and S. A. Shea, Stimulus-response characteristics of CO2-induced air hunger in normal subjects. Respir. Physiol. 103, 19–31 (1996).
6. S. H. Moosavi, E. Golestanian, A. P. Binks, R. W. Lansing, R. Brown, and R. B. Banzett, Hypoxic and hypercapnic drives to breathe generate equivalent levels of air hunger in humans. J Appl. Physiol. 94(1), 141–154 (2003).
7. A. Masuda, Y. Ohyabu, T. Kobayashi, C. Yoshino, Y. Sakakibara, T. Komatsu, and Y. Honda, Lack of positive interaction between CO_2 and hypoxic stimulation for Pco_2-VAS response slope in humans, Respir. Physiol. 121, 173–181 (2001).
8. L. Adams, R. Lane, A. Shea, A. Cockcroft, and A. Guz, Breathlessness during different forms of ventilatory stimulation: a study of mechanism in normal subjects and respiratory patients. Clin. Sci. 68, 663–672 (1985).
9. R. B. Banzett, R. W. Lansing, M. B. Reld, L. Adams, and R. Brown, 'Air hunger' arising from increased PCO_2 in mechanically ventilated quadriplegics. Respir. Physiol. 76, 53–67 (1989).
10. T. Nishino, I. Tohru, S. Tomoko, and S. Jiro, Inhaled furosemide greatly alleviates the sensation of experimentally induced dyspnea. Am. J. Respir. Crit. Care Med. 161, 1963–1967 (2000).

Frequency Response of the Input Reaching the Respiratory Centres During Moderate Intensity Exercise

Philippe Haouzi, Bruno Chenuel and Bernard Chalon

1. Introduction

It has been suggested that the intrinsic properties of the brainstem respiratory neurones, responsible for a short term potentiation phenomenon, could slow and magnify the effects of an immediate and steady centrally mediated stimulus to breathe during exercise[1,2,3]. Worth of note is that this phenomenom has been described mainly during and following stimulation of somatic afferent fibres or the sinus nerve[3]. Such a short term potentiation phenomenon was put forward to try to reconcile the traditional description of the dynamics of the $\dot{V}_E$ on and off-transient response to exercise, which has a 60 second time constant and follows the $\dot{V}O_2$ and $\dot{V}CO_2$ time course[4,5,6].

We sought to predict the behavior of the neural inputs reaching the brainstem respiratory neurones after extracting from available data in animals and in humans a short term potentiation phenomenom (STP). More specifically, our purpose was to determine if a temporal relationship exists between the $\dot{V}_E$ response, after "suppression" of a STP component, and the expected constant amplitude of motor activation during sinusoidal work rate changes at different periods of oscillations.

2. Methods

Two studies were re-examined: one reporting preliminary data obtained in trotting sheep and one previously published by Casaburi et al.[6].

Three sheep walked on a treadmill, breathing through a face mask. The speed of locomotion was changed sinusoidally from 50 to 100 m/min at period (T) ranging from 10 to 2 minutes. The fluctuating work rate was applied after 3 minutes of constant work rate at

---

Haouzi Philippe • Laboratoire de Physiologie, Faculté Médecine de Nancy, Avenue de la Forêt de Haye, B.P. 184, 54505 Vandoeuvre-lès-Nancy Cedex—France, Tel: 33 3 83 68 37 45, Fax: 33 3 83 68 37 39, E mail: p.haouzi@chu-nancy.fr

Post-Genomic Perspectives in Modeling and Control of Breathing, edited by Jean Champagnat, Monique Denavit-Saubié, Gilles Fortin, Arthur S. Foutz, Muriel Thoby-Brisson. Kluwer Academic/Plenum Publishers, 2004.

the mid point of the sinusoid (75 m/min). The phase lag (Ph) of the fundamental component (same frequency as the input function) of $\dot{V}E$, and $\dot{V}CO_2$ responses were computed as follows:

$$Ph = Tan^{-1}\ (Re/Im)$$

Where Re and Im are the real and imaginary parts of the response determined after second-by-second interpolation of the respiratory and locomotor (x) responses as:

$$Re = 2/NT \sum_{t=0}^{NT} (x(t)\text{-}Mx) \cos(2\pi ft) \text{ and } Im = 2/NT \sum_{t=0}^{NT} ((x(t)\text{-}Mx)) \sin(2\pi ft)$$

Where x (t) is the response value as time t (in seconds), Mx is the mean value of x for an integer number of cycle (N), T is the period of the input signal (in seconds) and f (=1/T) is its frequency in cycles per second.

In the study of Casaburi et al.[6], five healthy subjects performed cycloergometer exercise. Work rate was changed sinusoidally in the moderate intensity domain (below lactate threshold) at period (T) ranging from 10 to 0,75 minutes and in a stepwise manner (steady state).

According to the model proposed by Eldridge et al. [2], any signal from a central or peripheral source reaching the respiratory neurones is amplified and slowed due to some of their intrinsec properties. For moderate exercise, the gain of this component may represent half of the total gain of the response with a time constant which differs during the on and off transient (ranging from 15 to 50 seconds in the cat). We substracted the expected phase lag (Phst) of STP to the $\dot{V}E$ response using these parameters. Phst was computed as: $Phst = tg^{-1}\ (2\pi f\tau)$, where τ is the time constant of STP. τ values of 15 and 50 seconds have been used in this report.

3. Results

The phase lag between $\dot{V}E$ and the work rate input increased dramatically with the frequency of oscillations and was longer than $\dot{V}CO_2$ in sheep as already reported in humans. The increase in $\dot{V}E$ phase lag with the frequency of oscillations was more pronounced in humans than in sheep and was associated with slower pulmonary gas exchange dynamics.

The phase lag was corrected for a short term potentiation phenomenon[3] showing the expected temporal profile of the theoretical input to the respiratory neurones which would account for the observed ventilatory outcome (figure 1 and 2). This "signal" preceeded the changes in pulmonary gas exchange (shorter phase lag), but had a frequency response very different from the expected flat response of the motor control. Even, for the longest STP time constants (figure 2), the signal to breath is still decreasing with the frequency of oscillation lagging WR with a phase of about 45° at the period of oscillations of 2 minutes.

4. Discussion

The present results do not support a parallel activation of the systems controlling respiration and the locomotor or motor activity. As previously argued, the frequency response of the ventilatory response appears to be compatible with mechanisms linking ventilation

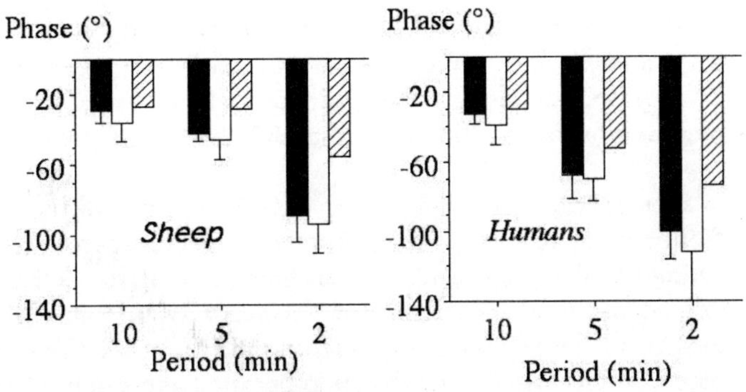

Figure 1. Evolution of the phase bewteen work rate oscillations and both CO_2 output (black bars) and ventilation (white bars) at 3 periods of cycling oscillations in sheep and humans. It is a convention that phases are shown with negative values. Note that phase lag of ventilation follows the changes in CO_2 output phase. The expected frequency response of the signal reaching the central nervous system after correction of the ventilatory response for a short tem potentiation phenomenon (STP) with a 15 sec time constant is shown in hatched bars. The frequency response of the signal reaching the central nervous system (without STP) is expected to be faster than the pulmonary gas exchange.

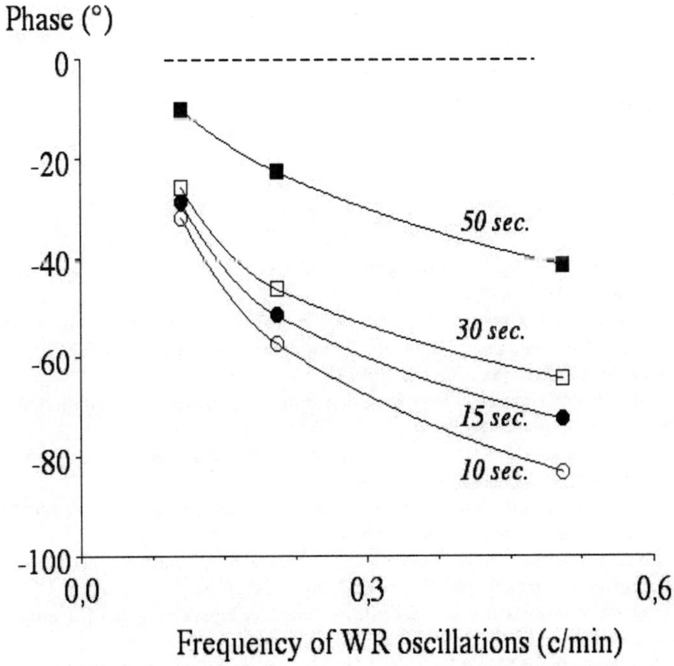

Figure 2. Evolution of the phase bewteen work rate oscillations and ventilation after extraction of a short term potentiation component of various time constants (ranging from 10 to 50 seconds). Note that the phase lag between work rate oscillations and this "signal" increased when the frequency oscillations increased even with the "slowest" short term potentiation phenomenom. The horizontal dotted line is the expected frequency response of a central command component which has no time constant.

to $\dot{V}CO_2$[4,10]. Results were very similar in humans and animals, suggesting commun control mechanisms of respiration during exercise in mammals coupling ventilation and $\dot{V}CO_2$[4,10]. The faster ventilatory response during trotting in sheep was always associated with faster $\dot{V}CO_2$ kinetics than in humans resulting in similar difference in phase lag between these two factors.

However, a short term potentiation phenomenon[2,3] would slow the ventilatory outcome resulting from any rapid signal coming from a peripheral source. This implies that the signal stimulating breathing has faster kinetics than those of gas exchange at the lung level. The expected dynamic characteristics of this feedforward signal is incompatible with a signal proportional to the motor act which should have a flat frequency response (figure 2). The nature of this signal is still open to question and can not be inferred from our analysis, but it is proportional to events associated with the change in metabolic rate (phase lag increased with the frequncey of oscillations) and is faster than $\dot{V}CO_2$.

The parameters used to describe STP were drawn from animal studies. The gain during moderate exercise was assumed to be half of the total gain as postulated for moderate exercise[1]. There are no data available on the characteristics of STP in humans or in large mammals in response to similar peripheral stimulations. However, the fact that $\dot{V}E$ phase lag is only slightly longer than $\dot{V}CO_2$ phase implies that, even with very fast STP dynamics, the signal stimulating breathing during exercise should lead $\dot{V}CO_2$.

In conclusion, the present analysis confirms pervious results which do not support the contention that a proportional adjustment of motor activity and minute ventilation exists during moderate intensity exercise. They support however the idea that the main component of the $\dot{V}E$ response is linked to the level of pulmonary gas exchange, but relying on signal(s) preceeding the changes in gas exchange in the lungs ($\dot{V}CO_2$).

References

1. Elridge FL, Milhorn DE and Waldrop TG. Exercise hyperpnea and locomotion: Parallel activation from the hypothalamus. *Science* **211**, 844–846 (1981).
2. Eldridge FL and Waldrop TG. *Neural control of breathing during exercise.* In lung Biology in Health and disease, vol. 52, Exercise pulmonary physiology and pathophysiology. Ed. Whipp B and Wasserman K, Dekker Inc., New-York, Basel, pp. 309–370 (1991).
3. Wagner P.G. and Eldridge F.L. Development of short-term potentiation of respiration. *Respir. Physiol.* **83**, 129–140 (1991).
4. Whipp B.J., Ward S.A., Lamara N., Davis J. and Wasserman K. Parameters of ventilatory and gas exchange. *J. Appl. Physiol.* **52**, 1506–1513 (1982).
5. Casaburi R., Barstow T.J., Robinson T. and Wasserman K. Influence of work rate on ventilatory and gas exchange kinetics. *J. Appl. Physiol.* **67**, 547–555 (1989).
6. Casaburi R., Whipp B., Wasserman K., Beaver W.L. and Koyal S.N. Ventilatory and gas exchange dynamics in response to sinusoidal work. J. Appl. Physiol. **42**, 300–311(1977).
7. Dejours P. La régulation de la ventilation au cours de l'exercice musculaire chez l'homme. *J. Physiol. Paris*, **51**, 163–261 (1959).
8. Bakker H.K., Struikenkamp R.S. and De Vries G.A. Dynamics of ventilation, heart rate, and gas exchange: sinusoidal and impulse work loads in man. *J. Appl. Physiol.* **48**, 289–301 (1980).
9. Fujihara Y., Hildebrandt J. and Hildebrandt J.R. Cardiorespiratory transients in exercising man. II. Linear models. *J. Appl. Physiol.* **35**, 68–76 (1973).
10. Whipp B.J. The control of exercise hyperpnea. In: *The Regulation of breathing*. Vol. 52, edited by T. Hornbein. New York: Dekker, pp. 1069–1139 (1981).

8

VARIABILITY AND PLASTICITY OF BREATHING

Effects of Resistive Loading on Breathing Variability

Non Linear Analysis and Modelling Approaches

S. Thibault, P. Calabrese, G. Benchetrit and P. Baconnier

1. Introduction

The ventilatory system as many biological systems is a complex dynamical one involving many controls and regulations. More over, in a reconstructed phase space respiratory data present a structure similar to a strange attractor (cf. Figure 1). In this way, considering the respiratory system as a deterministic chaotic one seems to be a good hypothesis to understand breathing and its variability. This hypothesis remains controversial. Donaldson (1992) shows that respiratory trajectories of resting human are not random but chaotic using the Largest Lyapunov Exponent (LLE). Unfortunately, Hughson *et al.* (1995) said that previous suggestion of Donaldson (1992) that human respiratory pattern was chaotic appears from the outcome of surrogate data analysis in their study to have been premature. Similarly, Fortrat *et al.* (1997), in human resting breathing could not conclude that ventilation derive from a deterministic chaotic system. On the contrary, Sammon *et al.* (1991, 1993a, 1993b) show that irregular inspiratory-expiratory phase switching and central respiratory pattern generator (CRPG) output in rats are consistent with low-dimensional chaos. Small *et al.* (1999) bring evidence that respiratory variability in infants during quiet sleep is deterministic and not random. Starting from the hypothesis that controlling the presence of chaos by changing a parameter of a system brings further evidence of the chaotic nature of the system we tried to modify, experimentally and on simulations, the chaotic dimension of the ventilatory system by changing the resistive load in resting breathing human.

S. Thibault, P. Calabrese, G. Benchetrit and P. Baconnier • Laboratoire TIMC/IMAG (UMR CNRS 5525), Université Joseph Fourier (UJF), Faculté de Médecine de Grenoble, 38706 La Tronche Cedex, France.

Post-Genomic Perspectives in Modeling and Control of Breathing, edited by Jean Champagnat, Monique Denavit-Saubié, Gilles Fortin, Arthur S. Foutz, Muriel Thoby-Brisson. Kluwer Academic/Plenum Publishers, 2004.

2. Method

2.1. Data Acquisition

We used data of Calabrese *et al.* (1998). Digitalisation frequency was 64 Hz. Airflow signal from 7 volunteers have been recorded with a pneumotachograph mounted on a face mask. Recording was made under 5 conditions: control and 4 different levels of resistance, 3.6, 5.8, 8.8 and 13.1 $cmH_2O.L^{-1}.s$. Flow resistive loads were added throughout the entire breath.

2.2. Analysis

We used the LLE which gives the average rate of divergence of two nearby trajectories as an index of global variability of the system. First, we have to reconstruct the phase space (cf. Figure 1) for applying the tools for chaos detection. Embedding dimension and a fixed lag are needed and are determined using, respectively, the algorithm of false nearest neighbours and the first minimum of mutual information. The LLE was then computed with the algorithm of Wolf (1985) which can be simply described as follows (cf. Figure 2): an initial length $L(t_0)$ is found by joining the initial point t_o to its nearest neighbour on a different trajectory. This one must occur later enough in the time series and, not for the first step but for the others, the orientation of joining vector compared to the last one (θ_1 and θ_2) must not be too different. The final length $L'(t_1)$ joins the points at the time t_1, and the procedure is repeated along the fiducial trajectory, in other words, until we have gone through the entire time series.

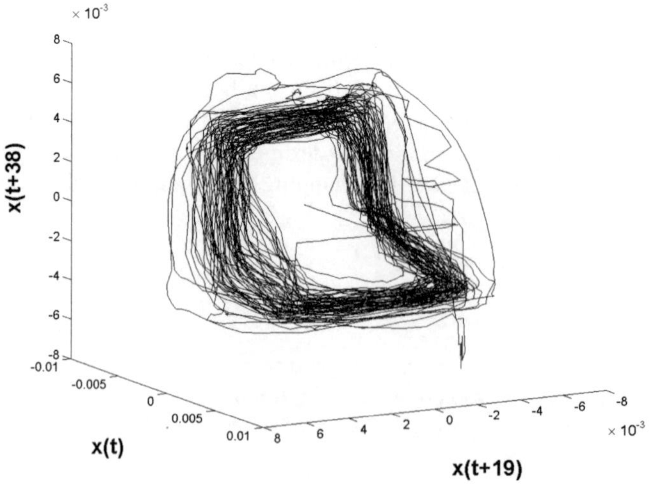

Figure 1. Reconstructed phase space from airflow data (subject #7). Embedding dimension and time lag are respectively 3 and 19 iterations (≈0.30 s).

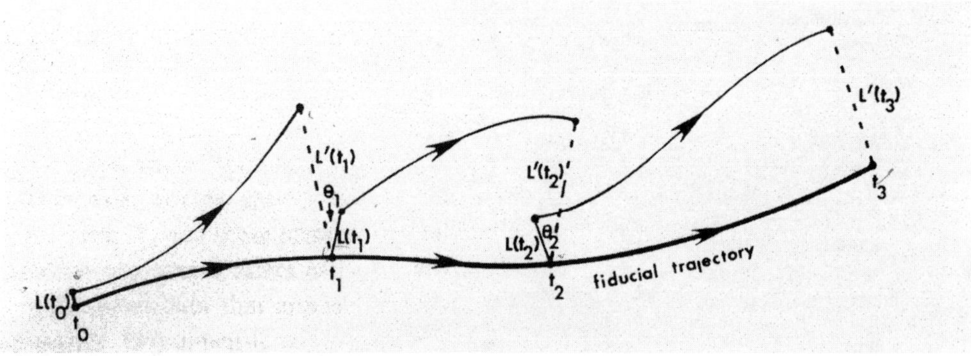

Figure 2. The figure represents the evolution and the replacement procedure used to estimate the LLE from experimental data. From Wolf (1985).

The LLE is computed such that:

$$\lambda = \frac{1}{t_{\mathrm{K}} - t_0} \sum_{\mathrm{k}=1}^{\mathrm{K}} \log_2 \frac{L'(t_{\mathrm{k}})}{L(t_{\mathrm{k}-1})}$$

where λ is LLE (bits.iteration^{-1}) and K is the total number of iterations (see Figure 2 with $\mathrm{K} = 3$).

2.3. Modelling

In parallel, we have built a mathematical model of the ventilatory system. It simulates the interactions between 2 objects: an active one, CRPG and a passive one, the mechanical respiratory system. As the CRPG has been shown to be able to exhibit a deterministic chaotic-like behaviour (Del Negro *et al.*, 2002), to model it, we use a Duffing oscillator (Dang-Vu *et al.*, 2000) which can exhibit chaotic behaviour. The passive mechanical ventilatory system is depicted as a single homogenously ventilated alveolar compartment of a fixed elastance (E) connected to a rigid airway of a fixed resistance (R) (Otis *et al.*, 1956). The pressure generated by the respiratory muscles is the result of the conversion by the muscles of the CRPG output (x) into a pressure; since only inspiration is active we represent that by a conditional function, $f(x)$. When the afferent pathways are functional, the CRPG activity can be modulated by the lung volume (z) variation, we represent that, as simply as possible by the term $\mathrm{A} \cdot \dot{z}$. Finally we obtain the following equation set:

$$\begin{cases} \ddot{x} = \mathrm{a} \cdot \dot{x} + \mathrm{b} \cdot x^3 + \alpha \cdot \sin(\omega \cdot t) + \mathrm{A} \cdot \dot{z} \\ \dot{z} = -\dfrac{\mathrm{E}}{\mathrm{R}} \cdot z + \dfrac{1}{\mathrm{R}} \cdot f(x) \\ \dot{t} = 1 \end{cases}$$

with

$$f(x) = \begin{cases} \mathrm{B} \cdot x & x > 0 \\ 0 & x < 0 \end{cases}$$

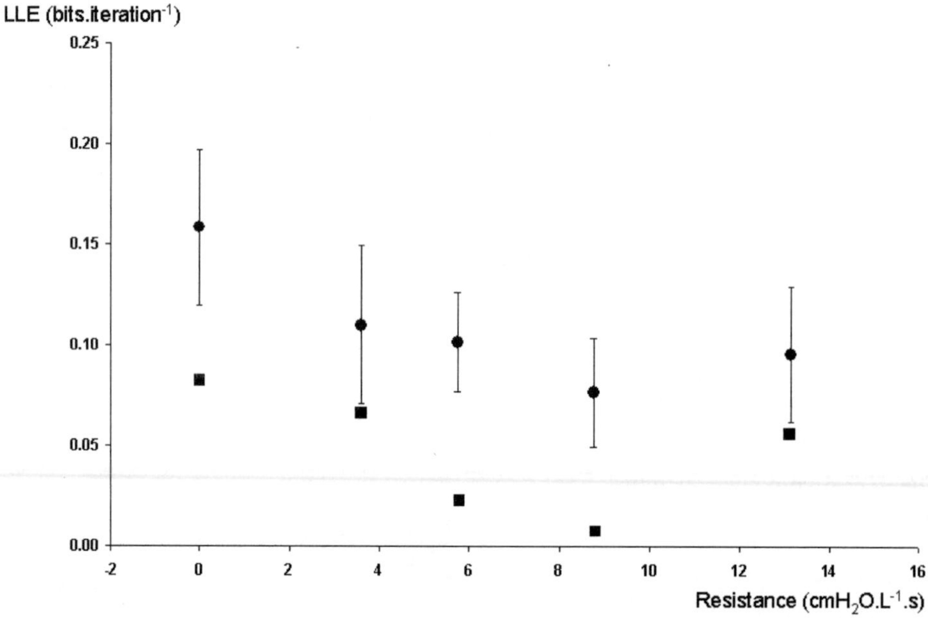

Figure 3. Value of LLE in function of resistive loads. (●) represents mean value (±SD) of LLE for experimental data and (■) value of LLE obtained by simulation at the same resistive loads.

t is the time. a, b, α and ω are Duffing oscillator parameters set in a way to present a chaotic behaviour. Only the value of resistance R changes, taken the same values as that of the experimental conditions.

3. Results

3.1. Experimental Data

Time lag increases with resistive load. This is in accordance with the decrease of respiratory frequencies as described by Calabrese *et al.* (1998). In fact, mutual information in airflow data is strongly linked to breath cycle duration. Value of embedding dimension is equal to 2 for subjects #3 and #5 and to 3 for the others. Mean values of LLE are 0.16 bits.iteration^{-1} for control, 0.11, 0.10, 0.08 and 0.10 bits.iteration^{-1} respectively for the four successive loads (cf. Figure 3). An ANOVA analysis shows a significant decrease of LLE with increase of resistance value ($p < 0.05$), and a post hoc Fischer's PLSD test shows a significant decrease in LLE values as compared with resting breathing at all levels of resistance. In five out of the seven subjects we observed an increase of LLE from the third level to the fourth level (cf. Table 1).

3.2. Simulations

Time lag, contrarily to experimental data does not present a significant increase with resistive load. The embedding dimension is equal to two in all cases which is in

Table 1. Value of LLE (bit.iteration^{-1}) for all subjects and for all levels of resistance. The last row presents means (±SD) for all levels of resistance.

	resistive loads ($cmH_2O.L^{-1}.s$)				
subjects	control	3.6	5.8	8.8	13.1
#1	0.19	0.12	0.14	0.05	0.08
#2	0.10	0.08	0.08	0.06	0.09
#3	0.18	0.12	0.11	0.10	0.07
#4	0.14	0.05	0.08	0.07	0.04
#5	0.21	0.11	0.07	0.03	0.11
#6	0.18	0.18	0.12	0.11	0.16
#7	0.11	0.10	0.11	0.10	0.12
mean (±SD)	0.16 (±0.04)	0.11 (±0.04)	0.10 (±0.02)	0.08 (±0.03)	0.10 (±0.03)

accordance with those found with experimental data. We observe with simulations the same tendency as for the experimental results namely a decrease in LLE with the increase of resistance until the value of 8.8 $cmH_2O.L^{-1}.s$. Moreover, simulation amazingly mimics the increase of LLE from the third level to the fourth level observed in five out of the seven subjects (cf. Figure 3). Simulations with extra resistive loads lead to a LLE value equal to −0.02 bit.iteration^{-1} for 85 $cmH_2O.L^{-1}.s$ indicating that we obtained chaos vanishing and resurgence of limit cycle or fixed point.

4. Discussion

First our global result indicates that adding resistance allows to reduce chaos intensity of breathing. Moreover, the increase of LLE from the third to the fourth level of added resistance obtained in five subjects is not an isolated observation. In an analysis of breathing variability using multilinear regression model, Brack et al. (1998) showed an increase of random uncorrelated fraction of breathing variability with high levels of resistance. This is likely to agree with our result on the effect of the highest level of added resistance as the random uncorrelated fraction of Brack et al. (1998) includes non linearity. In this way our results suggest that the use of LLE seems to be a good way to quantify breath to breath variability. The final increase of LLE questions the use of adding high resistive loads as a way to obtain chaos vanishing and thus as a tool to assess the chaotic nature of breathing. Despite this and the fact that a positive LLE does not systematically mean that a system is a deterministic chaotic one (Ruelle, 1990), simulations using our deterministic model present the same qualitative tendencies for LLE values than experimental ones. This supports the hypothesis of chaotic nature of respiratory system. That embedding dimension value stay the same for a subject along the load increase could reveal no other oscillators are involved in CRPG with adding resistive loads. Finally, respiratory variability may become pathological under assisted ventilation leading to instability. This situation may be seen as a stronger chaotic behaviour that should be important to control using a key parameter and/or predict their occurrence using our modified model adding ventilatory support.

References

Brack, T., A. Jubran and J. T. Tobin. 1998. Effect of Resistive Loading on Variational Activity of Breathing. *Am. J. Respir. Crit. Care Med.* 157: 1756–1763.

Calabrese, P., T. Pham Dinh, A. Eberhard, J.-P. Bachy and G. Benchetrit. 1998. Effects of Resistive Loading on the Pattern of Breathing. *Respiration Physiology.* 113: 167–179.

Dang-Vu H., C. Delcarte and M. Wysocki. 2000. A mathematical model of respiratory system exhibiting chaos. *Int. J. Diff. Equat. Appl.* 1: 399–408.

Del Negro, C. A., C. G. Wilson, R. J. Butera, H. Rigatto and J. C. Smith. 2002. Periodicity, Mixed-Mode Oscillations, and Quasiperiodicity in a Rhythm-Generating. *Biophysical Journal.* 82: 206–214.

Donaldson, G. C. 1992. The chaotic behaviour of resting human respiration. *Respir. Physiol.* 88: 313–321.

Fortrat J.-O., Y. Yamamoto and R. L. Hughson. 1997. Respiratory influences on non-linear dynamics of heart rate variability in humans. *Biol. Cybern.* 77: 1–10.

Hughson R. L., Y. Yamamoto and J.-O. Fortrat. 1995. Is the Pattern of Breathing at Rest Chaotic? A Test of the Lyapunov Exponent. *Adv. Exp. Med. Biol.* 393: 15–19.

Otis, A. B., C. B. Mc Kerrow, R. A. Bartlett, J. Mead, M. B. McIlroy, N. J. Selverstone and E. P. Radford 1956. Mechanical Factors in Distribution of Pulmonary Ventilation. *J. Appl. Physiol.* 8: 427–443.

Ruelle, D. 1990 Deterministic chaos: the science and the fiction. *Proc. R. Soc. Lond. A.* 427: 241–248.

Sammon M. P. and E. N. Bruce. 1991.Vagal afferent activity increases dynamical dimension of respiration in rats. *J. Appl. Physiol.* 70(4): 1748–1762.

Sammon M., J. R. Romaniuk and E. N. Bruce. 1993a. Bifurcations of the respiratory pattern associated with reduced lung volume in the rat. *J. Appl. Physiol.* 75(2): 887–901.

Sammon M., J. R. Romaniuk and E. N. Bruce. 1993b. Bifurcations of the respiratory pattern produced with phasic vagal stimulation in the rat. *J. Appl. Physiol.* 75(2): 912–926.

Small M., K. Judd, M. Lowe and S. Stick. 1999. Is breathing in infants chaotic? Dimension estimates for respiratory patterns during quiet sleep. *J. Appl. Physiol.* 86(1): 359–376.

Wolf, A., J. B. Swift, H. L. Swinney and A. Vastano. 1985. Determining Lyapunov Exponents From a Time Series. *Physica* 16D 285–317.

Effects of Intermittent Hypoxic Training and Detraining on Ventilatory Chemosensitive Adaptations in Endurance Athletes

Keisho Katayama, Kohei Sato, Hiroshi Matsuo, Koji Ishida, Shigeo Mori, and Miharu Miyamura

Key words: hypoxic ventilatory chemosensitivity, hypercapnic ventilatory chemosensitivity

1. Introduction

It is well known that chronic exposure to hypoxia and sojourns at high altitude lead to an increase in the hypoxic and hypercapnic ventilatory responses (HVR and HCVR), as indexes of ventilatory chemosensitivities to hypoxia and hypercapnia[1]. Similarly, recent studies[2,3] found that intermittent hypoxia at rest also induces an increase in HVR, whereas only a few studies have investigated the influence of intermittent hypoxic training, i.e., live low-train high, on ventilatory chemosensitivity; Levine et al.[4] and Benoit et al.[5] indicated that an enhanced HVR appeared after intermittent hypoxic training. We also have found that HVR tended to increase after intermittent hypoxic training in untrained subjects, but HCVR did not[6]. However, surprisingly, no study has attempted to investigate ventilatory chemosensitive adaptations to intermittent hypoxic training in endurance athletes who have blunted ventilatory chemosensitivity, despite the fact that intermittent hypoxic training is commonly used by endurance athletes.

Cardiorespiratory adaptations to hypoxic training have been reported by many investigators, but physiological responses after the cessation of altitude training have received little attention. To elucidate the change in ventilatory chemosensitivity after hypoxic training, we measured HVR and HCVR in untrained subjects before and after intermittent hypoxic training and again after detraining[6]. Consequently, it was found that the increased HVR after intermittent hypoxic training returned to the initial level after detraining. However,

---

Keisho Katayama • Research Center of Health, Physical Fitness and Sports, Nagoya University, Nagoya 464-8601, Japan.

Post-Genomic Perspectives in Modeling and Control of Breathing, edited by Jean Champagnat, Monique Denavit-Saubié, Gilles Fortin, Arthur S. Foutz, Muriel Thoby-Brisson. Kluwer Academic/Plenum Publishers, 2004.

to our knowledge, there are no available data concerning the influence of detraining after intermittent hypoxic training on ventilatory chemosensitivity in endurance athletes.

The purpose of the present study, therefore, was to elucidate the characteristics of ventilatory chemosensitive adaptations to intermittent hypoxic training and detraining in endurance athletes by comparing those observed in untrained subjects.

2. Methods

Five trained male endurance runners (athlete group), who belonged to a collegiate track team [peak maximal oxygen uptake ($\dot{V}O_{2peak}$) > 65 ml·kg^{-1}·min^{-1}], and five untrained male subjects (non-athlete group, $\dot{V}O_{2peak}$ < 55 ml·kg^{-1}·min^{-1}) participated in this study. The subjects were informed of the experimental procedures and their informed consent was obtained. This study was approved by the Human Research Committee of the Research Center of Health, Physical Fitness and Sports, Nagoya University.

Subjects were first familiarized with the equipment used in the experiment. Before the endurance exercise training, $\dot{V}O_{2peak}$ at sea level was determined for each subject. To determine a training intensity, $\dot{V}O_{2peak}$ was also measured in a hypobaric chamber, which simulated an altitude of 4,500 m (432 torr). On another day, the measurements of HVR and HCVR were performed for each subject. These measurements were performed at sea level at both the pre- (Pre) and post-training. The post-training test was performed twice, i.e., on the first day after 2 wk of intermittent hypoxic training (Post) and on the first day after 2 wk of detraining (De). The subjects in each group trained for 30 min/d, 5 d/wk for 2 wk on a mechanically braked bicycle ergometer at 70% of altitude $\dot{V}O_{2peak}$ in the hypobaric chamber, which was set to simulate an altitude of 4,500 m.

HVR was determined by using a progressive isocapnic hypoxic test[7], and a rebreathing system[3,6] was used. Inspired minute ventilation ($\dot{V}_I$), end-tidal CO_2 and O_2 fractions (F_{ETCO_2} and F_{ETO_2}), and arterial oxygen saturation (Sa_{O_2}) were measured continuously during rebreathing. The subjects breathed through a mouthpiece attached to a hot wire flowmeter. Sample gas was drawn continuously through a sampling tube connected to the mouthpiece in order to measure F_{ETCO_2} and F_{ETO_2} by means of a gas analyzer, and end-tidal partial pressure of CO_2 and O_2 (P_{ETCO_2} and P_{ETO_2}) were calculated from F_{ETCO_2} and F_{ETO_2}. P_{ETCO_2} was maintained within ±2 torr of the resting level during measurement. Sa_{O_2} was measured by using a finger pulse oximeter. HVR was estimated as the slope of the line calculated by the linear regression relating $\dot{V}_I$ to Sa_{O_2} ($\Delta\dot{V}_I/\Delta Sa_{O_2}$, l·min^{-1}·%$^{-1}$), and the slope was presented in positive numbers by convention.

HCVR was determined by two methods: the CO_2 rebreathing ($HCVR_R$) and the single breath CO_2 ($HCVR_{SB}$) methods. The measurements of $HCVR_R$ and $HCVR_{SB}$ were also similar to those of our previous studies[3,6]. In the rebreathing method, subjects rebreathed a gas mixture of 7% CO_2 in O_2 from a bag (5–6 liters) in a plastic box for 3–4 min[8]. $\dot{V}_I$ and P_{ETCO_2} were recorded in the same ways as in the HVR test. $HCVR_R$ was assessed as the slope of the line (S) determined by the linear regression relating P_{ETCO_2} to $\dot{V}_I$ ($\Delta\dot{V}_I/\Delta P_{ETCO_2}$, l·min^{-1}·torr^{-1}). On the other hand, the single breath CO_2 test was performed according to the protocol described by McClean et al.[9], i.e., application of a single CO_2 mixture composed of 13%CO_2-21%O_2-66%N_2 was repeatedly given to each subject: six to eight times, separated by 2- to 3-min intervals. The apparatus consisted of a bag-in-box circuit similar to that used for the CO_2 rebreathing test. The subjects were seated comfortably in a chair and began breathing room air through a mouthpiece with a noseclip. A T-valve was attached between the bag and the mouthpiece, and the port was connected to either room air or the bag containing the test gas. During testing, tidal volume (VT), $\dot{V}$I, F_{ETCO_2}, F_{ETO_2}, and inspiratory time (TI) were recorded continuously. When stable levels of F_{ETCO_2} and VT had been achieved, the inspiratory gas was switched from room air to the bag for a single tidal breath by turning the T valve during the expiratory phase of the previous breath. During the expiratory phase of the test breath, the T valve was turned back again to its initial position. Data for analyzing $HCVR_{SB}$ were limited to breaths within the first 20 sec following transients of hypercapnia. $HCVR_{SB}$ was quantified in a manner similar to that suggested by Khoo[10], units $[\Delta(V_T/T_I)/\Delta P_{ETCO_2}\ ml{\cdot}sec^{-1}{\cdot}torr^{-1}]$.

The maximal exercise test was conducted in the same way as in our previous studies[3,6]. An incremental protocol on an electromechanically-braked bicycle ergometer was used, i.e., starting at 60 W, the load was increased by 30 W every 2 min. Expired gases were collected into a Douglas bag during the last 30 sec of each intensity level until exhaustion. Expired gas volume ($\dot{V}$E) was measured with a wet-gas meter. Gas analysis was performed by means of an O_2 and CO_2 analyzer. Heart rate (HR) was continuously recorded using a three-lead electrocardiogram. The maximal ventilation per minute was estimated as $\dot{V}E_{peak}$, and the peak HR value was expressed as HR_{peak}. The highest values obtained for $\dot{V}O_2$ during the exercise protocol was used as $\dot{V}O_{2peak}$.

The values were expressed as means ± SD. The differential changes in the parameters during the experimental periods between the athlete and non-athlete groups were compared using a two-way analysis of variance (ANOVA) with repeated measurements. Differences in the parameters at each session (Pre, Post, and De) within each group were determined by using a Newman-Keuls test, and the comparison of parameters between groups at each session was achieved using the Mann-Whitney test. The level of significance was set at 0.05.

3. Results

Figure 1A indicates the changes in HVR at Pre, Post, and De in the two groups. HVR in the athlete group was significantly lower than that in the non-athlete group at Pre. In the athlete group, HVR increased significantly after intermittent hypoxic training. Similarly, the HVR in the non-athlete group tended to increase at Post, but the increase was not significant. Following 2 weeks of detraining, the increased HVR in both groups returned to the pre-training levels. There was no statistical difference in HVR between the groups during the experimental period. The magnitude of the changes in HVR (%ΔHVR) was calculated as the difference between those obtained before and after intermittent hypoxia (Pre-Post) and after intermittent hypoxic training and after detraining (Post-De). An increase in %ΔHVR after intermittent hypoxic training was significantly greater in the athlete group than in the non-athlete group (Fig. 1B). Reduction of %ΔHVR in the athlete group following detraining was also greater than that in the non-athlete group, but it was not statistically significant (Fig. 1B).

$HCVR_R$ in the athlete group at Pre and De were significantly lower than those in the non-athlete group (Table 1). There were no significant changes in $HCVR_R$ in either group after intermittent hypoxic training and detraining, and there was not a significant

Table 1. $HCVR_R$ and $HCVR_{SB}$ throughout the experimental period.

		Pre	Post	De
$HCVR_R$	A	1.17 ± 0.47 †	1.33 ± 0.50	1.01 ± 0.49 †
($l \cdot min^{-1} \cdot torr^{-1}$)	N	2.40 ± 0.53	2.29 ± 0.41	2.48 ± 0.72
$HCVR_{SB}$	A	8.8 ± 5.2 †	9.2 ± 3.2 †	8.3 ± 3.4
($ml \cdot sec^{-1} \cdot torr^{-1}$)	N	14.1 ± 6.2	14.8 ± 6.1	13.6 ± 6.4

A: athlete group, N: non-athlete group. † Significantly different from the non-athlete group ($P < 0.05$).

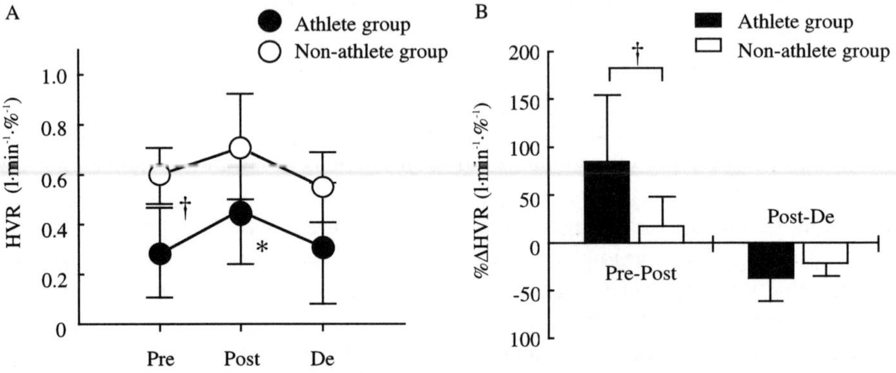

Figure 1. A: Changes in HVR throughout the experimental period. B: The magnitude changes in HVR (%ΔHVR) in both groups during intermittent hypoxic training and detraining. Values are means ± SD. *Significantly different from Pre ($P < 0.05$). †Significantly different between the groups ($P < 0.05$).

difference in $HCVR_R$ between the athlete and non-athlete groups during the experimental period. $HCVR_{SB}$ in the athlete group at Pre and Post was also significantly lower than that in the non-athlete group (Table 1). $HCVR_{SB}$ did not show any changes in the athlete and non-athlete groups after intermittent hypoxic training and detraining, and there was not a significant difference in $HCVR_{SB}$ between the groups during the experimental period.

$\dot{V}O_{2peak}$ in the athlete group tended to increase after intermittent hypoxic training, but it showed a decrease after detraining [67.0 ± 1.8 (Pre), 70.2 ± 3.2 (Post), 68.7 ± 3.2 $ml \cdot kg^{-1} \cdot min^{-1}$]. $\dot{V}O_{2peak}$ in the non-athlete group increased significantly after intermittent hypoxic training, and a significant loss of $\dot{V}O_{2peak}$ occurred after 2 weeks of detraining [52.8 ± 1.4 (Pre), 56.7 ± 3.9 (Post), 53.8 ± 2.4 (De) $ml \cdot kg^{-1} \cdot min^{-1}$]. No significant changes were observed in $\dot{V}E_{peak}$ and HR_{peak} in either group during the experimental period.

4. Discussion

In the present study, we found that 1) HVR in the athlete group increased significantly after intermittent hypoxic training for 2 weeks, and the increased HVR returned to the pre-training level after 2 wk of detraining. %ΔHVR after intermittent hypoxic training was greater in the athlete group than in the non-athlete group; 2) $HCVR_R$ and $HCVR_{SB}$ were unchanged in both groups after intermittent hypoxic training and detraining.

A limited number of investigators have examined that the effect of intermittent hypoxic training on ventilatory chemosensitive adaptations[4,5]: they have reported that an enhanced HVR appeared after intermittent hypoxic training. In our previous study[6], we also found that HVR tended to increase after intermittent hypoxic training. However, in these studies, the subjects were untrained individuals, and no study has attempted to investigate the ventilatory chemosensitive adaptations to intermittent hypoxic training in endurance athletes. HVR in the athlete group before intermittent hypoxic training was lower than that in the control group (Fig. 1A), and this result is consistent with previous study[11]. Following intermittent hypoxic training, the blunted HVR in the athlete group increased significantly, and HVR in the non-athlete group also tended to increase, although it was not significant. After detraining, the increased HVR in both groups returned to a level similar to that before training. %ΔHVR after intermittent hypoxic training and detraining was greater in the athlete group than in the non-athlete group, but it was not statistically significant after detraining (Fig. 1B). These results suggest that hypoxic ventilatory chemosensitivity is variable by intermittent hypoxic training and detraining, and that hypoxic ventilatory chemosensitivity are changeable in endurance athletes than in untrained individuals after intermittent hypoxic training and detraining.

It has hitherto been reported blunted $HCVR_R$, as an index of central hypercapnic chemosensitivity, in endurance trained athletes as compared with that in untrained subjects[11]. Similiary, $HCVR_R$ in the athlete group is significantly lower than that in the untrained subjects in this study. There was no change in $HCVR_R$ as a result of either endurance training or detraining (Table 1), and this indicates that central hypercapnic chemosensitivity does not increase readily due to intermittent hypoxic training. Since previous study in humans has shown an increase in the slope of $HCVR_R$ during chronic exposure to hypoxia[1], one of the reasons why there was no increase in $HCVR_R$ after intermittent hypoxic training may be attributable to the duration of the exposure to hypoxia. In addition, $HCVR_{SB}$ was also assessed before and after intermittent hypoxic training in the present study. Before training, $HCVR_{SB}$ in the athlete group was significantly lower than that in the non-athlete group (Table 1). Several investigators have proposed that a single breath of CO_2 is a useful method for evaluating sensitivities of peripheral chemosensitivity to hypercapnia[9]. Thus, the results from this study indicate that peripheral hypercapnic ventilatory chemosensitivity, as well as central hypercapnic ventilatory chemosensitivity, is lower in endurance athletes. As far as we know, this is the first study to assess $HCVR_{SB}$ in endurance athletes. However, no significant changes were found in $HCVR_{SB}$ after intermittent hypoxic training in either group. Previous studies have reported that peripheral hypercapnic ventilatory chemosensitivity increases during sojourns at high altitude[12] and after intermittent hypoxia (1h/day)[3]. Therefore, the duration of intermittent hypoxic training (30 min/day), as applied here, may have been insufficient to elicit an increase in $HCVR_{SB}$. In other words, it is possible to hypothesize that $HCVR_{SB}$ may increase after intermittent hypoxic training as training periods are prolonged.

In conclusion, HVR in the athlete group increased significantly after intermittent hypoxic training, and the increased HVR returned to the pre-training level after detraining. %ΔHVR after intermittent hypoxic training and detraining were greater in the athlete group than in the non-athlete group. $HCVR_R$ and $HCVR_{SB}$ in both groups were unchanged after intermittent hypoxic training and detraining. These results suggest that hypoxic ventilatory chemosensitivity is more sensitive to intermittent hypoxic training and detraining than hypercapnic ventilatory chemosensitivity, and that there is more variation in the changes in hypoxic ventilatory chemosensitivity in endurance athletes than in untrained individuals after intermittent hypoxic training and detraining.

5. Acknowledgement

This study was supported in part by a Grant-in-Aid for Scientific Research from the Japanese Ministry of Education, Science, Sports and Culture (grant no. 15700413) and the Uehara Memorial Foundation.

References

1. R. B. Schoene, R. C. Roach, P. H. Hackett, J. R. Sutton, A. Cymerman, and C. S. Houston. Operation Everest II: ventilatory adaptation during gradual decompression to extreme altitude. *Med. Sci. Sports Exerc.* **22**: 804–810 (1990).
2. N. Garcia, S. R. Hopkins, and F. L. Powell. Effects of intermittent hypoxia on the isocapnic hypoxic ventilatory response and erythropoiesis in humans. *Respir. Physiol.* **123**: 39–49 (2000).
3. K. Katayama, Y. Sato, Y. Morotome, N. Shima, K. Ishida, S. Mori, and M. Miyamura. Intermittent hypoxia increases ventilation and Sa_{O_2} during hypoxic exercise and hypoxic chemosensitivity. *J. Appl. Physiol.* **90**: 1431–1440 (2001).
4. B. D. Levine, D. B. Friedman, K. Engfred, B. Hanel, M. Kjaer, P. S. Clifford, and N. H. Secher. The effects of normoxic or hypobaric hypoxic endurance training on the hypoxic ventilatory response. *Med. Sci Sports Exerc.* **24**: 769–775 (1992).
5. H. Benoit, M. Germain, J. C. Barthelemy, C. Denis, J. Castells, D. Dormois, J. R. Lacour, and A. Geyssant. Pre-acclimatization to high altitude using exercise with normobaric hypoxic gas mixture. *Int. J. Sports Med.* **13**, Suppl.: S213–S216 (1992).
6. K. Katayama, Y. Sato, Y. Morotome, N. Shima, K. Ishida, S. Mori, and M. Miyamura. Ventilatory chemosensitive adaptations to intermittent hypoxic exposure with endurance training and detraining. *J. Appl. Physiol.* **86**: 1805–1811 (1999).
7. J. V. Weil, E. Byrne-Quinn, I. E. Sodal, W. O. Friesen, B. Underhill, G. F. Filley, and R. F. Grover. Hypoxic ventilatory drive in normal man. *J. Clin. Invest.* **49**: 1061–1072 (1970).
8. D. J. C. Read. A clinical method for assessing the ventilatory response to carbon dioxide. *Aust. Ann Med.* **16**: 20–32 (1967).
9. P. A. McClean, E. A. Phillipson, D. Martinez, and N. Zamel. Single breath of CO_2 as a clinical test of the peripheral chemoreflex. *J. Appl. Physiol.* **64**: 84–89 (1988).
10. M. C. K. Khoo. A model-based evaluation of the single-breath CO_2 ventilatory response test. *J. Appl. Physiol.* **68**: 393–399 (1990).
11. E. Byrne-Quinn, J. V. Weil, I. E. Sodal, G. F. Filley, and R. F. Grover. Ventilatory control in the athlete. *J. Appl. Physiol.* **30**: 91–98 (1971).
12. J. N. Pande, S. P. Gupta, and J. S. Guleria. Ventilatory response to inhaled CO_2 at high altitude. *Respiration* **31**: 473–483 (1974).

Effects of 5 Consecutive Nocturnal Hypoxic Exposures on Respiratory Control and Hematogenesis in Humans

Jon C. Kolb, Philip N. Ainslie, Kojiro Ide, and Marc J. Poulin

Introduction

Alterations in respiratory control during periods of either chronic or discontinuous hypoxic exposures seem to be facilitated by two main processes, which contribute to the increase in ventilation and a decrease in end-tidal P_{CO_2} ($P_{ET_{CO_2}}$). The first of these processes is an increase in the acute hypoxic ventilatory response (AHVR). An increase in the AHVR allows ventilatory acclimatization to altitude to proceed, despite respiratory alkalosis and a withdrawal of the stimulus to the chemoreceptors[1]. The second process is evident in a leftward shift of the acute hypercapnic ventilatory response (AHCVR), and also in an increase in the slope of this relationship, which is normally determined under euoxic or hyperoxic conditions. The leftward shift in the AHCVR has been suggested to be related to the degree of re-setting of the central chemoreceptors to start responding to a lowered P_{CO_2}[2].

Several studies have demonstrated a progressive increase in the AHVR during both natural altitude acclimatization[3,4] and intermittent hypoxic exposures[5,6]. During natural altitude acclimatization, a leftward shift of the AHCVR has been well documented[7,8]. However, the effect of intermittent exposure to hypoxia on AHCVR has received little investigation. In one study, subjects were exposed to a simulated altitude of 4,500 m for 30 min$\cdot$day^{-1} for 6 days[5]. Whilst there was a significant increase in AHVR there was no significant difference in the AHCVR[5].

The general aim of the present study was to establish an overnight hypoxic exposure that would elicit changes in ventilatory sensitivity to variations in O_2 and CO_2, which are similar to those observed during natural acclimatization. In particular we wished to address the following questions:

Jon C. Kolb, Philip N. Ainslie, Kojiro Ide, and Marc J. Poulin • Faculty of Kinesiology, and Departments of Physiology & Biophysics and Clinical Neurosciences, Faculty of Medicine, University of Calgary, 3330 Hospital Drive NW, Calgary, Alberta, T2N 4N1 Canada.

Post-Genomic Perspectives in Modeling and Control of Breathing, edited by Jean Champagnat, Monique Denavit-Saubié, Gilles Fortin, Arthur S. Foutz, Muriel Thoby-Brisson. Kluwer Academic/Plenum Publishers, 2004.

- Would five consecutive overnight exposures to a simulated altitude of 4300 m elicit changes in the slope and/or intercept of the AHCVR?
- Assuming that alterations in the AHVR are similar to that previously reported for both intermittent and chronic hypoxia, would changes in respiratory control persist following five days of normoxic recovery?
- Would the hematological variability during the nocturnal hypoxic exposures be similar to other intermittent and chronic hypoxic studies?

We tested the specific hypothesis that five consecutive nocturnal hypoxic exposures would elicit similar changes in respiratory control and hematogenesis to those reported during chronic altitude acclimatization.

Methods

The experiments were conducted in our laboratory located at 1103 m above sea level, and the average barometric pressure for the study days was 667 ± 14 (SD) Torr. Twelve male subjects (26.6 ± 4.1 yrs) slept 8 h/day overnight for 5 consecutive days in purpose-designed tents (Hypoxico, Inc) at a simulated altitude of 4300 m ($FiO_2 = \sim 13.8\%$). Subjects entered the tents at approximately 2300 hrs and exited at 0700 hrs the following morning. Between the nocturnal hypoxic exposures (~16 h normoxia), subjects maintained their normal daily activities. Throughout a control sleep (normoxia) and all overnight hypoxic exposures, pulse oximetry (Nellcor N-295, Nellcor Inc., Hayward, CA, USA) was used to continuously monitor arterial oxygen saturation (SaO_2). Prior to entering the tents, subjects were fitted with a fine nasal catheter which was connected to a CO_2 analyzer (Normocap Oxy Monitor, Datex-Ohmeda, Mississauga, Ontario, Canada) for determining overnight $P_{ET}CO_2$ values. $P_{ET}CO_2$ values were measured and averaged over a 5 minute period each hour throughout each night.

Blood measurements were conducted each night prior to entering the tents and following 8 h of hypoxia prior to exiting the tents. Micro blood samples (~200 μl) were obtained by finger tip penetration and immediately analyzed for capillary blood gases (P_cO_2 and P_cCO_2), hemoglobin, and hematocrit using an i-STAT Portable Clinical Analyzer (G6+ cartridge, i-STAT Corp., Princeton, NJ, USA). After the 5 nocturnal hypoxic exposures subjects were permitted to go home, but were asked to maintain their normal diet and physical activity levels.

Measurements of AHVR, ventilation in hyperoxia ($V_{E hyperoxia}$) and AHCVR were assessed twice prior to, immediately after, and 5 days following the hypoxic exposure using the technique of dynamic end-tidal forcing[9,10] according to the following protocol. After an initial 8 min lead in period of eucapnic euoxia ($P_{ET}O_2 = 88$ Torr), the hypoxic stimulus was varied by holding $P_{ET}O_2$ at 6 descending steps ($P_{ET}O_2 = 75.2$, 64.0, 57.0, 52.0, 48.2, and 45.0 Torr), each step lasting 90 sec. Throughout the hypoxic challenge, the $P_{ET}CO_2$ was held constant 1.5 mmHg above resting values. The hypoxic levels were calculated to provide equal steps in oxygen saturation of the arterial blood (SaO_2) via the relationship described by Severinghaus[11]. A linear regression was then performed between the mean values for V_E and 100-SaO_2 during the final 20 sec of each step. The slope of this relationship yielded numerical values for AHVR; this represents an index of the sensitivity to hypoxia[12]. Immediately following the measurement of AHVR, the $P_{ET}O_2$ was elevated to 300 Torr for 5 min while PET_{CO_2} remained constant at eucapnia. The last 2 min of this period was averaged and used to determine $VE_{hyperoxia}$. Then, PET_{CO_2} was raised rapidly over 1–2 breaths by an additional 7.5 Torr for 5 min whilst PET_{O_2} remained constant at 300 Torr. Combined with the results from the measurements of $VE_{hyperoxia}$, the last 2 min of the hypercapnic period was used to calculate the AHCVR.

Results

Mean SaO_2 values for all subjects during all 5 hypoxic exposures were significantly lower ($P < 0.001$) than SaO_2 measurements recorded during the control sleep (left panel, Figure 1). Mean SaO_2 on the 5th hypoxic exposure was significantly higher when compared to the 1st hypoxic night ($P < 0.001$) and 2nd and 3rd hypoxic exposures ($P < 0.05$). The mean PET_{CO_2} values for all subjects (right panel, Figure 1) during each of the overnight hypoxic exposures were significantly lower than control sleep values ($P < 0.001$). Furthermore, the overnight PET_{CO_2} group mean from the 5th hypoxic night was significantly lower than hypoxic nights 1, 2, and 3 ($P < 0.001$).

Figure 2 illustrates a decline in P_cCO_2 values which persisted between the intermittent hypoxic exposures (lower left panel), with a significant difference observed between evening values (normoxia) at the commencement of the control sleep and prior to exposure 4 ($P < 0.01$). Following 5 days of recovery, P_cCO_2 blood gas measurements had returned to near control values. The hypoxic intervention elicited responses for hemoglobin and hematocrit (Figure 2, upper and lower right panels), illustrated by increasing trends for both evening and morning values. Significant differences were revealed for morning hemoglobin and hematocrit on exposures 3 and 5 when compared with the control sleep ($P < 0.05$).

Immediately following the hypoxic exposure AHVR was increased by 1.6 ± 1.3 $l{\cdot}min^{-1}{\cdot}\%^{-1}$ ($P < 0.01$) when compared with the control values (Figure 3, left panel). When tested at 5 days following the hypoxic exposure, the AHVR had returned to control

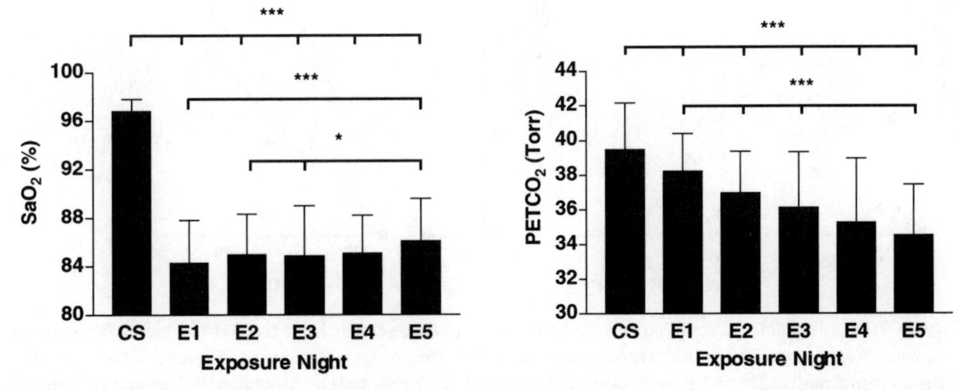

Figure 1. Changes in arterial oxygen saturation (SaO_2) and end-tidal P_{CO_2} (PET_{CO_2}) throughout the nocturnal hypoxic intervention. CS, control sleep ($F_IO_2 = 20.9\%$); E1–E5, hypoxic exposures ($F_IO_2 = 13.8\%$). Values represent overnight means ± SD. * Denotes significantly different than CS, or significantly different from E5 (* $P < 0.05$; *** $P < 0.001$).

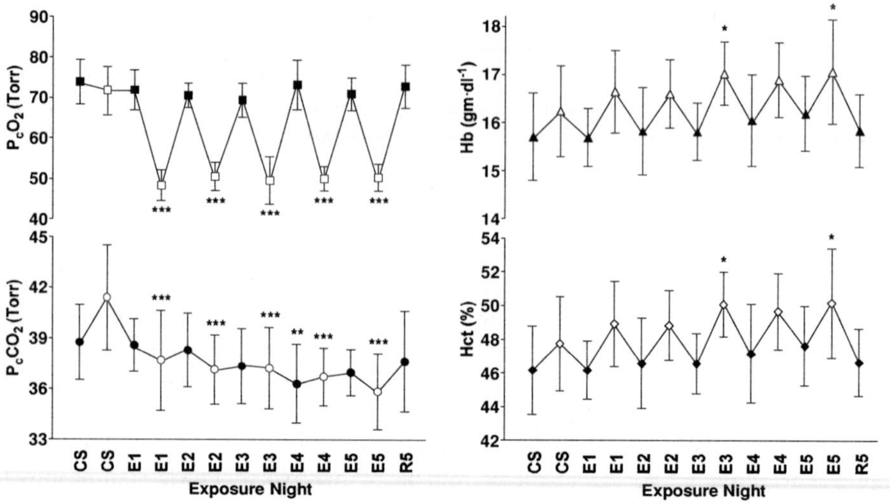

Figure 2. Hematological variability during each of the overnight sessions. CS, control sleep ($F_IO_2 = 20.9\%$), E1–E5, hypoxic exposures ($F_IO_2 = 13.8\%$). R5, recovery after five days post intervention. Closed symbols, evening measurements prior to entering hypoxic tent; Open symbols, morning measurements prior to exiting hypoxic tent. * Denotes significant difference between evening values against evening control sleep, or significant difference between morning values against morning control sleep. (* $P < 0.05$; ** $P < 0.01$; *** $P < 0.001$).

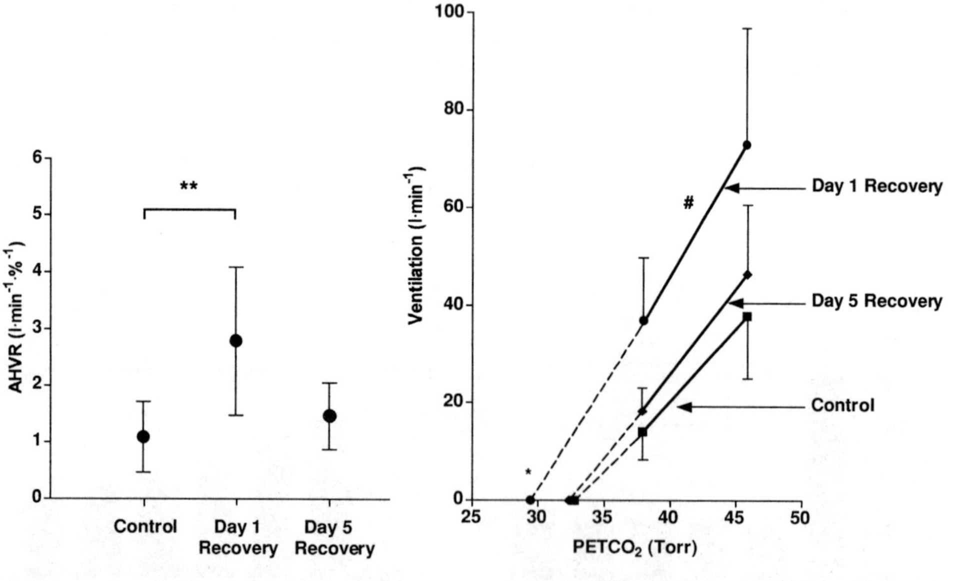

Figure 3. (Left panel): Acute hypoxic ventilatory response to isocapnic hypoxia (AHVR) before, immediately following 5 consecutive nocturnal hypoxic exposures, and after 5 days of recovery. Values are means ± SD. * Denotes significantly different from control measurements. (** $P < 0.01$). (Right panel): Ventilatory responses to increases in P_{ETCO_2} during hyperoxic conditions prior to, and following the 5 nights of normobaric hypoxia. Values are means ± SD. * Denotes significant difference ($P < 0.05$) in the intercept of the AHCVR at day 1 recovery from control values; # denotes significant difference ($P < 0.05$) in the slope of the AHCVR at day 1 recovery from control.

levels. There was a significant increase in ventilation when measured under conditions of hyperoxia ($P < 0.001$). This increase in hyperoxic ventilation returned to control values when measured following the 5 days of recovery. Mean values for both the slope and intercept of the AHCVR are illustrated in Figure 3 (right panel). There was a significant increase in the slope (1.5 ± 1.4 l·min^{-1}. Torr^{-1}; $P < 0.05$) and a decrease in the intercept (-2.7 ± 4.3 Torr; $P < 0.05$) of the AHCVR relationship. These changes in the AHCVR returned to baseline following 5 days of recovery. Collectively, the data indicate that 5 days recovery from the hypoxic exposure provided a suitable time frame during which AHVR, $VE_{hyperoxia}$, and AHCVR returned to pre exposure values. In other words, the time course for both ventilatory acclimatization and de-acclimatization were very similar.

Discussion

The present study has yielded a number of important and novel findings:

- Five nights of normobaric hypoxia elicited a pronounced increase in AHVR and leftward intercept shift of the AHCVR, together with an increase in the slope of this relationship.
- The significant decrease in PET_{CO_2} throughout the nocturnal hypoxic intervention correlated both with the leftward shift and increase in the slope of the AHCVR ($r = 0.55$; $P < 0.05$ and $r = 0.40$; $P < 0.05$ respectively).
- The hematology responses to discontinuous hypoxia were similar to those reported during chronic hypoxia[13–15].

Consistent with our hypothesis, these results indicate that 5 consecutive nights of normobaric hypoxia elicit similar modifications in respiratory control and hematogenesis as those reported in response to chronic altitude exposure. As far as we are aware, this study is the first to show a leftward intercept shift of the ventilatory response to an increase in PET_{CO_2}, together with an increase in the slope of this relationship, following an intermittent hypoxic exposure. The major reasons why previous intermittent studies[5] and chamber studies[16,17] have not shown changes in the AHCVR are most likely due to the level of hypoxia being too moderate and/or the duration of exposure being too short. For example, Schoene et al. demonstrated that AHCVR did not increase at barometric pressure of 452 Torr, while it did increase at 305 Torr[7]. The simulated altitude utilized in the present investigation may have been sufficient to elicit substantial respiratory alkalosis and subsequent changes in the AHCVR intercept and slope similar to observations made during natural altitude acclimatization.

Acknowledgements

This research study was approved by the Conjoint Health Research Ethics Board (University of Calgary) and was supported by the Alberta Heritage Foundation for Medical Research, the Heart and Stroke Foundation of Alberta, NWT, & Nunavut, the Olympic Oval Endowment Fund, and the Calgary Olympic Development Association. We extend our gratitude to Professor P. A. Robbins for his assistance in establishing the technique of dynamic end-tidal forcing in Calgary.

References

1. J.A. Dempsey, and H.V. Forster, Mediation of ventilatory adaptations. *Physiol.* Rev. **62**, 262–346 (1982).
2. D.J.C. Cunnningham, P.A. Robbins, and C.B. Wolf, Integration of respiratory responses to changes in alveolar partial pressures of CO_2 and O_2 and in arterial pH. In: Fishman, A.P. *Handbook of Physiology,* Section 3: The Respiratory System, Vol. II, Part 2: Control of Breathing. American Physiology Society, Bethesda, MD, 475–528 (1986).
3. J.T. Reeves, R.E. McCullough, L.G. Moore, A. Cymerman, and J.V. Weil, Sea-level PCO_2 relates to ventilatory acclimatization at 4,300 m. *J. Appl. Physiol.* **75**, 1117–1122 (1993).
4. M. Sato, J.W. Severinghaus, and P. Bickler, Time course of augmentation and depression of hypoxic ventilatory response at altitude. *J. Appl. Physiol.* **77**, 313–316 (1994).
5. K. Katayama, Y. Sato, K. Ishida, S. Mori, and M. Miyamura, The effects of intermittent exposure to hypoxia during endurance training on the ventilatory responses to hypoxia and hypercapnia in humans. Eur. *J. Appl. Physiol.* **78**, 189–194 (1998).
6. S. Mahamed, and J. Duffin, Repeated hypoxic exposures change respiratory chemoreflex control in humans. *J. Physiol.*, London, **534**, 595–603 (2001).
7. R.B. Schoene, R.C. Roach, P.H. Hackett, J.R. Sutton, A. Cymerman, and C.S. Houston, Operation Everest II: Ventilatory adaptation during gradual decompression to extreme altitude. *Med. Sci. Sports. Exerc.* **22**, 804–810 (1990).
8. D.P. White, K. Glesson, C.K. Pickett, A.M. Rannels, A. Cymerman, and J.V. Weil, Altitude acclimatization: influence on periodic breathing and chemoresponsiveness during sleep. *J. Appl. Physiol.* **63**, 401–412 (1987).
9. P.A. Robbins, G.D. Swanson, and M.G. Howson, A prediction correction scheme for forcing alveolar gases along certain time courses. *J. Appl. Physiol.* **52**, 1353–1357(1982).
10. M.G. Howson, S. Khamnei, M.E. McIntyre, D.F. O'Connor, and P.A. Robbins, A rapid computer controlled binary gas mixing system for studies in respiratory control. *J. Physiol.*, London, 394, 7P(1987).
11. J.W. Severinghaus, Simple, accurate equations for human blood O_2 dissociation computations. *J. Appl. Physiol.* **46**: 599–602 (1979).
12. X.B. Mou, L.S. Howard, and P.A. Robbins, A protocol for determining the shape of the ventilatory response to hypoxia in humans. *Respir. Physiol.* **101**: 139–43 (1995).
13. A. Ricart, H. Casas, M. Casas, T. Pages, L. Palacios, R. Rama, F.A. Rodriguez, G. Viscor, and J.L. Ventura, Acclimatization near home? Early respiratory changes after short-term intermittent exposure to simulated altitude. *Wilderness Environ Med* **11**: 84–88, (2000).
14. J.P. Richalet, J. Bittel, J.P. Herry, G. Savourey, J.L. Le Trong, J.F. Auvert, and C. Janin, Use of a hypobaric chamber for pre-acclimatization before climbing Mount Everest. *Int J Sports Med* **13**: S216–220, 1992.
15. M.V. Singh, S.B. Rawal, and A.K. Tyagi, Body fluid status on induction, reinduction and prolonged stay at high altitude on human volunteers. *Int J Biometeorol* **34**: 93–97, (1990).
16. L.S.G.E. Howard, and P.A. Robbins, Alterations in respiratory control during eight hours of isocapnic and poikliocapnic hypoxia in humans. *J. Appl. Physiol.* **78**, 1098–1107 (1995).
17. J.G. Tansley, M. Fatemian, L.S.G.E. Howard, M.J. Poulin, and P.A. Robbins, Changes in respiratory control during and after 48 h of isocapnic and poikliocapnic hypoxia in humans. *J. Appl. Physiol.* **85**, 2125–2134 (1998).

Memory, Reconsolidation and Extinction in *Lymnaea* Require the Soma of RPeD1

Susan Sangha, Nishi Varshney, Mary Fras, Kim Smyth, David Rosenegger, Kashif Parvez, Hisayo Sadamoto, and Ken Lukowiak

Abstract

The central pattern generator (CPG) that drives aerial respiratory behaviour in *Lymnaea* consists of 3 neurons. One of these, RPeD1—the cell that initiates activity in the circuit, plays an absolutely necessary role as a site for memory formation, memory reconsolidation, and extinction. Using an operant conditioning training procedure that results in a long-term non-declarative memory (LTM), we decrease the occurrence of aerial respiratory behaviour. Since snails can still breathe cutaneously learning this procedure is not harmful. Concomitant with behavioural memory are changes in the spiking activity of RPeD1. Going beyond neural correlates of memory we directly show that RPeD1 is a necessary site for LTM formation. Expanding on this finding we show that this neuron is also a necessary site for memory reconsolidation and 'Pavlovian' extinction. As far as we can determine, this is the first time a *single* neuron has been shown to be a necessary site for these different aspects memory. RPeD1 is thus a key neuron mediating different hierarchical aspects of memory. We are now in a position to determine the necessary neuronal, molecular and proteomic events in this neuron that are causal to memory formation, reconsolidation and extinction.

Introduction

The freshwater snail, *Lymnaea stagnalis*, serves as an excellent model in the study of associative learning and memory (Lukowiak et al., 2003). Aerial respiration is a simple, easily observable, and tractable behaviour that is driven by a 3-neuron CPG, whose necessity and sufficiency has been experimentally demonstrated (Syed et al., 1990; 1992; Figure 1). Since *Lymnaea* are bimodal breathers (cutaneous and aerial respiration) it is possible to teach the snail not to perform this behaviour in a hypoxic environment where this behaviour

---

Susan Sangha, Nishi Varshney, Mary Fras, Kim Smyth, David Rosenegger, Kashif Parvez, Hisayo Sadamoto, and Ken Lukowiak • Department of Physiology & Biophysics, Calgary Brain Institute, University of Calgary, Calgary, Alberta, Canada T2N 4N1.

Post-Genomic Perspectives in Modeling and Control of Breathing, edited by Jean Champagnat, Monique Denavit-Saubié, Gilles Fortin, Arthur S. Foutz, Muriel Thoby-Brisson. Kluwer Academic/Plenum Publishers, 2004.

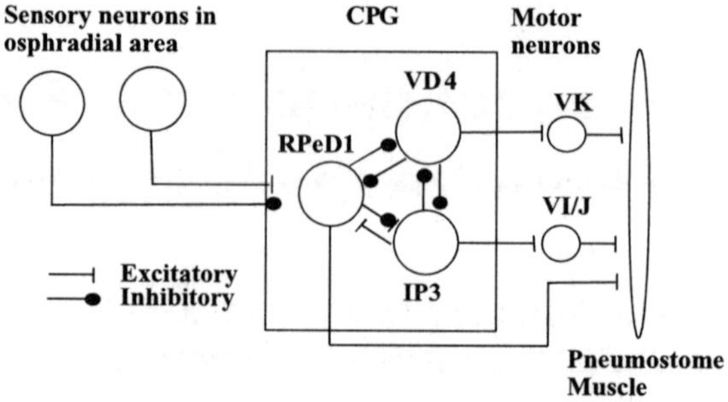

Figure 1. A schematic of the aerial respiratory neural network. The central pattern generator (CPG) drives aerial respiration; its activity is modified by sensory neuronal input to RPeD1. Sensory input (both mechano- and chemosensory) from the osphradial area can excite or inhibit CPG activity by altering RPeD1 activity. The VD4 (inspiration) excites VK cells, which are closer pneumostome motor neurons and IP3 (expiration) excites VI/J, which are opener motor neurons. RPeD1 also causes excitation of opener pneumostome muscles. Tactile stimulation of the pneumostome results in its closure and suppression of CPG rhythmogenesis. The 3-neuron CPG is both sufficient and necessary to produce the rhythmic neural output necessary for aerial respiration.

should predominate without causing harm to the snail (Lukowiak et al., 2000). Depending on the specific training procedure used snails exhibit varying lengths of memory persistence ranging from 2–3 hours (termed intermediate term memory, ITM) to 3–4 weeks (termed long-term memory, LTM; Lukowiak et al., 2000). Protein and RNA synthesis inhibitors differentially affect ITM and LTM; LTM requires *both* new protein and RNA synthesis whereas ITM only requires new protein synthesis (Sangha et al., 2003a). This learned behaviour and its memory are context-dependent and subject to extinction (McComb et al., 2002). Neural correlates of memory have in addition been found in one of the three CPG neurons, RPeD1 (Spencer et al., 1999, 2002).

While we are interested in coming to an understanding of the neuronal and molecular mechanisms of LTM, we are currently expending a great deal of time and energy attempting to determine the neuronal mechanisms of two related memory phenomena, reconsolidation and extinction. By doing so, we believe that we will be better able to determine how memory is formed.

What is reconsolidation of memory? The notion that the events experienced immediately prior to brain trauma are the ones most likely to be forgotten has been well-accepted since it was first proclaimed by Ribot in 1882 in the form of the "*Loi de Regression*". From this finding the *consolidation hypothesis* was formulated stating that memories initially exist in a fragile form and are strengthened over time, becoming less and less vulnerable to interference. More recently, there have been accounts that a memory re-enters a labile state upon memory reactivation or recall, and must be *re*consolidated before once again returning to a stable state. *Reconsolidation* therefore may provide a dynamic mechanism by which memories can be updated and changed (Nader, 2003). To what extent the entire post-acquisition cascade of intracellular events is recapitulated each time a memory is activated and reorganized has yet to be fully elucidated.

The notion that a classically conditioned memory could be extinguished is generally attributed to Pavlov. However, while just about every undergraduate psychology student knows that extinction occurs, the phenomenon is not really understood. First of all we have to appreciate that extinction is **not forgetting**, nor is it simply a case of unlearning. Rather, 'Pavlovian' extinction involves the formation of a new memory that *co-exists* alongside the original memory. The data presented here are consistent with the hypothesis that, during extinction, a new associative memory is being formed, specifically, an association between the behaviour and the absence of reinforcement. In addition the data support the view that, shortly after extinction training, there is a critical period, in which protein synthesis and gene transcription are required for its induction in order for extinction-memory to be stored alongside the original LTM.

Methods

Operant Conditioning Procedure

Briefly, individually labeled snails are placed in a 1-liter beaker containing 500 ml of room temperature hypoxic pond-water. The water is made hypoxic by bubbling N2 through it 20 minutes prior to and during training. A gentle tactile stimulus (using a sharpened wooden applicator) is applied to the pneumostome area (the respiratory orifice) every time the snail begins to open its pneumostome to perform aerial respiration. This tactile stimulus typically only evokes pneumostome closure. The time of each attempted opening is recorded and tabulated. Between training sessions snails are placed in eumoxic pond water where they were allowed to freely perform aerial respiration.

Operational Definitions of Learning and Memory

We operationally define memory as we have previously done (Spencer et al., 1999; 2002). Learning is present if the number of attempted pneumostome openings in the last training session is significantly less than the number of attempted openings in the 1st training session. In order to have memory, two criteria have to be met: i) the number of pneumostome openings in the memory test is significantly lower than in the first training session and, ii) the number of pneumostome openings in the memory session is not significantly higher than that of the last training session. If *both* these criteria are not met, memory for operant conditioning is not present. An ANOVA followed by the appropriate *post-hoc* statistical test is used to determine if the criteria are met. The form of memory tested in our experiments is known as *non-declarative* memory and the memory is stored within the same neural circuit that mediates the behaviour.

Soma Ablation Procedure

The soma of RPeD1 is required for LTM formation (Scheibenstock et al., 2002). The soma ablation procedure involves anesthetizing the animals with 1–3 ml of 50 mM $MgCl_2$ that was injected through the foot. This paralyzes the snail, allowing a dorsal midline incision to be made that exposes the animal's brain. Using a fine glass hand-held microelectrode, the RPeD1 soma is ablated by gently "poking" it. The incision is small enough to allow the animal to heal without suturing. Animals began to wake from the effects of the anesthetic within several hours of the surgery. To ensure that the proper cell's soma is ablated, a trained individual 'blinded to' the experiments attempts to visualize the cell that was ablated under the microscope at the conclusion of the experiment. In all cases the cell that had been ablated could not be found.

Results

RPeD1 is a Necessary Site of Memory Formation

One of our model system's experimental advantages is that it is possible to surgically remove the soma of RPeD1, whilst leaving behind a fully functional (i.e. all synaptic activity both pre- and post- continue to occur and *de novo* protein synthesis continues to occur) primary neurite. Ablation of the somata, however, removes the nucleus and thus transcription of new message *cannot occur*. If RPeD1's soma is ablated *before* operant conditioning training, learning occurs as does ITM (only dependent on new protein synthesis), but LTM formation cannot. Removal of RPeD1's soma *after* learning and memory consolidation occurs does not interfere with the ability to access the already formed LTM. Thus RPeD1 is a necessary site of LTM formation (Figure 2; Scheibenstock et al., 2002).

RPeD1 is a Necessary Site for Memory Reconsolidation

The notion that the events experienced immediately prior to brain trauma are the ones most likely to be forgotten (Ribot, 1882) led to the consolidation hypothesis, stating that memories initially exist in a fragile form and are strengthened over time (Nader, 2003). We now wish to explore what happens when a memory is made active. We hypothesize, as have others, that if the recently activated memory is to be retained, it is **necessary** for

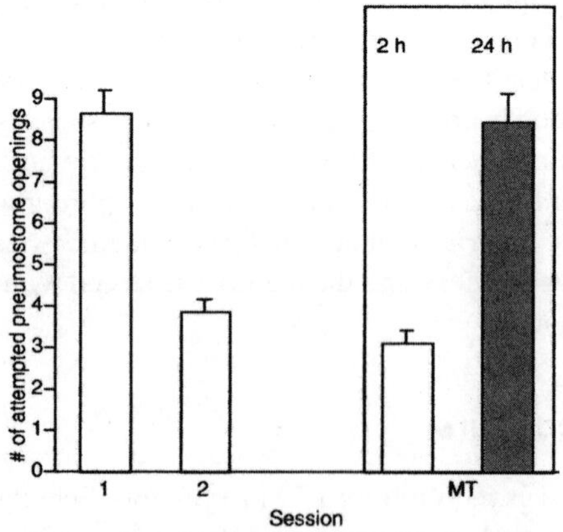

Figure 2. RPeD1 soma ablation and memory formation. Ablating RPeD1 somata 2 days before training does not affect the snails (N = 20) ability to learn. That is, there is a statistically significant difference between Session 1 and 2. In addition, these snails (N = 20) can form intermediate term memory (ITM). That is, in a memory test (MT) 2 h after the last training session the criteria for memory are met. However, if the memory is tested 24 h later in these same snails (N = 20), the criteria for memory are not met. We conclude that RPeD1 soma-ablated snails do not form LTM.

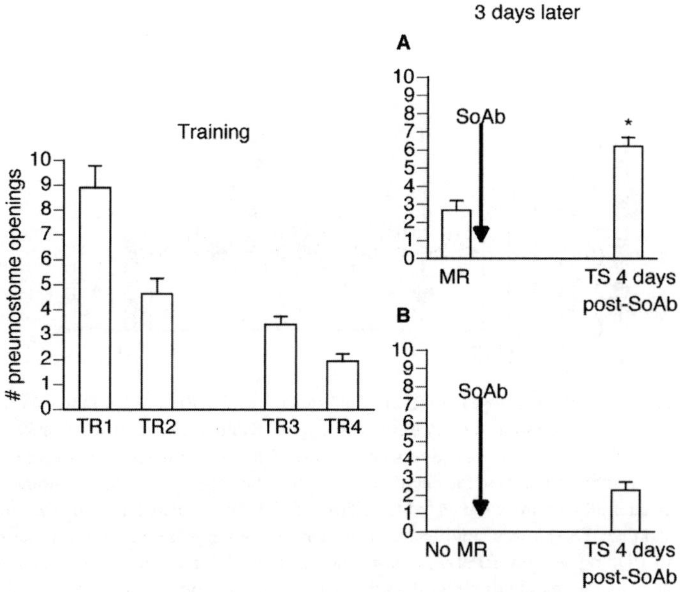

Figure 3. The soma of RPeD1 is required for reconsolidation. Operant training was administered (TR1–TR4) over the course of 2 days ($n = 19$). Three days later 9 animals received a memory recall session (MR in A) immediately followed by the ablation of the soma of RPeD1 (SoAb); while the remaining animals (B; $n = 10$) did not receive a MR session but still underwent the ablation procedure. Animals were tested for savings 4 days later. Memory was not observed in the group that received the MR session in conjunction with the ablation procedure (TS 4 days post RPeD1 soma ablation). Memory was observed in the other soma-ablated group, the one that did not receive a MR session. *$p < 0.01$ as compared to MR.

a *reconsolidation process to occur* (i.e. altered gene activity and *de novo* protein synthesis *are required* for 're-storage' of memory). This implies that memory re-enters a labile state when it is made active and must be *re*consolidated before once again returning to a stable state. Whether the reconsolidation process occurs within the same neurons, using the same processes as were necessary for the initial consolidation process, remains to be determined. Based on our findings concerning initial memory consolidation we hypothesize that if we remove the soma of RPeD1 *immediately after* a memory test *reconsolidation will not occur* and thus **there will be no memory**. That is, we will 'instantaneously' bring about forgetting by preventing the molecular processes necessary for memory formation from re-occurring. As can be seen (Figure 3), reconsolidation does not occur if we ablate RPeD1s somata immediately after memory recall. Thus, RPeD1 in addition to being a site for memory consolidation is also an *obligatory site* for memory reconsolidation.

RPeD1 is a Necessary Site for Extinction

The modern study of extinction began with Pavlov and has been shown to be ubiquitous across species (*C. elegans* to humans). Extinction, however, is **not** the destruction of the original memory. We hypothesize, as have others, that extinction is *new learning* which is consolidated into a *new memory* (dependent on altered gene activity and new

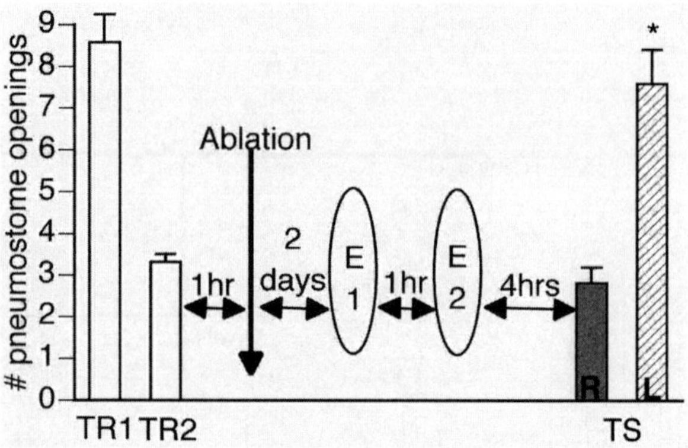

Figure 4. Extinction requires the soma of RPeD1. 25 snails received two 45-minute operant conditioning sessions. 1 hour later animals underwent surgery in which the soma of either LPeD1 ($n = 14$) or RPeD1 ($n = 11$) was ablated. 2 days later extinction training was administered: two 45-minute extinction sessions separated by 1 hour (snails did not receive a tactile stimulus to the pneumostome when they opened the pneumostome to perform aerial respiration). 4 hours later all animals were tested for LTM. LPeD1 soma-ablated animals demonstrated memory for the extinction training; TS (L) was significantly different than the last training session but not significantly different than the 1st training session. RPeD1 soma-ablated animals did not demonstrate memory for extinction training; TS (R) was not significantly different than the last training session and was significantly lower than the first training session.

protein synthesis) and this *new* memory coexists with and 'covers up' the old memory. Forgetting, on the other hand, is the **obliteration** of memory. What distinguishes extinction from forgetting is the phenomenon of *spontaneous recovery*, the reappearance of the original memory with time. While the behavioural phenotype of extinction and forgetting are similar, the neuronal mechanisms have to be different. The mechanistic details of extinction are of interest as extinction is used as a therapeutic tool in the treatment of fear disorders and substance addiction.

It is uncertain what the underlying neuronal mechanisms of extinction are. To begin to answer this question we operantly conditioned snails, allowed the memory to be consolidated and then ablated the soma of RPeD1 (Figure 4). We then attempted to extinguish the learned behaviour. Following the ablation of RPeD1 somata LTM could not be extinguished. These data are consistent with the hypothesis that: 1) The memory for extinction requires altered gene activity and new protein synthesis; and 2) That during extinction a new associative memory is being formed. Thus the soma of RPeD1 is a necessary site for extinction. These data show: 1) That the memory for extinction is stored along with the original LTM in RPeD1, and 2) Because spontaneous recovery occurs, show that the memory for extinction only *obscures* the original LTM.

Discussion

Together all these data support our general hypothesis that RPeD1 is a key neuron that plays important and necessary roles in different aspects of memory. We are now in a position to directly determine the molecular processes that cause memory formation, extinction and reconsolidation. We will also to begin to determine how one memory can temporarily obscure another (i.e. extinction) at the single cell level. We will also determine how the prevention of the reconsolidation process allows forgetting.

Aerial respiratory behaviour in *Lymnaea* is an important but stereotyped homeostatic behaviour. Only 3 neurons (RPeD1, VD4 and IP3) are capable of interacting to produce the appropriate rhythm. Other neurons when tested both *in vitro,* in culture and *in vivo,* following removal of the specific neuron from the CNS, with similar transmitter and electrophysiological phenotype cannot substitute for these 3 specific neurons. The obligatory rhythm needed to drive the aerial respiratory behaviour is an emergent property of the interplay between the 3 neurons. Yet this 'hard-wired' circuit possesses great plastic properties, in that specific training procedures are capable of adjusting the operation of the circuit, so that when the circuit should be active (i.e. in hypoxia) it is not. Embedded within the 'hard-wired' circuit are ample 'plastic-properties' (i.e. the ability to change the synaptic input to and from neurons; as well as the ability to change specific intrinsic membrane properties, etc.) that are competent to allow associative learning to come about and for the learning to be consolidated into LTM.

Knowing that a single neuron is a necessary site for memory formation, memory reconsolidation (Sangha et al 2003b) and extinction (Sangha et al 2003c) will enable us to direct our attention to the underlying molecular mechanisms of each of these different aspects of memory. As a start we can show that an inducer of gene activity, CREB, is both present in RPeD1 and is activated during the memory consolidation process (Sadamoto et al., 2004). We are also in the process of initiating studies directed to gaining an understanding of the proteomic basis of LTM in RPeD1. However, we need to emphasize that other neurons besides RPeD1 are probably necessary players in the mediation of LTM. We will therefore have to perform similar experiments on those neurons (e.g. VD4, IP3, RPeD11) as we have done for RPeD1 before we are able to completely specify the necessary and sufficient molecular mechanisms of LTM of this form of associative learning in *Lymnaea.*

References

Lukowiak, K., Adatia, N., Krygier, D., and Syed, N. (2000). Operant conditioning in Lymnaea: Evidence for intermediate and long-term memory. Learning and Memory 7: 140–150.

Lukowiak, K., Sangha, S., McComb, C., Varshney, N., Rosengger, D., Sadamoto, H. and Scheibenstock, A. (2003). Associative learning and memory in *Lymnaea stagnalis*: How well do they remember? J. Exp. Biol 206: 2097–2103

McComb, C., Meems, R., Syed, N., and Lukowiak, K. (2003). Electrophysiological differences in the neuronal circuit controlling aerial respiratory behaviour between juvenile and adult *Lymnaea.* J. Neurophysiol. 90: 983–992.

Muller, G.E. and Pilzecker, A. (1900). Experimentelle beitrage zur lehre vom gedachtnis. Z. Psychol. Suppl. 1.

Nader, K. (2003). Memory traces unbound. *Trends in Neurosciences* 26(2): 65–72.

Ribot, T. (1882). Diseases of memory. Appleton Century Crofts, New York, NY.

Sadamoto, H., Sato, H., Kobayashi, S., Murakami, J., Aonuma, H., Ando, H., Fujito, Y., Hamano, K., Awaji, M., Lukowiak, K., Urano, A., and Ito, E. (2004). CREB in the Pond Snail *Lymnaea stagnalis*: Cloning, Gene Expression and Function in Identifiable Neurons of the Central Nervous System. J. Neurobiol. 58: 455–66.

Sangha, S., Morrow, R., Smyth, K., Cooke, R., and Lukowiak, K. (2003a). Cooling blocks ITM and LTM Formation and preserves memory. Neurobiol. Learning & Memory. 80 130–139.

Sangha, S., Scheibenstock, A., and Lukowiak, K. (2003b). Reconsolidation of a long-term memory in *Lymnaea* requires new protein and RNA synthesis and the soma of RPeD1. J. Neuroscience 23: 8034–8040.

Sangha, S., Scheibenstock, A., Morrow, R., and Lukowiak, K. (2003c). Extinction requires new RNA and protein synthesis and the soma of the cell RPeD1 in *Lymnaea stagnalis* J. Neuroscience 23: 9842–9851.

Scheibenstock, A., Krygier, D., Haque, Z., Syed, S., and Lukowiak, K. (2002). The soma of RPeD1 must be present for LTM formation of associative learning in *Lymnaea*. J. Neurophysiol 88: 1584–1591.

Spencer, G., Syed, N., and Lukowiak, K. (1999). Neural changes following operant conditioning of aerial respiratory behaviour in *Lymnaea stagnalis*. J. Neuroscience 19, 1836–1843.

Spencer, G., Kazmi, M., Syed, I., and Lukowiak, K. (2002). Changes in the activity of a central pattern generator neuron following the reinforcement of an operantly conditioned behavior in *Lymnaea* J. Neurophysiol 88: 1915–1923.

Syed, N. I., Bulloch, A. G. M. and Lukowiak, K. (1990). *In vitro* reconstruction of the respiratory central pattern generator of the mollusk *Lymnaea*. *Science* **250**: 282–285.

Syed, N. I., Ridgway, R., Lukowiak, K., and Bulloch, A. G. M. (1992). Transplantation and functional integration of an identified respiratory interneuron in *Lymnaea stagnalis*. *Neuron* **8**: 767–774.

9

Conclusions and Perspectives

Modeling and Control of Breathing: Perspectives from Pre- to Post-Genomic Era

Opening Remarks for IX Oxford Conference

Chi-Sang Poon

Three years ago on Cape Cod, Massachusetts in the U.S., the VIII Oxford Conference celebrated a new millennium of frontiers in modeling and control of breathing. Today, we are thrilled to be here to once again try and push the envelope on this important field with a brand new post-genomic perspective.

I wish to first of all take this opportunity to recognize those who are past participants of the Oxford Conference series. Thanks to your continuing contributions and support, you are now part of the legacy of this legendary event. And for those who are joining us for the first time: congratulations and welcome onboard! This is really a historic moment for new membership as this particular meeting (which is about to take place) will mark the silver jubilee of the Oxford Conference series.

Yes, it was some 25 years ago, in September 1978, when the first meeting of this kind began at Oxford University as a brain child of two pioneers: Dan Cunningham at Oxford and Richard Herczynski at the Polish Academy of Sciences. The idea back then was to bring together people like the two of them—a physiologist and a mathematician—in order to talk about control of breathing from both ends of the spectrum. That first meeting proved to be a smashing success and that prompted a sequel meeting four years later at Lake Arrowhead, CA, which was organized by (among others) Dr. Brian Whipp and Dr. Susan Ward, then at UCLA (I am glad both of them are here with us today). From then on, just one thing after another and the "Oxford Conference" has continued every three years in various places around the globe. The thrust of each Oxford Conference might vary somewhat from time to time but one thing has never changed: a call for both modeling and experimental approaches to the study of control of breathing.

Chi-Sang Poon • Chi-Sang Poon, Harvard-MIT Division of Health Sciences and Technology, Massachusetts Institute of Technology, Cambridge, MA, USA.

Post-Genomic Perspectives in Modeling and Control of Breathing, edited by Jean Champagnat, Monique Denavit-Saubié, Gilles Fortin, Arthur S. Foutz, Muriel Thoby-Brisson. Kluwer Academic/Plenum Publishers, 2004.

Physiology, of course, is an experimental science and we cannot overemphasize the importance of experiments. But then—why "modeling"? Well, the answer is very simple indeed: believe it or not, modeling is arguably one of the "oldest professions" in science! In the history of physiology, modeling has been used to advantage in many fundamental discoveries by people like Darwin, Mendel, Heymans, Watson and Crick, and Hodgkin and Huxley, etc. And yet, in the pre-genomic era through the end of the 20th Century, modeling had somehow fallen by the wayside. Indeed, almost all of the Nobel awards after Hodgkin and Huxley have been in areas of cellular/molecular biology and genetics, and none had anything to do with physiological modeling whatsoever. Ironically, the only cardiorespiratory related award during this period (aside from the award on the development of beta-blocker, which was attributed to the associated principles for drug treatment) was about cell-to-cell signaling by nitric oxide, which is best known nowadays for its treatment of erectile dysfunction. (It was no coincidence that this award came just shortly after the wonder drug Viagra hit the market in 1998!)

So, what has gone wrong? Make no mistake about it: Nobel or not, cardiorespiratory physiology as a major scientific discipline is going strong. But who knows what will be the next breakthrough in this field, seven decades after Heymans' landmark discovery of the carotid bodies? There are certainly many, many exciting leads right now, some of which will be found at this meeting. But one thing almost for certain is: the next big break (I mean really big break) would probably not simply come from a new cell or molecule or gene alone, but would likely involve some kind of model or theory as well. The post-genomic era has been a real awakening for biomedical research: now that the genome bubble is over, science's "oldest profession" is coming back full circle! All of a sudden the new buzzwords these days are like bioinformatics, systems biology, computational biology, in-silico experiments, etc, which are all code names for computational modeling.

Let me illustrate... The genome is kind of like a Chinese abacus, the world's first computer. Obviously, in order for an abacus to work we must have all the beads in place. But even then it still won't do us any good if we don't know how to do the math on those beads. The founding fathers of the Oxford Conference had the wisdom to realize early on that in order to fully understand such a complex phenomenon as control of breathing, one will need not only good experimental data but also good computational modeling in order to make sense of those data.

Finally, in commemoration of the silver jubilee of the Oxford Conference I had the pleasure of making a little mascot out of a mini-abacus. Nine numbers (1–9) have been entered on the abacus to signify the nine Oxford Conferences. Now, on behalf of the XIII Oxford Conference I have the honor to present this mascot to Dr. Jean Champagnat, the organizer of the IX Oxford Conference, as a symbol of the passing of the torch.

Thank you very much.

Post-Genomic Perspectives in Modeling and Control of Breathing

Jean Champagnat

Robustness and flexibility of neuro-biological systems are important parameters to consider with respect to possible gene-behavior links. On the one hand, respiration is an exceptionally reliable and continuous rhythmic behaviour, so that the rhythmic neuronal network requires strict biological specifications because it becomes vital after birth. Thus, rhythm generation is a robust mechanism that persists in a variety of reduced states and preparations. Many contributors in the present volume illustrate the efficiency of such an *in vitro* strategy. This has provided extremely powerful approaches for examining many issues of interest at the cellular and molecular levels such as rhythm generation, network configuration, synaptic transmission, neuromodulation and signal transduction. On the other hand, the respiratory behaviour has to be highly flexible and to continuously adapt so that it remains appropriate in the face of ever changing conditions. One of the major successes of the Oxford Conference series has been to include a balanced contribution from molecular and cellular biologists, modelers and human physiologists, which together show that respiratory physiology has developed a unique experience in examining complex, integrated behaviors such as the ventilatory responses to exercise, hypoxia or sleep in animals and humans.

By the beginning of this Millennium, Physiology has probably reached a major milestone in its history. Sequencing projects in humans and laboratory animals have renewed biology because we have, for the first time, quantitative and statistical views of the genome. Initial draft DNA sequences of the entire human genome are easily accessible. Strategies aimed at the systematic identification of genes, inter-specific comparisons and studies of human polymorphisms are under way on a large scale. This is a very critical time for physiologists. Questions are now arising concerning the function of these 28 000 genes that seem to be included in the human genome. Identifying all the genes and proteins in an organism is like providing a catalog of individual components: we need to know how these parts are assembled to form functional organs or regulatory pathways. The respiratory system is highly suitable to understand physiological compensations as well as complex disease etiology following experimental or pathological gene inactivation. However, it is unclear

---

Jean Champagnat • Neurobiologie Génétique et Intégrative, Institut de Neurobiologie Alfred Fessard, CNRS, 91196, Gif-sur-Yvette, France.

Post-Genomic Perspectives in Modeling and Control of Breathing, edited by Jean Champagnat, Monique Denavit-Saubié, Gilles Fortin, Arthur S. Foutz, Muriel Thoby-Brisson. Kluwer Academic/Plenum Publishers, 2004.

yet how respiratory physiologists and neurobiologists will tackle new perspectives offered by systematic explorations of gene and protein expression patterns. Whilst the transition from genetically oriented work to successful biological achievements is occurring relatively slowly, one of the greatest challenges in integrative physiology and biomedical research today is the elucidation of the underlying genetic architecture of complex behaviors. This is particularly true in the field of respiratory physiology and neurobiology. One of the topics newly introduced into the 9th Oxford Conference is a first inquiry into this question. Indeed, many of the contributors have oriented their presentation in relation to a possible link with gene expression, or in contrast, with functional flexibility, the major property of breathing control during everyday life.

Animal models are now available to investigate the neurobiological basis of the large variety of normal and pathological breathing patterns. Mechanisms underlying central and obstructive sleep apneas, exercise-induced hyperpnea, apneustic respiration, periodic, cluster type, and other irregular breathing are considered in this volume from the *in vivo, in vitro* and modeling approaches. Sections 2 to 4 of this volume are devoted to the core function of the respiratory network, sensory afferent control and elementary mechanisms for generating an inspiratory rhythm and burst pattern. Several contributors show that the respiratory controller can be isolated in a variety of experimental situations that maintain rhythmic inspiratory activity. In these preparations, individual rhythmic cells are now recorded electrophysiologically and viewed by direct optical imaging of intracellular calcium levels; they can be modelled, characterized immunochemically and their voltage-dependent and synaptic properties are currently analysed to eventually reveal the link between protein expression and generation of the rhythm.

Accumulation of new data on molecular mechanisms of respiratory control provides an ideal situation for the use of genetically engineered models. Mice in which molecular regulations are affected by the deletion of specific genes were actively generated during the past ten years and they are widely used for functional studies in a post-genomic perspective. Since the 2000 Oxford Conference, the transgenic model strategy has in fact invaded the field of respiratory physiology. Examples can be found in this volume and many other on-going studies were presented in Paris. For example, because multiple neurotransmitters are involved in adaptive respiratory responses, loss-of-function studies in sections 2 and 4 of this volume, have concentrated on transgenic mice in which genes involved in the biosynthesis of neurotransmitters (glutamic acid decarboxylase, acetylcholinesterase) and ligand/receptor interactions (purinergic P2X receptors, pituitary adenylate cyclase-activating polypeptide) are inactivated. Analysis of these mutants obviously requires the entire panel of physiological skills together with a comparative analysis of multiple or conditional mutants to understand compensatory behavioral regulations. It is now clear that targeted gene inactivation is highly effective in analyzing the central control of breathing. Importantly, this approach provides together with pharmacology, a unique opportunity to investigate relevant molecular signaling in intact animals *in vivo*. Oxygen sensing initiates another set of integrative respiratory regulations that interest molecular geneticians. Progress is hindered due to the complexity of regulations and the lack of appropriate pharmacological tools: studies using transgenic mice might help to resolve the role of specific metabolic or transduction pathways in central and peripheral chemoreceptors (section 2 of this volume). Different sites may vary in their sensitivity and threshold as well as in the stimulus specificity for CO_2 or pH. Heme and/or redox sensitive proteins such as mitochondrial cytochromes are oxygen sensors. Important potassium channels include the K(ATP) and small-conductance Ca2+− activated, as well as the background K+ channel subunit TASK-1, which comprises four transmembrane segments and two pore domains involved in both oxygen- and acid-sensing. All these molecules are actively studied in transgenic mice.

Gene microarrays and proteomics are tools now available and currently used to address cellular signaling. The aim is to follow genomic expression under specified physiological or pathological conditions, or after down-regulation, inactivation or over-expression of specific genes. From a clinical point of view, the identification of differentially expressed transcripts in normal versus pathological human tissues, coupled to the growing availability of single-nucleotide polymorphism maps add to the process of gene discovery. Many aspects of human respiratory pathology are considered in the present volume, while only a few diseases are related to inherited disorders. One example is the mutation in the gene encoding succinate deshydrogenase D causing carotid body tumors (this volume, page 71). For an extensive review on the genetic basis of breathing parameters, the reader should consider the recent issue of Respiratory Physiology and Neurobiology on "Functional genomics and proteomics in control of breathing" (vol. 135, nos 2–3, pages 107–268, 2003). Otherwise, the genetic analysis of aberrant control of ventilation using human twins and familial aggregation strategies remains difficult to differentiate from confounding environmental effects. Variability of respiratory parameters in human populations, although provisionally important to approach gene control underlying biological diversity in complex respiratory regulations, is highly dependent on physiological conditions including cardiovascular regulations and cerebral blood flow (this volume, section 6). Furthermore, despite an extensive history of physiological investigation and modelling analysis, understanding breathing patterns during sleep (section 5) or exercise (section 7) remains a challenge in integrative respiratory neurosciences. While some combination of neurogenic, chemoreflex and circulatory-coupled processes are likely to contribute, the precise details of the control process remain unresolved. There is growing agreement that the ventilatory control process during exercise contains elements of substantial redundancy or occlusive interactions, so that synaptic plasticity provides novel perspectives in modeling ventilatory control (section 8 of this volume). In this respect, as stressed by C.S. Poon (page 321), the extensive modelling effort has led to many new concepts and methodologies in respiratory control, including the dynamics of the chemical control of breathing, the re-breathing technique for CO_2 response measurement, the systems analysis of respiratory instability or exercise hyperpnea, the optimization of ventilatory pattern and the nonlinear analysis of breathing pattern variability.

Central to discussions at the 9th Oxford Conference has been the new and growing interest in transcription factors that are required for the control of breathing patterns. For example, recent investigations provided evidence that a single factor, the hypoxia-inducible factor HIF-1, (pages 53 to 64 of the present volume) plays a major role in essential adaptive responses to hypoxia, including brain regulation of catecholamine synthesis. Manipulation of HIF-1 either pharmacologically or by gene therapy, may represent a novel approach to the treatment of cerebral and myocardial ischemia. Among transcription factors, those involved in development are important to consider. Developmental aspects of the respiratory rhythm have gained recent importance because results obtained in vitro are mostly from neonates. Attempts to isolate the respiratory network at the different stages of development until adulthood have been stimulated and results have accumulated on the mechanisms underlying embryonic maturation leading to the onset of fetal breathing, the prenatal evolution of the rhythm, birth-related events and postnatal development. Transgenic mice are widely used in developmental studies and important respiratory defects are induced at birth by the inactivation of, for example, *Hox* or *Krox-20* genes that are important in the patterning of the rhombencephalic neural tube (pages 143). Furthermore, development and plasticity of respiratory control is also critically dependent on peptide neurotrophic factors that mediate survival, proliferation and differentiation of different neuronal populations as well as the control of mature respiratory rhythm generation (page 115). Studies of early stages of network development in vertebrate evolution indicate that the respiratory circuit may be under the same genetic constraints and factors that are involved in early hindbrain patterning and control respiratory-like rhythmic movements that are among the earliest detectable behaviors of the mammalian fetus. These valuable insights open new avenues as to why the genetic codes underlying a vital function such as breathing have been selected, conserved and optimized during evolution. Genetic approaches offer great potential to improve our understanding of the physiology and pathology of respiratory control. It is our hope that this volume of Advances in Experimental Medicine and Biology will stimulate continued discussion of the major challenges facing respiratory neuroscience in the post-genomic perspective.

Abbreviations

A1	adenosine receptors
A1C1, A2C2, A5	catecholaminergic cell groups
ACh	acetylcholine
AChE	acetylcholinesterase
AHCVR	acute hypercapnic ventilatory response
AHVR	acute hypoxic ventilatory response
AMPA	alpha-amino-3-hydroxy-5-methylisoxazole-4-propionate
4AP	4-amino-pyridine
AP-4	2-amino-4-phosphonobutyric acid
AP-5	2-amino-5-phosphonopentanoic acid
ATP	adenosine tri-phosphate
aug-E	augmenting expiratory neurons
BChE	butyrylcholinesterase
BDNF	Brain derived neurotrophic factor
BötC	Bötzinger complex
CB	carotid bodies
CBF	cerebral blood flow
CBFV	cerebral blood flow velocity
CCR	carotid chemoreflex,
CNQX	6-cyano-7-nitroquinoxaline-2,3-dione
CNS	central nervous system
COPD	chronic obstructive pulmonary disease
CPT	8-cyclopentylmtheophilline
CPX	ciclopirox olamine
CREB	cyclic AMP response element binding protein
CVRG	caudal ventral respiratory group
DFO	desferrioxamine
DLH	D,L-homocysteic acid
DT	electroencephalogram dipole tracing
EAA	excitatory amino acids
early-I	early inspiratory neurons
ECG	electrocardiogram
EEG	electroencephalogram
EOF	expiratory off-switching
EPSCs	excitatory postsynaptic currents
fMRI	functional magnetic resonance imaging
GABA	gamma-aminobutyric acid
GAD	glutamic acid decarboxylase

HB, HBR	Hering-Breuer reflex
HCVR	hypercapnic ventilatory response
HIF-1α	hypoxia inducible factor 1α
5HT	serotonin
5-HT_{1B}, 5-HT_{2A}	serotoninergic receptors
HVD	hypoxic ventilatory depression
HVR	hypoxic ventilatory response
I	inspiratory neurons
Ih	hyperpolarization-activated current
I_{NaP}	persistent sodium current
IOS	inspiratory off-switch
IPSCs	inhibitory postsynaptic currents
IVRG	intermediate ventral respiratory group
KF	Kölliker-Fuse
KYN	kynurenic acid
late-I	late inspiratory neurons
LC	locus coeruleus nucleus
LTD	long term depression
LTP	long term potentiation
MABP, MAP	blood pressure
MAC	minimal alveolar concentration
α,β-meATP	α,β-methylene ATP
MOS	metal-oxide-silicon (transistor circuits)
NaCN	sodium cyanide
NHE-3	Na^+/H^+ exchanger type 3
NK1	neurokinin 1 receptors
NMDA	N-methyl-D-aspartate
NPBM	nucleus parabrachialis medialis
NRA	nucleus retroambigualis
NREM	non-rapid eye movement (sleep)
NTS	nucleus tractus solitarius
$P2X_1$-$P2X_6$	purinergic P2X ionotropic receptor subunits
P2Y	purinergic metabotropic receptor
PAC1	PACAP receptor
PACAP	pituitary adenylate cyclase-activating polypeptide
PAG	periaqueductal gray
PBC	pre-Bötzinger complex
PBr	parabrachial nucleus
PET	positron emission tomography
$P_{ET_{CO_2}}$	end-tidal partial pressure of CO2
$P_{ET_{O_2}}$	end-tidal partial pressure of O2
PGL1	paraganglioma type I
PPADS	pyridoxal-5′-phosphate-6-azophenyl-2′4′-disulphonic acid
pre-BötC	pre-Bötzinger complex
pre-I	pre-inspiratory neurons
PSR	pulmonary stretch receptors

PvCO2	venous CO2 partial pressure
ramp-I,	ramp inspiratory neurons
Raw	airway resistance
REM	rapid eye movement (sleep)
ROS	reactive oxygen species
RP	raphe pallidus nucleus
RT/PCR	reverse transcription/polymerase chain reaction
RVRG	rostral ventral respiratory group
SaO_2	arterial oxygen saturation
SAP-SP	saporin conjugated to substance P
SA-PSR	slowly adapting pulmonary stretch receptors
SDHD	succinate dehydrogenase D
SIDS	sudden infant death syndrome
SIFT	stroboscopic interferometric filtering technique
SSB/DT	dipole tracing method of the scalp-skull-brain head model
STP	short term potentiation
TASK	two-pore domain acid-sensitive potassium channel
TEA	tetraethylammonium
TH	tyrosine hydroxylase
TP	tracheal pressure
TrkB	tyrosine protein kinase receptors B
VAH	ventilatory acclimatization to hypoxia
VEGF	vascular endothelial growth factor
VLM	ventro-lateral medulla
VLSI	very large scale integration
VRC	ventral respiratory column
VRG	ventral respiratory group

Author Index

Subject Index

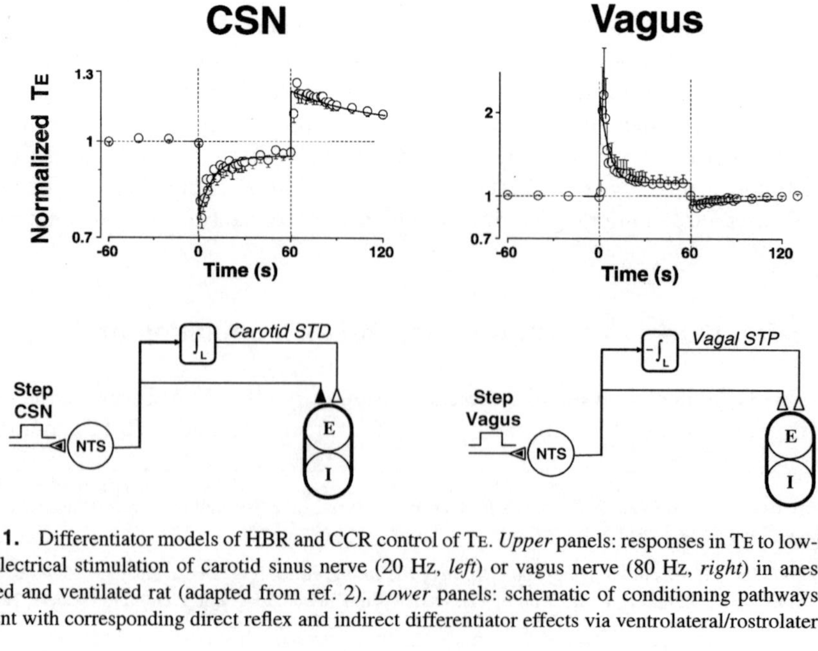

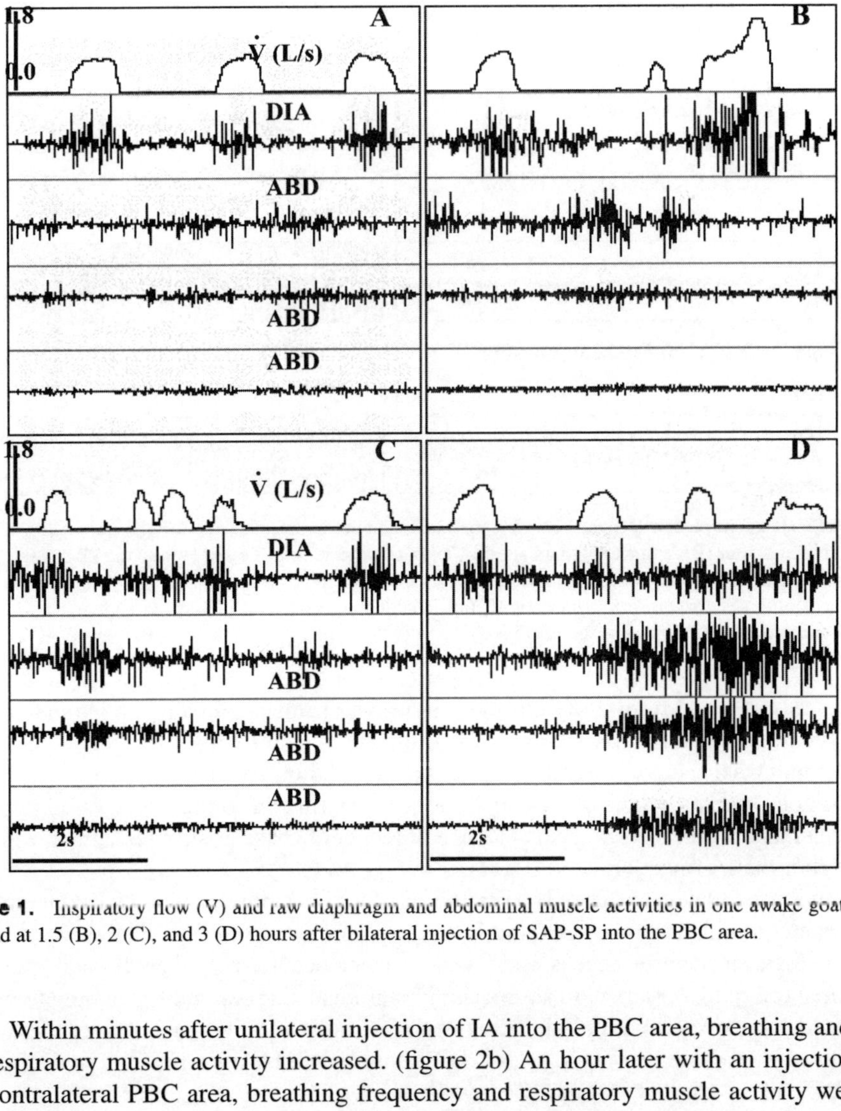

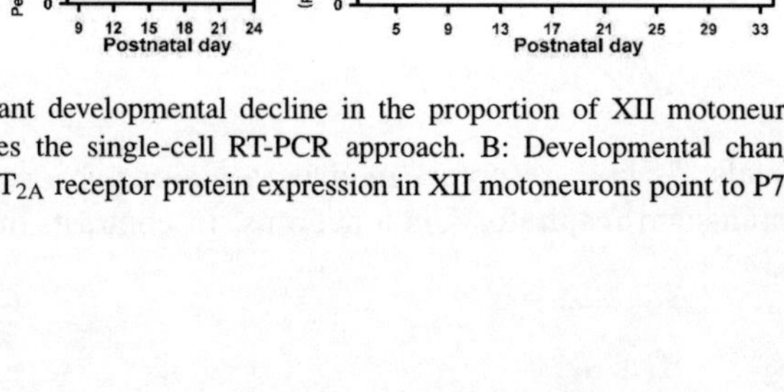

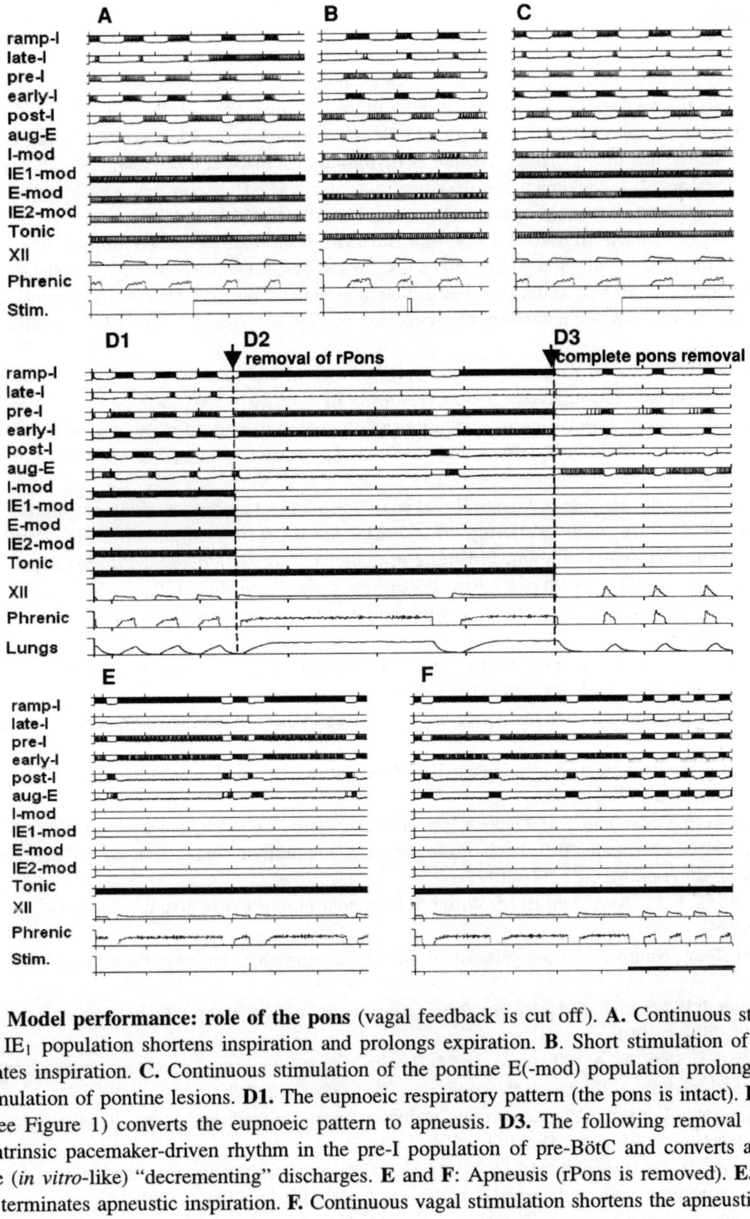

Table 1. Ventilatory measurements during normoxia and the sensitivities. Vent, Freq, V_T, T_i/T_{tot}, and, V_i/T_i calculated during the last two minutes of normoxia. Freq sens, V_T sens, T_i/T_{tot} sens, and T_i/T_i sens are the change in the measurement from normoxia (averaged over minute 3–4) to hypoxia (averaged over minute 7–8) divided by the change in saturation. * = $P < 0.01$.

	AHR ($l \cdot min^{-1} \cdot \%sat^{-1}$)	Vent (l/min)	Freq (min^{-1})	Freq sens ($l \cdot min^{-1} \cdot \%sat^{-1}$)	V_T (mL)	$V_T$sens (ml/%sat)	T_i/T_{tot}	T_i/T_{tot}sens ($\%sat^{-1}$)	V_i/T_i (l/sec)	V_i/T_i sens ($l \cdot sec^{-1} \cdot \%sat^{-1}$)
AVS	0.53	12.9	17.9	0.07	733	26.8	0.40	−1.52	0.58	0.020
	±0.35	±3.6	±2.3	±0.14	±146	±15.2	±0.04	±1.91	±0.13	±0.013
Pain	0.36	10.5	14.8	0.01	735	24.5	0.38	−2.59	0.48	0.012
	±0.27	±2.7	±2.7	±0.12	±200	±19.1	±0.04	±1.94	±0.10	±0.008
Rest	0.45	9.8	14.2	0.07	732	26.8	0.38	−1.65	0.46	0.018
	±0.33	±1.7	±2.9	±0.18	±189	±20.4	±0.05	±2.86	±0.09	±0.018
Test sig.	*	*	*	NS	NS	NS	P = 0.06	NS	*	*
Sex sig.	NS	*	NS	P = 0.04	*	NS	NS	NS	*	NS

The stimulatory effect of AVS on minute ventilation occurred during both normoxia and hypoxia. During normoxia, this was attributed to both increased frequency of ventilation and increased inspiratory drive (V_i/T_i). For AVS, the AHR was increased because of an increase in respiratory frequency without a change in the tidal volume sensitivity (eq. 2) or by an increase in V_i/T_i without a change in the V_i/T_i sensitivity (eq. 3). For Pain, the AHR was not significantly increased. However, analysis of the components of AHR showed a significant decrease in the V_i/T_i sensitivity.

There did not seem to be any important sex differences in the results. The significant difference in the normoxic ventilation and tidal volume could be because of the difference in body mass between men and women.

4. Discussion

The effects of pain on respiration are not simple. Clinical observation confirms that pain is often a respiratory stimulant and this is supported by the literature.[9] The interaction of pain and hypoxia, however, has not been clearly elucidated. Studies examining the respiratory effects of post-operative pain relief have shown the most common agents to be associated with significant hypoxemic events[10–12] that are not always predictably based on patient or drug characteristics. Furthermore, the pathways that modulate these responses may not involve chemoreflexes[13] and may even bypass cortical structures.[14]

In this study, painful thermal stimulation did not significantly increase normoxic ventilation, nor did it augment the acute hypoxic response. This agrees with previous experiments done by Sarton et al.[1] in which a painful electrical stimulus (to a VAS of 4.5–5.5) did not augment the AHR. However, in contrast to our study, Sarton's painful stimulus did cause a significant increase in normoxic minute ventilation (with non-significant increases in normoxic

